T0281021

# Lecture Notes in Computer Science 14486

Founding Editors

Gerhard Goos
Juris Hartmanis

## Editorial Board Members

The series Lecture Notes in Computer Science (LNCS), including its subseries Lecture Notes in Artificial Intelligence (LNAI) and Lecture Notes in Bioinformatics (LNBI), has established itself as a medium for the publication of new developments in computer science and information technology research, teaching, and education.

LNCS enjoys close cooperation with the computer science R & D community, the series counts many renowned academics among its volume editors and paper authors, and collaborates with prestigious societies. Its mission is to serve this international community by providing an invaluable service, mainly focused on the publication of conference and workshop proceedings and postproceedings. LNCS commenced publication in 1973.

Dmitry I. Ignatov · Michael Khachay ·
Andrey Kutuzov · Habet Madoyan ·
Ilya Makarov · Irina Nikishina ·
Alexander Panchenko · Maxim Panov ·
Panos M. Pardalos · Andrey V. Savchenko ·
Evgenii Tsymbalov · Elena Tutubalina ·
Sergey Zagoruyko
Editors

# Analysis of Images, Social Networks and Texts

11th International Conference, AIST 2023
Yerevan, Armenia, September 28–30, 2023
Revised Selected Papers

Springer

*Editors*

Dmitry I. Ignatov (ID)
National Research University Higher School
of Economics
Moscow, Russia

Andrey Kutuzov (ID)
University of Oslo
Oslo, Norway

Ilya Makarov (ID)
Artificial Intelligence Research Institute
Moscow, Russia

Alexander Panchenko (ID)
Skolkovo Institute of Science and Technology
Moscow, Russia

Panos M. Pardalos (ID)
University of Florida
Gainesville, FL, USA

Evgenii Tsymbalov (ID)
Apptek
Aachen, Germany

Sergey Zagoruyko (ID)
MTS AI
Moscow, Russia

Michael Khachay (ID)
Krasovskii Institute of Mathematics
and Mechanics of Russian Academy
of Sciences
Yekaterinburg, Russia

Habet Madoyan
American University of Armenia
Yerevan, Armenia

Irina Nikishina (ID)
University of Hamburg
Hamburg, Germany

Maxim Panov (ID)
Mohamed bin Zayed University of Artificial
Intelligence
Abu Dhabi, United Arab Emirates

Technology Innovation Institute, UAE
Abu Dhabi, United Arab Emirates

Andrey V. Savchenko (ID)
National Research University Higher School
of Economics
Nizhny Novgorod, Russia

Elena Tutubalina (ID)
Kazan Federal University
Kazan, Russia

ISSN 0302-9743 ISSN 1611-3349 (electronic)
Lecture Notes in Computer Science
ISBN 978-3-031-54533-7 ISBN 978-3-031-54534-4 (eBook)
https://doi.org/10.1007/978-3-031-54534-4

This Springer imprint is published by the registered company Springer Nature Switzerland AG
The registered company address is: Gewerbestrasse 11, 6330 Cham, Switzerland

Paper in this product is recyclable.

# Preface

This volume contains the refereed proceedings of the 11th International Conference on Analysis of Images, Social Networks and Texts (AIST 2023)[1]. The previous conferences (during 2012–2021) attracted a significant number of data scientists, students, researchers, academics, and engineers working on interdisciplinary data analysis of images, texts, and networks. The broad scope of AIST makes it an event where researchers from different domains, such as computer vision and natural language processing, exploiting various data analysis techniques, can meet and exchange ideas. As the test of time has shown, this leads to the cross-fertilisation of ideas between researchers relying on modern data analysis machinery.

Therefore, AIST 2023 brought together all kinds of applications of data mining and machine learning techniques. The conference allowed specialists from different fields to meet each other, present their work, and discuss both theoretical and practical aspects of their data analysis problems. Another important aim of the conference was to stimulate scientists and people from industry to benefit from knowledge exchange and identify possible grounds for fruitful collaboration.

The conference was held during September 28–30, 2023. The conference was organized with the support of Zaven and Sonia Akian College of Science and Engineering, American University of Armenia.

This year, the key topics of AIST were grouped into five tracks:

1. Data Analysis and Machine Learning chaired by Evgenii Tsymbalov (Apptek, Germany) and Maxim Panov (Mohamed bin Zayed University of Artificial Intelligence and Technology Innovation Institute, UAE)
2. Natural Language Processing chaired by Andrey Kutuzov (University of Oslo, Norway) and Elena Tutubalina (Kazan Federal University, Russia)
3. Network Analysis chaired by Irina Nikishina (Universität Hamburg, Germany) and Ilya Makarov (HSE University and AIRI, Russia)
4. Computer Vision chaired by Sergei Zagoruyko (MTS AI, Russia), and Andrey Savchenko (HSE University, Russia)
5. Theoretical Machine Learning and Optimization chaired by Panos Pardalos (University of Florida, USA) and Michael Khachay (IMM UB RAS and Ural Federal University, Russia)

The Program Committee and the reviewers of the conference included 137 well-known experts in data mining and machine learning, natural language processing, image processing, social network analysis, and related areas from leading institutions of many countries including Armenia, Austria, the Czech Republic, Finland, France, Georgia, Germany, Greece, India, Ireland, Italy, Mexico, Montenegro, the Netherlands, Norway, Portugal, Russia, Slovenia, Spain, the UAE, the UK, and the USA. This year, we received 106 submissions, mostly from Russia but also from Armenia, China, Finland, France,

---

[1] https://aistconf.org.

Georgia, Germany, India, Iraq, Kyrgyzstan, Saudi Arabia, Singapore, Switzerland, the UAE, and the USA.

Out of the 106 technical submissions, 13 were poster submissions, and 76 submissions remained after desk rejects and withdrawals. For the remaining papers, only 24 were accepted into this main volume. In order to encourage young practitioners and researchers, we included 21 papers in the companion volume published in Springer's Communications in Computer and Information Science (CCIS) series. Thus, the acceptance rate of this LNCS volume is 32%. Each submission was double-blind reviewed by at least three reviewers, experts in their fields, in order to supply detailed and helpful comments.

The conference featured several invited talks and tutorials dedicated to current trends and challenges in the respective areas.

The invited talks from academia covered a wide range of machine learning and artificial intelligence areas:

- Narine Sarvazyan (George Washington University and American University of Armenia): "Decoding Hyperspectral Imaging: From Basic Principles to Medical Application"
- Hakim Hacid (Technology Innovation Institute): "Towards Edge AI: Principles, current state, and perspectives"
- Samuel Horvath (MBZUAI): "Towards Real-World Federated Learning: Addressing Client Heterogeneity and Model Size"
- Artem Shelmanov (MBZUAI): "Safety of Deploying NLP Models: Uncertainty Quantification of Generative LLMs"
- Muhammad Shahid Iqbal Malik (HSE University): "Threatening Content and Target Identification in low-resource languages using NLP Techniques"

We would like to thank the authors for submitting their papers and the members of the Program Committee for their efforts in providing exhaustive reviews.

According to the track chairs, and taking into account the reviews and presentation quality, the Best Paper Awards were granted to the following papers:

- Data Analysis and Machine Learning: "Ensemble Clustering with Heterogeneous Transfer Learning" by Vladimir Berikov;
- Natural Language Processing: "Benchmarking Multi-Label Topic Classification in Kyrgyz Language" by Anton Alekseev, Sergey Nikolenko, and Gulnara Kabaeva;
- Network Analysis: "Limit Distributions of Friendship Index in Scale-Free Networks" by Sergei Sidorov, Sergei Mironov, and Alexey Grigoriev;
- Computer Vision: "DeepLOC: Deep Learning-based Bone Pathology Localization and Classification in Wrist X-ray Images" by Razan Dibo, Andrey Galichin, Pavel Astashev, Dmitry V. Dylov, and Oleg Y. Rogov;
- Theoretical Machine Learning and Optimization: "Is Canfield Right? On the Asymptotic Coefficients for the Maximum Antichain of Partitions and Related Counting Inequalities" by Dmitry Ignatov.

We would also like to express our special gratitude to all the invited speakers and industry representatives. We deeply thank all the partners and sponsors, especially the hosting organization: Zaven and Sonia Akian College of Science and Engineering, American University of Armenia, with special thanks to Habet Madoyan and Amalya Hambardzumyan, as local organizers. Our special thanks go to Springer for their help, starting from the first conference call to the final version of the proceedings. Last but not least, we are grateful to the volunteers, whose endless energy saved us at the most critical stages of the conference preparation.

Here, we would like to mention that the Russian word "aist" is more than just a simple abbreviation as (in Russian) it means "stork". Since it is a wonderful free bird, a symbol of happiness and peace, this stork gave us the inspiration to organize the AIST conference series. So we believe that this conference will likewise bring inspiration to data scientists around the world!

December 2023

Dmitry Ignatov
Michael Khachay
Andrey Kutuzov
Habet Madoyan
Ilya Makarov
Irina Nikishina
Alexander Panchenko
Maxim Panov
Panos Pardalos
Andrey Savchenko
Evgenii Tsymbalov
Elena Tutubalina
Sergey Zagoruyko

# Organization

## Organizing Institutions

- Zaven and Sonia Akian College of Science and Engineering, American University of Armenia (Yerevan, Armenia)
- Skolkovo Institute of Science and Technology (Moscow, Russia)
- Krasovskii Institute of Mathematics and Mechanics, Ural Branch of the Russian Academy of Sciences (Yekaterinburg, Russia)
- School of Data Analysis and Artificial Intelligence, HSE University (Moscow, Russia)
- Laboratory of Algorithms and Technologies for Networks Analysis, HSE University (Nizhny Novgorod, Russia)
- Laboratory for Models and Methods of Computational Pragmatics, HSE University (Moscow, Russia)

## Program Committee Chairs

| | |
|---|---|
| Michael Khachay | Krasovskii Institute of Mathematics and Mechanics of Russian Academy of Sciences & Ural Federal University, Russia |
| Andrey Kutuzov | University of Oslo, Norway |
| Ilya Makarov | HSE University, Moscow and Artificial Intelligence Research Institute, Russia |
| Irina Nikishina | Universität Hamburg, Germany |
| Maxim Panov | Mohamed bin Zayed University of Artificial Intelligence and Technology Innovation Institute, UAE |
| Panos Pardalos | University of Florida, USA |
| Andrey Savchenko | HSE University, Nizhny Novgorod, Russia |
| Elena Tutubalina | HSE University, Moscow and Kazan Federal University, Russia |
| Evgenii Tsymbalov | Apptek, Germany |
| Sergey Zagoruyko | MTS AI, Russia |

## Proceedings Chairs

| | |
|---|---|
| Dmitry I. Ignatov | HSE University, Moscow, Russia |
| Evgenii Tsymbalov | Apptek, Germany |

## Steering Committee

| | |
|---|---|
| Dmitry I. Ignatov | HSE University, Moscow, Russia |
| Michael Khachay | Krasovskii Institute of Mathematics and Mechanics of Russian Academy of Sciences, Russia & Ural Federal University, Russia |
| Alexander Panchenko | Skolkovo Institute of Science and Technology, Russia |
| Andrey Savchenko | HSE University, Nizhny Novgorod, Russia |

## Program Committee

| | |
|---|---|
| Anton Alekseev | St. Petersburg Department of V.A. Steklov Institute of Mathematics of the Russian Academy of Sciences, Russia |
| Ammar Ali | ITMO University, Russia |
| Vladimir Arlazarov | Smart Engines Service LLC, Federal Research Center "Computer Science and Control" of Russian Academy of Sciences, Russia |
| Ekaterina Artemova | HSE University, Moscow, Russia |
| Yulia Badryzlova | HSE University, Moscow, Russia |
| Jaume Baixeries | Universitat Politecnica de Catalunya, Spain |
| Amir Bakarov | HSE University, Moscow, Russia |
| Artem Baklanov | International Institute for Applied Systems Analysis, Austria |
| Jeremy Barnes | University of the Basque Country, Spain |
| Nikita Basov | University of Manchester, UK |
| Vladimir Batagelj | University of Ljubljana, Slovenia |
| Tatiana Batura | A.P. Ershov Institute of Informatics Systems SB RAS; Novosibirsk State University, Russia |
| Vladimir Berikov | Sobolev Institute of Mathematics SB RAS, Russia |
| Malay Bhattacharyya | Indian Statistical Institute, India |
| Mikhail Bogatyrev | Tula State University, Russia |
| Elena Bolshakova | Moscow State Lomonosov University, Russia |
| Pavel Braslavski | Ural Federal University, Russia and Nazarbaev University, Kazakhstan |
| Aram Butavyan | American University of Armenia, Armenia |
| Alexey Chernyavskiy | Philips Innovation Labs, Russia |
| Vera Davydova | Sber AI, Russia |
| Radhakrishnan Delhibabu | Vellore Institute of Technology, India |
| Oksana Dereza | National University of Ireland, Ireland |

| | |
|---|---|
| Michael Diskin | HSE University, Russia |
| Anna Dmitrieva | University of Helsinki, Finland |
| Ekaterina Dmitrieva | HSE University, Russia |
| Boris Dobrov | Lomonosov Moscow State University, Russia |
| Ivan Drokin | Prequel App, USA |
| Pavel Efimov | ITMO University, Russia |
| Anton Eremeev | Omsk Branch of Sobolev Institute of Mathematics SB RAS, Russia |
| Elena Ericheva | botkin.ai, Russia |
| Anna Ermolayeva | Peoples' Friendship University of Russia, Russia |
| Kirill Fedyanin | Technology Innovation Institute, UAE |
| Alena Fenogenova | SberDevices, Russia |
| Alexander Fishkov | Skolkovo Institute of Science and Technology, Russia |
| Georgii Gaikov | MTS AI, Russia |
| Yuriy Gapanyuk | Bauman Moscow State Technical University, Russia |
| Olga Gerasimova | HSE University, Russia |
| Edward Kh. Gimadi | Sobolev Institute of Mathematics, SB RAS, Russia |
| Petr Gladilin | Huawei and ITMO University, Russia |
| Maksim Glazkov | Neuro.net, Russia |
| Anna Glazkova | University of Tyumen, Russia |
| Alexander Gneushev | Moscow Institute of Physics and Technology, Russia |
| Elizaveta Goncharova | Artificial Intelligence Research Institute, Russia |
| Amalya Hambardzumyan | American University of Armenia, Armenia |
| Anastasia Ianina | Moscow Institute of Physics and Technology, Russia |
| Dmitry Ilvovsky | HSE University, Russia |
| Vladimir Ivanov | Innopolis University, Russia |
| Ilia Karpov | HSE University, Russia |
| Mikhail Khachay | IMM URAN, Russia |
| Vladimir Khandeev | Sobolev Institute of Mathematics, Siberian Branch of the Russian Academy of Sciences, Russia |
| Javad Khodadoust | Tecnológico de Monterrey, Mexico |
| Daniel Kireev | VisionLabs, Russia |
| Denis Kirjanov | HSE University, Russia |
| Dmitrii Kiselev | HSE University and Sber, Russia |
| Yury Kochetov | Sobolev Institute of Mathematics, Russia |
| Sergei Koltcov | HSE University, Russia |
| Jan Konecny | Palacký University Olomouc, Czech Republic |

| Anton Konushin | HSE University, Russia |
| Andrei Kopylov | Tula State University, Russia |
| Nikita Kotelevskii | Skolkovo Institute of Science and Technology, Russia |
| Evgeny Kotelnikov | Vyatka State University, Russia |
| Anastasia Kotelnikova | Vyatka State University, Russia |
| Angelina Kudriavtseva | MTS AI, Russia |
| Maria Kunilovskaya | University of Saarland, Germany |
| Anvar Kurmukov | IITP RAS, Russia |
| Andrey Kutuzov | University of Oslo, Norway |
| Dmitri Kvasov | University of Calabria, Italy |
| Florence Le Ber | icube, France |
| Alexander Lepskiy | HSE University, Russia |
| Tatyana Levanova | Sobolev Institute of Mathematics SB RAS, Omsk Branch, Russia |
| Natalia Loukachevitch | Moscow State University, Russia |
| Olga Lyashevskaya | HSE University, Russia |
| Habet Madoyan | American University of Armenia, Armenia |
| Jaafar Mahmoud | ITMO University, Russia |
| Ilya Makarov | Artificial Intelligence Research Institute, Russia |
| Alexey Malafeev | HSE University, Russia |
| Muhammad Shahid Iqbal Malik | HSE University, Russia |
| Vladislav Mikhailov | HSE University, Russia |
| Olga Mitrofanova | St. Petersburg State University, Russia |
| Petter Mæhlum | University of Oslo, Norway |
| Amedeo Napoli | LORIA Nancy (CNRS - Inria - Université de Lorraine), France |
| Irina Nikishina | University of Hamburg, Germany |
| Damien Nouvel | INaLCO, France |
| Victor Ohanyan | American University of Armenia, Armenia |
| Walaa Othman | ITMO University, Russia |
| Evgeniy M. Ozhegov | HSE University, Russia |
| Alexander Panchenko | Artificial Intelligence Research Institute, Russia |
| Maxim Panov | Mohamed bin Zayed University of Artificial Intelligence and Technology Innovation Institute, UAE |
| Panos Pardalos | University of Florida, USA |
| Stefan Pickl | University of the Bundeswehr Munich, Germany |
| Dina Pisarevskaya | Queen Mary University of London, UK |
| Lidia Pivovarova | University of Helsinki, Finland |
| Vladimir Pleshko | RCO, Russia |

| | |
|---|---|
| Arnak Poghosyan | Vmware, Inc.; Institute of Mathematics; American University of Armenia; Yerevan State University, Armenia |
| Aleksandr Rubashevskii | Skolkovo Institute of Science and Technology, Russia |
| Yuliya Rubtsova | University of Bonn, Germany |
| Alexey Ruchay | Chelyabinsk State University, Russia |
| Nicolay Rusnachenko | Bauman Moscow State Technical University, Russia |
| Alexander Sapin | ClickHouse B.V., The Netherlands |
| Andrey Savchenko | HSE University, Russia |
| Oleg Seredin | Tula State University, Russia |
| Tatiana Shavrina | HSE University, Russia |
| Denis Sidorov | Energy Systems Institute SB RAS, Russia |
| Henry Soldano | Laboratoire d'Informatique de Paris Nord, France |
| Alexey Sorokin | Moscow State University, Russia |
| Andrey Sozykin | Krasovskii Institute of Mathematics and Mechanics, Russia |
| Dmitry Stepanov | Program System Institute of Russian Academy of Sciences, Russia |
| Tatiana Tchemisova | University of Aveiro, Portugal |
| Mikhail Tikhomirov | Lomonosov Moscow State University, Russia |
| Yulya Trofimova | HSE University, Russia |
| Christos Tryfonopoulos | University of the Peloponnese, Greece |
| Magda Tsintsadze | Iv.Javakhishvili Tbilisi State University, Georgia |
| Evgenii Tsymbalov | Apptek, Germany |
| Elena Tutubalina | Kazan Federal University, Russia |
| Alsu Vakhitova | MTS AI, Russia |
| Daiana Vavilova | Kalashnikov Izhevsk State Technical University, Russia |
| Petr Vytovtov | Kalashnikov Izhevsk State Technical University, Russia |
| Dmitry Yashunin | Harman International, USA |
| Varduhi Yeghiazaryan | American University of Armenia, Armenia |
| Sergey Zagoruyko | MTS AI, Russia |
| Alexey Zaytsev | Skolkovo Institute of Science and Technology, Russia |
| Nikolai Zolotykh | University of Nizhny Novgorod, Russia |

## Organizing Committee

| | |
|---|---|
| Habet Madoyan (Organizing Chair) | American University of Armenia, Armenia |
| Amalya Hambardzumyan | American University of Armenia, Armenia |
| Dmitry Ignatov | HSE University, Moscow, Russia |
| Irina Nikishina | Universität Hamburg, Germany |
| Alexander Panchenko | Skolkovo Institute of Science and Technology and Artificial Intelligence Research Institute, Russia |
| Maxim Panov | Mohamed bin Zayed University of Artificial Intelligence and Technology Innovation Institute, UAE |
| Evgenii Tsymbalov | Apptek, Germany |

## Additional Reviewers

Ali, Ammar
Filatov, Andrei
Kireev, Daniel
Kovalenko, Aleksandr
Kurkin, Maxim

Kuvshinova, Ksenia
Musaev, Ruslan
Romanov, Vitaly
Severin, Nikita
Zhevnenko, Dmitry

## Sponsoring Institutions

American University of Armenia, Armenia
Artificial Intelligence Research Institute, Russia

# Invited Talks

# Towards Edge AI: Principles, Current State, and Perspectives

Hakim Hacid 🆔

Technology Innovation Institute, United Arab Emirates
hakim.hacid@tii.ae

**Abstract.** The artificial intelligence (AI) community has invested heavily in developing techniques that can digest very large amounts of data to extract valuable information and knowledge. Most techniques, particularly deep learning models, require large amounts of computing and storage power, making them suitable for cloud-based environments. The intelligence is therefore remote from the end user, raising concerns about, for example, data privacy and latency. Edge AI addresses some of the problems inherent in the cloud and focuses on best practices, architectures and processes for extending data AI outside the cloud. Edge AI brings AI closer to the end user and uses, for example, fewer communication resources, as processing is performed directly on the edge device. This presentation will introduce edge AI and give an overview of existing work and potential future contributions.

**Keywords:** Artificial intelligence · Edge AI · Cloud computing · Knowledge extraction

# Towards Real-World Federated Learning: Addressing Client Heterogeneity and Model Size

Samuel Horvath🆔

MBZUAI, United Arab Emirates
samohorvath11@gmail.com

**Abstract.** In this talk, I will introduce federated learning and discuss two recent approaches for addressing the challenges of client heterogeneity and model size in federated learning. In the first part of the talk, I will introduce federated learning. I will discuss the motivation for federated learning, the key challenges, and some of the existing approaches. In the second part of the talk, I will discuss the FjORD framework. FjORD is a framework for addressing the problem of client heterogeneity in federated learning. FjORD uses Ordered Dropout to gradually prune the model width without retraining, enabling clients with different capabilities to participate by tailoring the model width to the clients capabilities. In the third part of the talk, I will discuss the Maestro framework. Maestro is a framework for addressing the problem of model size in federated learning. Maestro uses a technique called trainable low-rank layers to compress the model without sacrificing accuracy. I will conclude the talk by discussing the future of federated learning.

**Keywords:** Federated learning · Client heterogeneity · Ordered dropout · Low-rank layers

# Decoding Hyperspectral Imaging: From Basic Principles to Medical Applications

Narine Sarvazyan ⓘD

George Washington University, USA
phynas@gwu.edu

**Abstract.** Over the past few decades, the application of hyperspectral imaging (HSI) has significantly expanded, finding widespread use in areas such as satellite imaging, agriculture, the food industry, and medicine. What sets HSI apart is its capacity to acquire complete spectral data from every pixel of an image. Each HSI dataset is a collection of individual images across numerous spectral bands and/or varied lighting conditions. A distinctive element of HSI is that, unlike grayscale or color images – where each pixel contains one or three to four color channels respectively – HSI captures hundreds of spectral bands for every pixel. Hence, the output from HSI is essentially a three- or four-dimensional dataset, with two dimensions representing spatial axes and the rest providing spectral values. Each dimension typically encompasses hundreds of individual values, so the massive amount of information collected by HSI hardware presents a great opportunity to apply ML and AI tools for data analysis. We overview the fundamental principles of HSI technology including examples from our own projects. The goal is to illustrate both the immense promise of HSI in revealing previously unseen surgical targets, as well as challenges posed by the high-dimensionality of HSI data. The key considerations for automatic processing and analysis of HSI data for medical use will also be touched upon. These include preserving the original spectral detail of an image to prevent the loss of information, ensuring processing efficiency for real-time application in a clinical environment, and managing the demands on processing power to ensure broader implementation.

**Keywords:** Hyperspectral imaging · High-dimensional data · Surgical targets · Clinical environment

# Safety of Deploying NLP Models: Uncertainty Quantification of Generative LLMs

Artem Shelmanov🆔

Technology Innovation Institute, United Arab Emirates
artem.shelmanov@mbzuai.ac.ae

**Abstract.** When deploying a machine learning (ML) model in practice, care should be taken to look beyond prediction performance metrics such as accuracy or F1. We should ensure also that it safe to use ML-based applications. This entails that applications should be evaluated along other critical dimensions such as reliability and fairness. The widespread deployment of large language models (LLMs) has made ML-based applications even more vulnerable to risks of causing various forms of harm to users. While streamline research effort has been devoted to the alignment via various forms of fine-tuning and to fact checking of the generated output, in this talk, we focus on uncertainty quantification as an effective approach to another important problem of LLMs. Models often hallucinate, i.e., fabricate facts without providing users an apparent means to discern the veracity of their statements. Uncertainty estimation (UE) methods could be used to detect unreliable generations unlocking the safer and more responsible use of LLMs in practice. UE methods for generative LLMs are a subject of bleeding-edge research, which is currently quite scarce and scattered. We systemize these efforts, discuss common caveats, and provide suggestions for the development of novel techniques in this area.

**Keywords:** Large language models · Fine-tuning · Model hallucinations · Uncertainty estimation

# Contents

**Computer Vision**

**Data Analysis and Machine Learning**

**Network Analysis**

**Theoretical Machine Learning and Optimization**

# Invited Paper

# Threatening Expression and Target Identification in Under-Resource Languages Using NLP Techniques

Muhammad Shahid Iqbal Malik[✉]

Department of Computer Science, National Research University Higher School of Economics, 11 Pokrovskiy Boulevard, Moscow 109028, Russian Federation
mumalik@hse.ru

**Abstract.** In recent decades, hate speech on social media platforms has been on the rise. It is highly desired to control this kind of material because it initiates unrest and harms to the society. Literature describes several forms of the hate speech and it is quite challenging to differentiate between these forms and to design an automated detection system, especially for under-resource languages. In this study, we propose a robust framework for threatening expressions and its target identification in Urdu (Nastaliq style) language. The proposed methodology presents each step in detail like data collection & annotation, cleaning & pre-processing step, and fine-tuning of Robustly Optimized Bidirectional Encoder Representations from Transformer (Urdu-RoBERTa) with grid search technique for hyper-parameters optimization. The study exploits the strength of a pre-trained Urdu-RoBERTa as a transfer learning technique with grid search fine-tuning. The proposed framework is compared with state-of-the art baseline and ten comparable models and it outperformed all for both tasks (threatening expression and target identification). Furthermore, the proposed framework obtained benchmark performance and improved the f1-score with substantial margin.

**Keywords:** Natural language processing · threatening expression · low-resource · target identification · RoBERTa · hyper-parameters

## 1 Introduction

A large number of users who use social media platforms is escalating at a very high speed according to the recent statistics, and various social media platforms are commonly used for sharing views and opinions. As social media is open for everyone by providing freedom of speech, it is being used for the spreading of positive views as well as for the propagation of hate speech and negative content. Therefore, it is highly desired to control this kind of material because it initiates unrest and harms to individuals and affect society by arousing violence, terrorist activities, aggression, etc. Two examples of hate speech expressions are presented in Fig. 1. In the left part (a), a tweet was posted in 2014 motivating killings of Jews for fun and in the right part (b), a leader is giving a threat to the United States.

© The Author(s), under exclusive license to Springer Nature Switzerland AG 2024
D. I. Ignatov et al. (Eds.): AIST 2023, LNCS 14486, pp. 3–17, 2024.
https://doi.org/10.1007/978-3-031-54534-4_1

(a)                                                                                    (b)

**Fig. 1.** Two examples of hate speech expressions on social media [1] (the sensitive information is patched by blue boxes by the editors) (Color figure online)

There is a consensus among the researchers on the definition of hate speech that "it is a language used to attack/target an individual or a group on the basis of ethnicity, race, gender, or religion etc." [2]. In literature, several state-of-the-art definitions of hate speech are presented and their sources are mainly from scientific studies and popular social media platforms [3–5]. Some examples are:

- "Hate speech attack others dependent on racism, ethnicity, public start, sexual bearing, sex, character, age, handicap, or genuine illness" [6]
- "Hate Speech is a purposeful attack on a specific social occasion of people motivated by the pieces of the group's character" [7]
- "Hate Speech is a toxic speech attack on a person's individuality and likely to result in violence when targeted against groups based on specific grounds like religion, race, place of birth, language, residence, caste, community, etc." [1]

These benchmark definitions describe the hate speech in several perspectives and researchers address hate speech according to their understanding, knowledge, and thinking prospective. In addition, literature described several forms of hate speech like toxicity [8], profanity [9], discrimination [10], Cyberbullying [11] etc., and these forms of hate speech are presented in the Fig. 2. The definitions of some of the forms and their differences compared to hate speech are presented below:

- Profanity vs Hate Speech: "Hostile or indecent words or expressions but hate speech can use profane words but not always"
- Toxicity vs Hate Speech: "Conveying content that is disrespectful, abusive, unpleasant, and harmful but Not all toxic comments contain hate speech"
- Discrimination vs Hate Speech: "Interaction via a distinction and afterward utilized as the premise of unreasonable treatment but Hate speech is a virulent form of discrimination"

Threatening text or violent threat is one of the form of hate speech. Few studies handled the task of threatening expression detection in high-resource languages like English [12–14] etc., but under-resource languages have very limited such approaches. Furthermore, target identification from threatening expression is almost ignored for

**Fig. 2.** Various forms of hate speech

under-resource languages. A large variety of languages are spoken worldwide. The landscape of world's popular spoken languages is presented in the Fig. 3. We can analyze the proportion of population speaking under-resource languages like Arabic, Urdu, Hindi, Chinese, Bengali and Russian in the Asian subcontinent. Urdu is the national language of Pakistan and it is being spoken by approximately 300 million people in Canada, USA, UK and India. Furthermore, it is an indigenous language of approximately 170 million speakers in the Asian region. Urdu language is identified as an under-resource language because several content processing toolkits and other resources are not available [15].

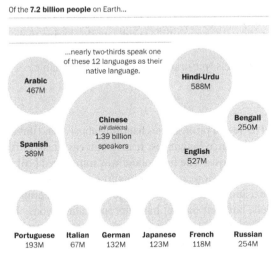

**Fig. 3.** The landscape of world languages in seven maps and charts (according to Washington post, 2022)

There are two writing styles in Urdu language, one is Nastaliq and other is Roman. In this study, we address Urdu language in Nastaliq writing style and introduced a model for threatening expression identification and then target identification from threatening expression. It is a kind of hierarchical classification task. Three objectives are planned to design an automated and effective threatening expression identification model.

- To develop an accurate system for the identification of threatening expression in Urdu (Nastaliq writing style) language.
- Regarding threatening expression; design a target identification framework to distinguish between individual and group.
- The proposed framework should be based on an automated feature-generation technique in contrast to hand-crafted features.

The remaining part of the paper is organized as follows; challenges with under-resource languages are presented in Sect. 2, followed by Sect. 3, in which related work is described. Section 4 presents the description of proposed model and experiments are discussed and analyzed in Sect. 5. Section 6 provides the conclusion of the research work and future prospective in this domain.

## 2 Challenges with Under-Resource Languages

While dealing with resource-poor languages, we encounter the following challenges.

- Lack of annotated datasets.
- Several essential resources and accurate text processing toolkits are not available, especially for Urdu, and Bengali languages etc.
- Some languages have multiple scripts, like Urdu has Nastaliq/Arabic style or Latin/Roman style.
- Social media users usually use multiple scripts while sharing their opinions. Issue of code-mixing.
- Pertinent language models (pre-trained) are scarcely available.

## 3 Related Work

In this section, a brief review of prior studies handling abusive and threatening expression identification in under-resource languages. It is not an easy task to filter out unwanted content from social media posts. The first model was introduced [16] in 2021 to address the task of threatening expression and target identification in Urdu. The word and char n-grams, and FastText are combined with some Machine Learning (ML) and Deep Learning (DL) models, but their dataset has improper annotations. Then another study [17] proposed a detection model for Urdu threat records but there was an issue of highly imbalanced dataset. Similarly, another detection model is developed by Das et al. [18] to handle abusive and threatening text detection in Urdu. They utilized transformer model with XGboost and obtained 54% f1-score for threat record detection. Then another detection model was proposed for abusive and threatening expression detection in Urdu [19] by exploiting word n-grams with word2vec model but their propose solution only achieved 49.31% f1-score. A recent study [20] proposed an ensemble model-based system for threat records detection and obtain 73.99% f1-score.

Some research studies focused on the design of identification models related to abusive expression detection in Urdu. Hussain et al. [21] introduced an offensive expression identification model in Urdu language using Facebook posts. They incorporated ensemble model with word2vec and obtained 88.27% accuracy on balanced dataset. Likewise, another framework [22] is introduced for Twitter platform by utilizing char and word n-grams, and FastText embeddings and obtained 82.68% f1-score. For Roman and Nastaliq Urdu, a significant framework is proposed for abusive expression detection. They utilized bag-of-words with ML and DL models and got 96% accuracy with Convolutional Neural Network (CNN) model. A recent study [23] used char and word n-grams, and BERT model for the detection of offensive content and their proposed system obtained 86% f1-score.

In literature, we found lack of annotated corpus for threatening content and target identification task. Furthermore, majority of approaches are based on hand-crafted features and lacking in automated feature engineering. The comparison between ML and DL models are rarely handled in the literature.

## 4  Proposed Model

In this section, proposed framework is described in detail. The pipeline for the design of proposed framework is presented in Fig. 4. Here, the complete flow of the process adopted in the design of proposed framework is defined.

### 4.1  Problem Definition

We address the task of threatening expression detection as a binary classification problem. Here, two-level of classification is performed; for the first level, text is categorized into threatening or not-threatening. For the second level, the threatening expressions are further categorized into individual or group category.

### 4.2  Dataset Collection and Annotation

Twitter platform was chosen to collect the tweets from Pakistani Twitter accounts. There is an annotation issues with the prior dataset [16] available for this task. Therefore, we designed a new dataset and crawled the tweets using Twitter API. The time period is chosen from August 2020 to August 2022 due to uncertainty and un-stability in the politics of Pakistan. At first, we designed a lexicon of seed words containing 250 keywords in total. This lexicon helped us to identify the relevant tweets from Pakistani Twitter accounts. Some example keywords are listed in the Table 1.

After that cleaning process is employed to the crawled data and the steps of cleaning process are described in the left part of Fig. 5. After cleaning the dataset, it was shared to three annotators for the annotation purposes. Two level of annotations are performed. In the first level, threatening vs not-threatening and in the second level, threatening tweets are further categorized into "individual or group" for target tagging. Some example guidelines are presented in the Table 2. The annotators are chosen by following some criteria described by the study [24].

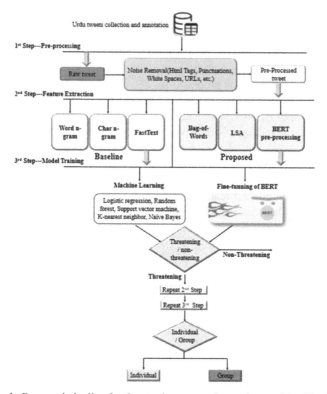

**Fig. 4.** Proposed pipeline for threatening expression and target identification

**Table 1.** Examples of Seed words

| Urdu | Translation | Urdu | Translation | Urdu | Translation |
|---|---|---|---|---|---|
| ٹکرے | Pieces | افسوسناک موت | Sad death | خون کے آنسو | Tears of blood |
| بلاک | Killed | ٹانگیں | Break the legs | عبرت کا نشان | Sign of lesson |
| لاشیں | Corpses | چھورا | Stabbing | خون پی جاؤں | Drink blood |
| دھمکی | Threat | چیر دوں | Tear | جان سے مار دو | Kill with soul |
| جنازہ | Funeral | الٹا لٹکا | Hang upside | سر تن سے جدا | Head separated from body |

## 4.3 Pre-processing

Following pre-processing steps are applied on the dataset:

- Removal of punctuations, mentions, hashtags, numbers, HTML tags, and URLs.
- Emoji/Emoticons are replaced with relevant text.
- Stop words removal (not for transformers)

An example of all pre-processing steps are presented in the Table 3 for the understanding of the readers.

**Table 2.** Examples (guidelines) for annotating the tweets

| S# | Urdu | Translation | Level 1 | Level 2 |
|----|------|-------------|---------|---------|
| 01 | لعنتی کتے عـــوام کا حـــال دیکھو لوڈشیڈنگ کو روکو خدا کی قسم ہم تم کو زندہ جلا دیں گے | Cursed dogs, look at the condition of the people, stop the load shedding, by God we will burn you alive | Threatening | Individual |
| 02 | دونوں کی آنکھیں نکال | I will take out the eyes of both | Threatening | Group |
| 03 | ابھی کچھ کچھ امید زندہ ہو رہی ہے | There is some hope now | Non-threatening | NA |
| 04 | تم کتے کی وہ دم ہو جو بارہ سال بند رکھی پھر بھی ٹیڑی یہ کام تمہارہ مرشد ہی کرتا تھا | You are the dog's tail that was tied for twelve years, yet your mentor used to do this | Non-threatening | NA |

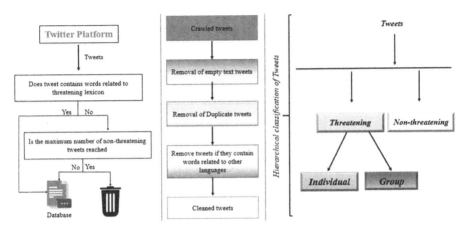

**Fig. 5.** Process of data collection, cleaning and classification

## 4.4 Urdu-RoBERTa Model

In this study, we exploited the strength of Urdu-RoBERTa model for the identification of threatening expression and target from the tweets. The RoBERTa model has already proved its effectiveness for several NLP tasks [25, 26]. There are several benefits of RoBERTa including fast development, fewer data requirements, and contextual feature generation, etc. It is a pre-trained transformer model proposed by the researchers [27] and mainly based on the BERT transformer model. Here, we used the Urdu-RoBERTa model (https://huggingface.co/urduhack/roberta-urdu-small) by fine-tuning some important hyperparameters for the threatening expression identification task. The word count and cloud representation of the annotated dataset is presented in Fig. 6.

**Fine-Tuning Process:** Some steps are usually required in the fine-tuning of any transformer model. The input data must be transformed into a pre-defined format to make it compatible for the RoBERTa architecture. After pre-processing, we need to apply data transformation step.

**Table 3.** Demonstration of pre-processing steps

| Urdu Text Preprocessing |
| --- |

| **1. Punctuations Removal** | |
| --- | --- |
| Before | After |
| ہم آنکھیں نکال لیں گے نواز شریف کی.....پھر نابینا باپ کو لے کر بیٹھی رہو گی جھوٹی اور مکار عورت | ہم آنکھیں نکال لیں گے نواز شریف کی پھر نابینا باپ کو لے کر بیٹھی رہو گی جھوٹی اور مکار عورت<br>We will take out the eyes of Nawaz Sharif then you will sit with the blind father a lying and deceitful woman |

| **2. Stop word Removal** | |
| --- | --- |
| دشمن کے دانت کھٹے اور جسم مردہ ہونگے انشاءاللہ<br>The teeth of the enemy will be sour and the body will be dead, God willing | دشمن دانت کھٹے جسم مردہ انشاءاللہ<br>enemy's teeth sour, body dead, God willing |

| **3. Replace Emoji with Corresponding Text** | |
| --- | --- |
| 😡 تم خبیث لوگ بچ نہیں سکو گے چاہے کچھ بھی کر لو<br>You evil people will not escape no matter what you do | مشتعل  تم خبیث لوگ بچ نہیں سکو گے چاہے کچھ بھی کر لو<br>You evil people will not be saved no matter what you do, enraged |

| **4. Hashtag, Numbers and Link Removal** | |
| --- | --- |
| چن چن کے ماریں گے تمھارے گھروں میں گھس کر ماریں گے 128514#&http://t.co/cOU2WQ5L4q; | چن چن کے ماریں گے تمھارے گھروں میں گھس کر ماریں گے |

**Fig. 6.** Word count and cloud representation of the annotated dataset

**Data Transformation:** It is a mandatory step for fine-tuning process. First of all, uncased RoBERTa tokenizer is applied to each tweet to break it into word tokens. After that, each sentence is appended by the [CLS] token at the start and the [SEP] token at the end. Then each token of the tweet is mapped to an index. As maximum length of a tweet is 280 characters, therefore we chose 64 and 128 sequence lengths for the experimental setup. The process of addition of attention masks is performed to fed input to RoBERTa base classifier.

**Classification:** We applied 80–20 data split on the dataset. The 20% is used for testing and the 80% part is further divided into 90–10, in which 90% is actually used for training and 10% is used for validation. A single layer is appended on the top of Urdu-RoBERTa base model for the binary classification task. For fine-tuning, we chose Grid Search technique to find the optimum values of hyper-parameters. The list of hyper-parameters and their values are presented in the Table 4. The optimizer function is utilized for updating all the parameters of each epoch.

**Catastrophic Forgetting and Overfitting:** As RoBERTa is pre-trained in a generic prospective on a big corpus and it needs appropriate fine-tuning with important hyper-parameters for a new learning. The new learning could encounter the issues of Catastrophic Forgetting, and Overfitting. Every transformer is prone to catastrophic forgetting. We dealt this issue by exploring a range of learning rates and concluded that higher learning rates encounter convergence failures and best results are obtained with learning rate of 2e−5. To deal with the issue of overfitting, we monitor loss value on the validation dataset and found that 5 epochs are appropriate to save the fine-tuning process from over and under fitting.

**Table 4.** List of hyper-parameters and their ranges

| Hyperparameters | Grid Search |
|---|---|
| Sequence length | 64, 128 |
| Batch size | 8, 16, 32 |
| Learning rate | 1e−4, 1e−5, 2e−5, 3e−4, 3e−5, 5e−5 |
| Weight decay | 0.01–0.1 |
| Warmup ratio | 0.06–0.1 |
| Hidden dropout | 0.05, 0.1 |
| Attention dropout | 0.05, 0.1 |
| Epochs | 1–10 |

### 4.5 Experimental Setup

The detail of benchmark and the comparable models are presented below. We chose following ML models because they demonstrated state-of-the-art performance in several NLP tasks [28, 29].

**Benchmark:** Amjad et al. study [16].

#### Comparable Models

- Latent Semantic Analysis (100) + Logistic Regression
- Latent Semantic Analysis (100) + Random Forest
- Latent Semantic Analysis (100) + Support Vector Machine

- Latent Semantic Analysis (100) + Naive Bayes
- Latent Semantic Analysis (100) + K-nearest neighbor
- Bag-of-words + Logistic Regression
- Bag-of-words + Random Forest
- Bag-of-words + Support Vector Machine
- Bag-of-words + Naive Bayes
- Bag-of-words + K-nearest neighbor

In total, two feature engineering techniques (latent semantic analysis and bag-of-words) and five ML models are utilized. In addition, the classifiers' performance is evaluated using standard accuracy, precision, recall and macro f1-score measures.

## 5  Results and Analysis

In this section, two types of experiments are performed to fine-tune the Urdu-RoBERTa model for threatening expression and target identification tasks. The proposed framework is compared with a baseline and ten comparable models.

### 5.1  Fine-Tuning Urdu-RoBERTa and Comparison with Baselines (Threatening Identification)

Six RoBERTa classifiers are trained, validated and tested using fine-tuning process employed to design an effective identification model for threatening expression. Two sequence lengths (64 and 128) and three batch sizes are tried with other hyper-parameters (listed in Table 4) for fine-tuning. The training and validation loss obtained by applying sequence lengths of 64, and 128 with batch sizes of 8 and 16 are presented in Fig. 7. It is clearly visible that training loss decreases continuously from epoch 1 to 5, indicates continuous learning of Urdu-RoBERTa in the training phase, but Fig. 7 shows that validation loss started decreasing up to the 3rd epoch and then started increasing up to 5th epoch. This signals that further training and validation may leads to overfitting.

Next, the performance of fine-tuned RoBERTa is compared with a baseline [16] and comparable models. From the applied ML models, logistic regression presented best performance as compared to random forest, support vector machine, naïve bayes and k-nearest neighbor. Therefore, we added only best results from the comparable models as shown in Table 5. The performance is presented in accuracy, precision, recall, and macro f1-score. Among the baseline, char 5-g presented highest metric values. i.e. f1-score is 85.83% and accuracy is 86.25%. In contrast, the proposed fine-tuned RoBERTa (with sequence length of 64 and batch size of 8) outperformed the baseline and the comparable models, by providing 87.8% f1-score and 87.5% accuracy. This leads to 2% improvement in f1-score and 1.25% in accuracy. Thus fine-tuned Urdu-RoBERTa proved its effectiveness for the identification of threatening expression on the Twitter network. The performance of FastText embedding in baseline is worst here, although it demonstrated state-of-the-art performance for other NLP tasks.

(a)                                          (b)

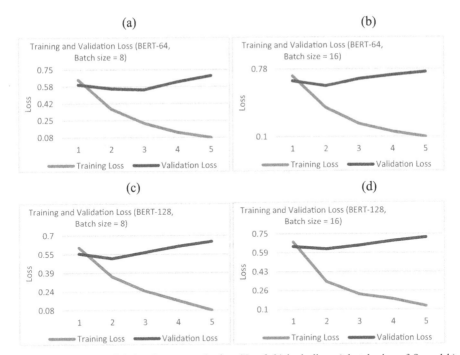

**Fig. 7.** Training & validation losses employing SL of 64 including a) batch-size of 8, and b) batch-size of 16, and SL of 128 including c) batch-size of 8, and d) batch-size of 16

### 5.2  Fine-Tuning Urdu-RoBERTa and Comparison with Baselines (Target Identification)

Here, we performed fine-tuning of Urdu-RoBERTa for the target identification task and compared it with baseline [16] and comparable models. Again six classifiers are trained, validated and tested using grid search fine-tuning process for five epochs and results are presented in Table 6. We employed the same procedure and range of parameters as presented in Sect. 5.1. The sequence lengths of 64 and 32 and batch size of 8, 16 and 32 are explored and results are demonstrated. The performance of baseline and comparable models are also added in the Table 6.

For target identification task, all evaluation metrics indicate that fine-tuned Urdu-RoBERTa with sequence length of 128 and batch size of 8 outperformed the baseline and comparable models by achieving the benchmark accuracy of 82.5% and 83.20% f1-score. The Urdu-RoBERTa with sequence length of 64 and batch size of 8 did not demonstrated highest performance for target identification task but presented comparable performance to bag-of-words + logistic regression model. Among the baseline, combined word (1–2-3) grams demonstrated better performance as compared to other features for target identification task. Again FastText embeddings did not perform well. As a whole, proposed framework improves accuracy by 0.58% and f1-score by 0.41% as compared to state-of-the-art baseline.

**Table 5.** Performance comparison of fined-tuned RoBERTa with baseline and comparable models (Threatening vs non-Threatening)

| Type | Features | Accuracy | Precision | Recall | F1-score |
|---|---|---|---|---|---|
| Baseline [16] | Word uni-gram | 83.83 | 81.51 | 80.17 | 80.83 |
| | Word bi-gram | 82.08 | 81.97 | 82.64 | 82.30 |
| | Word tri-gram | 75.42 | 80.39 | 67.77 | 73.54 |
| | Word combined (1-2-3) | 82.92 | 80.30 | 87.60 | 83.79 |
| | Char 1-g | 68.75 | 69.16 | 68.59 | 68.87 |
| | Char 2-g | 79.58 | 82.14 | 76.03 | 78.97 |
| | Char-3-g | 80.00 | 84.11 | 74.38 | 78.94 |
| | Char 4-g | 84.16 | 87.38 | 80.16 | 83.62 |
| | Char 5-g | **86.25** | **89.28** | **82.64** | **85.83** |
| | Char 6-g | 85.83 | 87.82 | 83.47 | 85.59 |
| | Char 7-g | 82.50 | 83.19 | 81.81 | 82.50 |
| | Char 8-g | 82.08 | 82.50 | 81.81 | 82.15 |
| | Char combined (1–8) | 80.00 | 74.82 | 90.90 | 82.08 |
| | FastText | 59.17 | 63.64 | 54.69 | 58.82 |
| Proposed | Bag of words | 83.75 | 81.54 | 87.60 | 84.46 |
| | Latent Semantic Analysis | 81.25 | 79.23 | 85.12 | 82.07 |
| | BERT-64 (8) | **87.5** | **86.4** | **89.26** | **87.8** |
| | BERT-64 (16) | 83.75 | 85.34 | 81.82 | 83.54 |
| | BERT-64 (32) | 84.58 | 87.5 | 80.99 | 84.12 |
| | BERT-128 (8) | **87.5** | **89.57** | **85.12** | **87.29** |
| | BERT-128 (16) | 84.17 | 86.73 | 80.99 | 83.76 |
| | BERT-128 (32) | 82.92 | 80.3 | 87.6 | 83.79 |

In the end, we conclude our experiments by summarizing three advantages of the proposed framework. First, it improves the identification performance for both tasks (threatening and target identification) in comparison with benchmark with substantial margin. Second, the proposed pipeline is based on automated feature generation method in contrast to hand-crafted features (baseline). Third, the state-of-the-art Urdu-RoBERTa language model is capable to capture the threatening language context realistically in the Urdu and experiments proved its effectiveness.

**Table 6.** Performance comparison of fined-tuned RoBERTa with baseline and comparable models (Target Identification)

| Type | Feature Set | Accuracy | Precision | Recall | F1-score |
|------|-------------|----------|-----------|--------|----------|
| Baseline [16] | Word uni-gram | 80.33 | 81.51 | 80.17 | 80.83 |
| | Word bi-gram | 82.08 | 81.97 | 82.64 | 82.30 |
| | Word tri-gram | 75.42 | 80.39 | 67.77 | 73.54 |
| | Word combined (1-2-3) | 81.92 | 80.30 | 87.60 | 82.79 |
| | Char 1-g | 60.83 | 68.08 | 50.00 | 57.65 |
| | Char 2-g | 69.16 | 72.13 | 68.75 | 70.40 |
| | Char 3-g | 70.83 | 72.30 | 73.43 | 72.86 |
| | Char 4-g | 70.83 | 73.77 | 70.31 | 72.00 |
| | Char 5-g | 69.16 | 71.42 | 70.31 | 70.86 |
| | Char 6-g | 67.50 | 71.18 | 65.62 | 68.29 |
| | Char 7-g | 60.83 | 66.66 | 53.12 | 59.13 |
| | Char 8-g | 61.66 | 68.75 | 51.56 | 58.92 |
| | Char combined (1–8) | 72.50 | 79.24 | 65.62 | 71.79 |
| | FastText | 54.17 | 51.67 | 54.39 | 52.99 |
| Proposed | Bag-of-words | 80.33 | 81.51 | 80.17 | 80.83 |
| | Latent Semantic Analysis | 68.33 | 68.57 | 75.00 | 71.64 |
| | BERT-64 (8) | 79.17 | 80 | 81.25 | 80.62 |
| | BERT-64 (16) | 75 | 81.48 | 68.75 | 74.58 |
| | BERT-64 (32) | 75 | 79.31 | 71.88 | 75.41 |
| | BERT-128 (8) | **82.5** | **85.25** | **81.25** | **83.2** |
| | BERT-128 (16) | 79.17 | 81.97 | 78.13 | 80 |
| | BERT-128 (32) | 76.67 | 83.33 | 70.31 | 76.27 |

# 6  Conclusion

This study addressed the task of threatening expression and target identification for an under-resource language. First, hate speech is described in detail with several definitions presented by the literature. Then, various forms of hate speech are summarized and threatening expression (type of hate speech) is defined. After that, challenges dealing with under-resource languages are discussed. Urdu language with Nastaliq style is chosen to demonstrate the steps of the proposed methodology for threatening expression and target identification task. The process of dataset collection and then annotation are described with examples. After that, data cleaning and pre-processing steps are demonstrated on real tweets. The Urdu-RoBERTa model is used with grid search fine-tuning to capture the actual context of threatening expression and their target identification. The issues of catastrophic forgetting and overfitting are highlighted and their solutions are discussed. After that, implementation detail of fine-tuning process is described and results are compared with a benchmark and ten comparable models. The proposed system outperformed the benchmark and comparable models for both tasks (threatening expression and target identification). Thus, the proposed system and its findings may assist law and enforcement organizations to detect and filter-out this kind of material from social media platforms.

Regarding future prospects, researchers will encounter following challenges: First, there will be an issue of interpretability because each low-resource language has different way of creating context to describe an opinion. Second, appropriate categorization of various types of hate speech is challenging in low-resource languages as their definitions overlap. Third, designing an efficient code-mixed content identification framework for low-resource language is not an easy task.

**Acknowledgments.** This article is an output of a research project implemented as part of the Basic Research Program at the National Research University Higher School of Economics (HSE University). Moreover, this research was supported in part by computational resources of HPC facilities at HSE University.

# References

1. Chhabra, A., Vishwakarma, D.K.: A literature survey on multimodal and multilingual automatic hate speech identification. Multimed. Syst. 1–28 (2023)
2. Schmidt, A., Wiegand, M.: A survey on hate speech detection using natural language processing. In: Proceedings of the Fifth International Workshop on Natural Language Processing for Social Media (2017)
3. Delgado, R., Stefancic, J.: Images of the outsider in American law and culture: can free expression remedy systemic social ills. Cornell L. Rev. **77**, 1258 (1991)
4. Fortuna, P., Nunes, S.: A survey on automatic detection of hate speech in text. ACM Comput. Surv. (CSUR) **51**(4), 1–30 (2018)
5. Youtube. YouTube hate policy. https://support.google.com/youtube/answer/2801939?hl=en. 2019
6. Twitter. Twitter_Hate Definition. https://support.twitter.com/articles/.2017
7. De Gibert, O., et al.: Hate speech dataset from a white supremacy forum. arXiv preprint arXiv: 1809.04444 (2018)
8. Andročec, D.: Machine learning methods for toxic comment classification: a systematic review. Acta Universitatis Sapientiae, Informatica **12**(2), 205–216 (2020)
9. Malmasi, S., Zampieri, M.: Challenges in discriminating profanity from hate speech. J. Exp. Theor. Artif. Intell. **30**(2), 187–202 (2018)
10. Thompson, N.: Social Problems and Social Justice. Bloomsbury Publishing (2017)
11. Chen, Y., et al.: Detecting offensive language in social media to protect adolescent online safety. In: 2012 International Conference on Privacy, Security, Risk and Trust and 2012 International Conference on Social Computing. IEEE (2012)
12. Ashraf, N., et al.: Individual vs. group violent threats classification in online discussions. In: Companion Proceedings of the Web Conference 2020 (2020)
13. Jiang, L., et al.: Intelligent control of building fire protection system using digital twins and semantic web technologies. Autom. Constr. **147**, 104728 (2023)
14. Mazari, A.C., Boudoukhani, N., Djeffal, A.: BERT-based ensemble learning for multi-aspect hate speech detection. Cluster Comput. 1–15 (2023)

15. Nawaz, A., et al.: Extractive text summarization models for Urdu language. Inf. Process. Manag. **57**(6), 102383 (2020)
16. Amjad, M., et al.: Threatening language detection and target identification in Urdu tweets. IEEE Access **9**, 128302–128313 (2021)
17. Kalraa, S., Agrawala, M., Sharmaa, Y.: Detection of Threat Records by Analyzing the Tweets in Urdu Language Exploring Deep Learning Transformer-Based Models (2021)
18. Das, M., Banerjee, S., Saha, P.: Abusive and threatening language detection in Urdu using boosting based and BERT based models: a comparative approach. arXiv preprint arXiv:2111.14830 (2021)
19. Humayoun, M.: Abusive and threatening language detection in Urdu using supervised machine learning and feature combinations. arXiv preprint arXiv:2204.03062 (2022)
20. Mehmood, A., et al.: Threatening URDU language detection from tweets using machine learning. Appl. Sci. **12**(20), 10342 (2022)
21. Hussain, S., Malik, M.S.I., Masood, N.: Identification of offensive language in Urdu using semantic and embedding models. PeerJ Computer Science **8**, e1169 (2022)
22. Amjad, M., et al.: Automatic abusive language detection in Urdu tweets. Acta Polytechnica Hungarica 1785–8860 (2021)
23. Saeed, R., et al.: Detection of offensive language and its severity for low resource language. ACM Trans. Asian Low-Resour. Lang. Inf. Process. **22**, 1–27 (2023)
24. Malik, M.S.I., Cheema, U., Ignatov, D.I.: Contextual embeddings based on fine-tuned Urdu-BERT for Urdu threatening content and target identification. J. King Saud Univ.-Comput. Inf. Sci. 101606 (2023)
25. Malik, M.S.I., et al.: Multilingual hope speech detection: a robust framework using transfer learning of fine-tuning RoBERTa model. J. King Saud Univ.-Comput. Inf. Sci. **35**(8), 101736 (2023)
26. Rehan, M., Malik, M.S.I., Jamjoom, M.M.: Fine-tuning transformer models using transfer learning for multilingual threatening text identification. IEEE Access (2023)
27. Liu, Y., et al.: RoBERTa: a robustly optimized BERT pretraining approach. arXiv preprint arXiv:1907.11692 (2019)
28. Younas, M.Z., Malik, M.S.I., Ignatov, D.I.: Automated defect identification for cell phones using language context, linguistic and smoke-word models. Expert Syst. Appl. **227**, 120236 (2023)
29. Malik, M.S.I., Imran, T., Mamdouh, J.M.: How to detect propaganda from social media? Exploitation of semantic and fine-tuned language models. PeerJ Comput. Sci. **9**, e1248 (2023)

# Natural Language Processing

# Benchmarking Multilabel Topic Classification in the Kyrgyz Language

Anton Alekseev[1,2,3,4]($\boxtimes$)(iD), Sergey Nikolenko[1,2,3](iD), and Gulnara Kabaeva[4](iD)

[1] Steklov Mathematical Institute at St. Petersburg, St. Petersburg, Russia
anton.m.alexeyev@gmail.com
[2] St. Petersburg State University, St. Petersburg, Russia
[3] Kazan (Volga Region) Federal University, Kazan, Russia
[4] Kyrgyz State Technical University n. a. I. Razzakov, Bishkek, Kyrgyzstan

**Abstract.** Kyrgyz is a very underrepresented language in terms of modern natural language processing resources. In this work, we present a new public benchmark for topic classification in Kyrgyz, introducing a dataset based on collected and annotated data from the news site *24.KG* and presenting several baseline models for news classification in the multilabel setting. We train and evaluate both classical statistical and neural models, reporting the scores, discussing the results, and proposing directions for future work.

**Keywords:** Topic classification · Kyrgyz language · Multi-label classification · Low-resource languages

## 1 Introduction

Kyrgyz is an agglutinative Turkic language spoken in several countries, notably China and Tajikistan in addition to Kyrgyzstan; it is by no means an endangered language, and several millions of people call it their mother tongue [42]. However, despite a large amount of linguistic work, including computational linguistics (see Sect. 2), it is certainly a *low-resource* language, with a very modest number of tools and datasets available in the open for Kyrgyz language processing[1]. A recent publication [42], following the taxonomy proposed in [25], labels Kyrgyz with the "Scraping By" status, defined as follows: "With some amount of unlabeled data, there is a possibility that they could be in a better position in the 'race' in a matter of years. However, this task will take a solid, organized movement that increases awareness about these languages, and also sparks a strong effort to collect labelled datasets for them, seeing as they have almost none". Therefore, we believe that a meaningful effort to construct open manually annotated text collections or other reliable resources for Kyrgyz language

---

[1] For a list of tools, corpora, and other language resources for Turkic languages including Kyrgyz, see e.g. https://github.com/alexeyev/awesome-kyrgyz-nlp and http://ddi.itu.edu.tr/en/toolsandresources.

D. I. Ignatov et al. (Eds.): AIST 2023, LNCS 14486, pp. 21–35, 2024.
https://doi.org/10.1007/978-3-031-54534-4_2

processing is in great demand; modern NLP, while shifting towards universal models, is still hard to imagine without at least evaluation data.

Text topic classification is a core task in natural language processing and information retrieval [41]; it is one of the most popular in practice, with applications in advertising [83], news aggregation, and many other industries. Very often, topic categorization is posed as a multilabel classification problem since the same text can touch upon multiple topics [40,55,81].

In this work, we present virtually the first labeled dataset for text classification in the Kyrgyz language based on the *24.kg* news portal. Moreover, we propose several baseline models and evaluate their results; in this evaluation, we see that multilingual models do help to process Kyrgyz, using even primitive stemming and passing from word $n$-grams to symbol $n$-grams quite expectedly help, and deep learning models that we have considered perform better than the best linear models with virtually no hyperparameter search. Thus, the contributions of our work are threefold: (1) a novel manually labeled dataset for texts in the Kyrgyz language, (2) several approaches to multi-label Kyrgyz text classification, (3) proof of concept for the feasibility of multilingual LLMs for Kyrgyz language processing in supervised tasks. The paper is organized as follows: Sect. 2 discusses related work, Sect. 3 introduces our dataset, Sect. 4 shows baseline models and experimental setup used in our experiments, Sect. 5 discusses experimental results, and Sect. 6 concludes the paper.

## 2   Related Work

*Topic Classification.* Text topic classification is one of the oldest and best known tasks in information retrieval and natural language processing [41,84]. It is straightforwardly defined as a supervised learning (classification) task, often expanding into multilabel classification since longer texts are hard to fit into a single topic [40,55,81], a problem that is still attracting attention in the latest deep learning context [38,74]. Many approaches have been developed for datasets of different nature: (i) news article datasets such as BBC News [20], Reuter [36], 20 Newsgroups [34], or WMT News Crawl [35]; (ii) scientific texts such as arXiv abstracts [35], patents [68], or clinical texts [51,65,78]; (iii) social media posts where topics are usually represented by [hash]tags [15], and more. We note especially prior efforts related to text classification for low-resource languages [2,13,16,18,21].

*Kyrgyz Language Processing.* There already exists a large corpus of linguistic research papers dedicated to various aspects of the Kyrgyz language: (1) grammar, syntax and morphology modeling [6,9,22–24,26,28–30,32,59,60,69,76,77,80], including a recent release of 780 dependency trees [7] as part of the Universal Dependencies initiative [50], (2) text-related statistics and construction of corpora/dictionaries [4,5,28–30,45,61], (3) computer-aided language learning and other educational systems [11,27], (4) machine translation [31,54,70,76], (5) lexicons and thesauri [9,64], (6) computational linguistics in general [47], and more. Kyrgyz also appears in multiple works as a part of multilingual research studies, e.g. on multiway machine translation [43,44] and even text categorization [37], although the latter uses a different (Arabic) script than our work.

News articles represent a traditional and widely used text domain, traditionally a valuable source of data both for information retrieval and natural language processing. News-based datasets have found many applications including such non-traditional ones as relation classification [58]. In this work, we concentrate on the news domain primarily because to the best of our knowledge, for the Kyrgyz language only fiction, news, and Wikipedia articles are readily available online. Collecting social media content, which may also be useful for numerous NLP tasks [3,10,33,46,48,49,71–73], is also possible but requires significant extra effort for preprocessing; in particular, preprocessing steps for social media sources would have to include language detection since oftentimes people writing in Kyrgyz also publish posts in Russian and other languages.

Several research efforts on Kyrgyz open corpora and dictionaries are currently in progress (see, e.g., [28–30]), but as of 2023, there are still very few manually annotated datasets useful for Kyrgyz language processing. We hope to start filling this gap with this work.

## 3 Dataset

### 3.1 Annotation

With permission of *24.kg*[2] editors, we collected 23 283 news articles in Kyrgyz, dated from May 2017 to October 2022. The portal does not provide any topical tags for articles in Kyrgyz, hence we had to either match collected articles with possibly available articles in Russian, which are tagged, or annotate them with our own topical categories. The original rubrics used at *24.kg* include: (1) Власть (government, politics and law), (2) Общество (society), (3) Экономика (economics), (4) Происшествия (accidents, current events), (5) Агент 024 (current events), (6) Спорт (sports), (7) Техноблог (tech), (8) Спецпроекты (special projects), (9) Кыргызча (articles in Kyrgyz), (10) English (articles in English), (11) Бизнес (business). Some of the rubrics are clearly not topical ("English", "Кыргызча"), some are multi-topic ("Спецпроекты", "Агент 024"), and some other topics also turned out to cover very diverse information. Therefore, we had to introduce our own topical labels.

While some general-purpose taxonomies for content classification do exist and are used, e.g., in advertising, including *dmoz*[3] and IAB[4] taxonomies, the label sets there are too broad for the purpose of news classification. Our preliminary experiments with translated news titles zero-shot classification with IAB Tier 1 tags (in the *label-fully-unseen* setting [79]) yielded poor prediction quality. However, we still consider this direction very promising from the practical point of view and leave it for future research.

---

[2] https://24.kg/.

[3] https://www.dmoz-odp.org/; previously https://www.dmoz.org/.

[4] https://iabtechlab.com/standards/content-taxonomy/.

**Table 1.** Sample cluster (#16).

| Title | Proposed labels |
| --- | --- |
| The presidential candidate who violated traffic rules paid... | law/crime, politics |
| Cars of drivers who do not pay fines on time will be... | law/crime |
| 44 percent of the 108,000 fines imposed for violating traffic... | law/crime |
| Party candidate was fined 7,500 soms for holding a concert... | law/crime, politics |
| Fines for garbage thrown from cars have been increased... | law/crime, ecology |

To motivate the introduction of a custom set of labels, we have automatically translated[5] titles of Kyrgyz articles into English, randomly sampled 500 out of 23 284 of them (a subset small enough to annotate in reasonable time yet hopefully large enough to derive meaningful conclusions regarding the topics), and obtained their embeddings via the SentenceBERT model [57] (`all-mpnet-base-v2`, the best-performing[6] fine-tuned MPNet model [66]). Then, we have grouped the resulting embeddings using agglomerative clustering (Euclidean distance, Ward linkage [75], other hyperparameters left at default as provided by *scikit-learn* version 1.0 [53]) into 100 clusters. Note that the exact clustering procedure and chosen parameters are not of significant importance here; the main idea is to group texts into hopefully small clusters of very similar titles to speed up annotation and, most importantly, to be able to easily invent topic names that are neither too general nor too specific. Note also that we had to translate the titles and apply the model trained on English language data not for the annotation itself but only because there are no good sentence embeddings models for Kyrgyz with reported quality. Where it was impossible to deduce the topics of the article from the title, we made decisions based on the original Kyrgyz news texts. A sample cluster is presented in Table 1.

The exploratory annotation task was defined as follows: for each cluster, invent a topic name that best describes most if not all titles and use it as the class label. Then correct the label for titles in the cluster that do not fit the invented topic. If multiple topics apply to some of the titles, add more tags where necessary. After that, we make another pass over all 500 titles since some of the labels were not "available" at the beginning, i.e., a certain "general" label might be added to the label set after some of the texts that would be appropriately labeled with it had already been annotated. As a result, we obtained a refined list of 20 labels shown in Table 2. To validate the label set by comparing label distributions with each other, we have annotated 500 more English translations of the titles using the same set of labels. We found that the difference in label count distributions in the two sets of 500 were relatively small, which showed that our annotation was consistent. Finally, we have annotated 500 more texts following the same procedure. We do not disclose the exact distribution of the

---

[5] Google Translate: https://translate.google.com/?sl=ky&tl=en&op=docs.
[6] As of 24.08.2023: https://www.sbert.net/docs/pretrained_models.html.

**Table 2.** Topical tags for *24.kg*, total number of tags in the first two annotation batches and their descriptions. The counts do not sum up to 500 since this is a multilabel task.

| Class label | 500-1 | 500-2 | Description |
|---|---|---|---|
| politics | 127 | 174 | Mentions of politicians and political decisions |
| law/crime | 126 | 128 | Judiciary and penitentiary sys., legislature, trials, crime |
| foreign affairs | 84 | 91 | Any non-Kyrgyzstan-related news |
| health | 68 | 63 | Health and medicine-related news (mostly COVID19) |
| local | 43 | 43 | Traffic rerouting, events scheduling |
| accidents | 41 | 31 | Disasters, fires, road accidents, etc. |
| econ/finance | 37 | 50 | Money, import-export and labour-related news |
| society | 36 | 49 | Local initiatives, protests, other citizen-related news |
| culture | 32 | 29 | Cultural events and initiatives, celebrity news |
| citizens abroad | 17 | 11 | Migration questions and Kyrgyz people abroad |
| sports | 16 | 15 | Awards, announcements, famous sportspeople mentions |
| natural hazards | 13 | 5 | Inconveniences and threats due to natural reasons |
| development | 12 | 25 | Realty, land use and infrastructural development |
| religion | 11 | 13 | Religion-related news |
| science/tech | 9 | 7 | Everything related to science and technology |
| border | 8 | 9 | Kyrgyzstan's borders-related conflicts and resolutions |
| education | 7 | 22 | News on educational procedures/events/institutions |
| weather | 6 | 4 | Weather forecasts and reports |
| ecology | 4 | 4 | Ecological initiatives, laws and reports |
| natural resources | 2 | 0 | Issues related to natural resources |

**Table 3.** Dataset statistics; sentences counted via the `sent_tokenize` method from NLTK [8]. Per-text values show mean value and standard deviation.

| | # texts | # sent. | Sent/text | # tokens | Unique tokens | Tok/text | Unique stems |
|---|---|---|---|---|---|---|---|
| Train | 1 000 | 7 319 | $7.32 \pm 5.36$ | 107 556 | 18 958 | $107.56 \pm 74.02$ | 9 872 |
| Test | 500 | 4 025 | $8.05 \pm 8.78$ | 57 414 | 12 885 | $114.83 \pm 101.15$ | 6 924 |

labels in the final batch for the sake of fairness in possible future competitions: knowing the exact number of texts with a certain label might be used as a test data leak to improve results.

## 3.2 Data Description

The dataset consists of 1 500 texts, annotated in three sessions as described above. Since the dataset is relatively small, we split it in only two parts: the first two batches of 500 (i.e., *training* set has 1 000 texts) and the last batch (i.e. the *test* set has 500 texts).

For further application of models based on the bag-of-ngrams approach, the texts had to be split into tokens and, possibly, stemmed/lemmatized. For tokenization, we used the splitting mechanism provided by the Apertium Project morphological analyzer [19,77]; to the best of our knowledge, this is the only open source engine for Kyrgyz morphology. Similarly, for *word normalization* we used the Apertium-Kir [77] FST's token segmentation; since prefixes are uncommon in Kyrgyz, the first segment was used as the stem. Overall dataset statistics are presented in Table 3.

# 4    Models and Experimental Setup

The resulting dataset is far too small to be used for training, especially for classical models that heavily depend on frequency estimates of various ratios of tokens and n-grams, e.g., models based on the bag-of-words assumption. However, we can use the dataset in cross-validation to make comparisons across models that perform transfer learning. Still, we include classical approaches into the benchmark as well, since several works have demonstrated that word/character n-gram baselines are sometimes surprisingly competitive, e.g., in entity linking [1,62], so they should not be ignored even for a relatively small dataset.

We have used grid search to find the best parameters. Since the training set is small and imbalanced in terms of labels, we used 2-fold validation for hyperparameter search with a stratified split into two subsets preserving the label distribution[7]. Below we show the considered values and ranges of hyperparameters in addition to the models themselves.

## 4.1    Approaches Based on the Bag-of-ngrams Assumption

To provide a classic baseline, we have considered several sparse text representations (essentially bags-of-ngrams) and several corresponding models.

*Text Preprocessing.* We tested several text representations. First, we tokenized text (Sect. 3.2) into unigrams, 1-2-grams, 1-2-3-grams, and 2-3-grams. For frequency cutoffs we retained tokens with maximum document frequency (maxdf) of 40%, 60%, 80%, and 100% and minimum occurrence (mincount) in 2, 5, or 10 documents. We also set the maximum number of features (maxfeat) equal to 2 000 or 10 000. In another set of experiments, we used character n-grams: 2-3-grams, 3-4-grams, and 5-6-grams; maxdf for a character n-gram was set to 40%, 70%, or 100%, mincount was 4, 10, and 15, and maxfeat was 2 000 or 10 000. Then, having stemmed the texts as in Sect. 3.2, we have run experiments with the same "vectorization" parameters.

---

[7] Specifically, we used the *IterativeStratification* algorithm from the *scikit-multilearn* library [67].

*Independent Classifiers ("Independent").* In this set of baselines, we train a separate model for every label, using models that are known to perform well for sparse features: logistic regression (both LBFGS and SGD optimization methods), a linear model with hinge loss (linear SVM), and a linear model with Huber loss (usually preferred for regression tasks). In the search for the best model, we treated log-loss, hinge loss, and Huber loss as hyperparameters. Experiments with LBFGS were carried out separately, which is reflected in the results table in Sect. 5. For logistic regression with the LBFGS optimizer (that includes $L_2$-regularization), we have tested the performance with regularization strength $C = \frac{1}{\lambda} \in \{0.7, 0.9, 1.0\}$ and limited the number of iterations to 1 000 or 10 000 steps. For other models, apart from the loss function, we have tested the averaging mechanism (enabling/disabling it), $L_1$, and $L_2$-regularizers, and limited the number of iterations to either 20 000 or 100 000 steps.

*Binary Classifiers Chain ("Chain").* In this approach, the base models (which are the same as in the previous paragraph) make predictions in a sequence; the training task for every label in a chain includes predictions for previous labels as features. Apart from the hyperparameters listed in the previous paragraph, we have tried different orderings of the prediction chain.

*Multilabel k-Nearest Neighbors.* We have added two models based on k-nearest neighbors to the grid search: (1) the model *ML-kNN* introduced in [82], which uses Bayesian inference to assign labels to test classes based on the standard kNN output, (2) a binary relevance kNN classifier (*BR-kNN*), a similar method introduced in [17] that assigns the labels that have been assigned to at least half of the neighbors. Although nearest neighbors classifiers are known to perform poorly for high-dimensional vectors (which bags-of-ngrams are), we added them to the task due to them being multi-label *by design*. We have tested $k \in \{1, 2, 3, 5, 10\}$ neighbors; for *ML-kNN*, we tried different values of the smoothing parameter, $s \in \{0.1, 0.5, 0.7, 1.0\}$. As a reliable implementation, we used a combination of models from the *scikit-learn* and *scikit-multilearn Python* libraries [53,67].

### 4.2   Neural Baseline

As a modern approach to fine-tuning neural networks, we have intentionally selected the most standard method, which is not necessarily state of the art for other languages. Among multilingual pretrained large language models, one of the most popular ones is XLM-RoBERTa (large)[8], which is essentially a RoBERTa model [12] trained on a 2.5TB segment of *CommonCrawl* data containing 100 languages, including Kyrgyz. We used XLM-RoBERTa in the multilabel classification fine-tuning setting, using a "classification head" with two linear feedforward layers with dropout and a binary cross-entropy loss. We have used the same split as before as the train-development split to find the best number of epochs (14 out of 15) based on the Jaccard score metric (see below). We used

---

[8] Available on HuggingFace: https://huggingface.co/xlm-roberta-large.

the AdamW optimizer [39] (the AMSGrad version [56]), with weight decay set to 0.01, learning rate set to 0.00002, and exponential learning rate scheduling with $\gamma$ coefficient set to 1.0; $\beta_1 = 0.9$, $\beta_2 = 0.999$, $\epsilon = 10^{-8}$. Batch size was set to 4 mostly due to the equipment-related constraints. Also note that the pretrained `xlm-roberta-large` checkpoint uses byte-pair encoding (bpe) [63] as the tokenizer (provided with the model).

### 4.3  Evaluation Metrics

Each prediction is a set of labels represented as a vector of 0 s and 1 s, where 1 means that the corresponding label has been predicted. Several metrics from "regular" binary and multi-class classification can also be applied for multilabel classification. The counterpart of accuracy here is the fraction of exact matches ("Exact"). The $F_1$ measure (a harmonic mean of precision and recall) can be computed for every sample if we treat each binary vector representation of the label set as binary prediction results and then averaged ("F1-sample" in Table 4). Besides, the $F_1$ measure can be computed for each label and, e.g., micro-averaged ("F1-micro"). We also used metrics unique to the multilabel setting: share of samples where at least one label is predicted correctly ("@l1") and the Hamming loss ("Hamm"), i.e., the Hamming distance between binary vectors of labels. Finally, we report the metric we have used for model selection: the sample-averaged Jaccard similarity computed for each pair of predicted and ground truth label sets; Jaccard similarity between sets $A$ and $B$ is defined as $\frac{|A \cap B|}{|A \cup B|}$.

## 5  Results

Results of our computational experiments are presented in Table 4. It clearly demonstrates that employing the multilingual models for supervised tasks with Kyrgyz text data is feasible, since a fine-tuned *XLM-RoBERTa*-based classifier (without any hyperparameter search) outperforms all other approaches. Note that this result has been far from obvious, since, in our preliminary experiments, fine-tuning another popular model `bert-base-multilingual-cased` (in our case essentially BERT [14] with an added feedforward layer and dropout) did not bring any meaningful results. Another interesting observation is that while (i) Apertium-Kir does not consider the contexts of words, (ii) it is not a lemmatizer, and (iii) the selected stemming method is very far from perfect, even this kind of text normalization does bring improvements compared to the basic bag-of-ngrams approach. Moving from word ngrams to character ngrams also improves the results in most cases, which one could expect since the Kyrgyz language is morphologically rich.

**Table 4.** Evaluation results: @ll—"at-least-one", Hamm—Hamming loss, JaccCV—mean Jaccard score in cross-validation, ↑—more is better, ↓—less is better.

| Configuration | JaccCV↑ | @ll↑ | Jaccard↑ | Exact↑ | Hamm↓ | F1-micro↑ | F1-sample↑ |
|---|---|---|---|---|---|---|---|
| **Bag-of-Token-Ngrams** | | | | | | | |
| Independent, LBFGS, 1-gram | .390 | .59 | .43 | .29 | .06 | .54 | .48 |
| Chain, LBFGS, 1-gram | .405 | .63 | .46 | .31 | .06 | .56 | .51 |
| Independent, SGD, hinge loss, 1-2-gram | .465 | .68 | .47 | .29 | .06 | .56 | .53 |
| Chain, SGD, hinge loss, 1-gram | .474 | .68 | .49 | .32 | .06 | .56 | .55 |
| ML-kNN, 1 neighbor, 0.1-smoothing, 1-gram | .276 | .45 | .30 | .19 | .10 | .33 | .35 |
| BRML-kNN, 1 neighbor,1-gram | .276 | .45 | .30 | .19 | .10 | .33 | .35 |
| **Bag-of-Token-Character-Ngrams** | | | | | | | |
| Independent, LBFGS, 2-3-grams | .491 | .69 | .49 | .33 | .06 | .58 | .55 |
| Chain, LBFGS, 2-3-grams | .494 | .69 | .49 | .33 | .06 | .58 | .55 |
| Independent, SGD, hinge loss, 3-4-grams | .521 | .70 | .46 | .26 | .07 | .55 | .54 |
| Chain, SGD, hinge loss, 3-4-ngrams | .524 | .71 | .48 | .28 | .07 | .55 | .55 |
| ML-kNN, 1 neighbors, 0.1-smoothing, 2-3-gram | .412 | .65 | .42 | .24 | .08 | .48 | .49 |
| BRML-kNN, 1 neighbor, 2-3-gram | .412 | .65 | .42 | .24 | .08 | .48 | .49 |
| **Bag-of-Stem-Ngrams** | | | | | | | |
| Independent, LBFGS, 1-gram | .451 | .67 | .50 | .34 | .05 | .59 | .55 |
| Chain, LBFGS, 1-gram | .463 | .68 | .51 | .35 | .05 | .60 | .56 |
| Independent, SGD, log loss, 1-gram | .514 | .74 | .52 | .33 | .06 | .61 | .59 |
| Chain, SGD, 1-gram | .516 | .74 | .54 | .36 | .06 | .61 | .60 |
| ML-kNN, 1 neighbor, 0.1-smoothing, 1-gram | .345 | .55 | .36 | .21 | .09 | .41 | .42 |
| BRML-kNN, 1 neighbor,1-gram | .345 | .55 | .36 | .21 | .09 | .41 | .42 |
| **Bag-of-Stem-Character-Ngrams** | | | | | | | |
| Independent, LBFGS, 2-4-grams | .494 | .71 | .52 | .35 | .06 | .61 | .58 |
| Chain, LBFGS, 5-6-grams | .490 | .70 | .51 | .35 | .06 | .60 | .57 |
| Independent, SGD, hinge loss, 3-4-grams | .522 | .70 | .49 | .32 | .06 | .58 | .55 |
| Chain, SGD, hinge loss, 3-4-grams | .524 | .69 | .50 | .33 | .06 | .58 | .56 |
| ML-kNN, 1 neighbor, 0.1-smoothing, 3-4-grams | .425 | .65 | .42 | .25 | .08 | .46 | .49 |
| BRML-kNN, 1 neighbor, 3-4-grams | .425 | .65 | .42 | .25 | .08 | .46 | .49 |
| XLM-RoBERTa (with bpe tokenization) | | **.88** | **.66** | **.46** | **.04** | **.72** | **.73** |

# 6  Conclusion

In this work, we have introduced a new annotated text collection in the Kyrgyz language for multilabel topic classification and evaluated several baseline models[9]. This is one of the first open datasets for the low-resource Kyrgyz language. As for baselines, we have found that while classical baselines can achieve acceptable results, especially after (even primitive) stemming, a straightforward neural baseline achieves significantly better results even with virtually no hyperparameter search.

In the future, we plan to further improve the current labeling scheme, additionally expanding and validating current annotations; we plan to ask multiple experts to label the texts using models trained on currently presented data to speed up the labeling. Then, we plan to increase the dataset size by annotating

---

[9] The dataset, baselines, and evaluation code will be released after a Kyrgyz-language-related competition we plan to hold, at the following URL: https://github.com/alexeyev/kyrgyz-multi-label-topic-classification.

more news texts. Afterwards, we plan to hold a competition that should uncover state of the art multilabel classification models for the Kyrgyz language news domain.

Also, to enhance the benchmark with an arguably even more fair comparison, we plan to: (1) translate original texts to English via *Google Translate* and report the scores of the relevant neural models that employ English LLMs as backbones or the scores of zero-shot classification via prompting state of the art generative models such as, e.g., GPT-4 [52]; (2) add the results of the *fastText* supervised classification model trained on our data to the benchmark after publication; (3) study whether using data from a similar domain in other Turkic languages can help improve classification quality. In general, we hope that the presented dataset will be able to serve as the basis for these and other experiments and become a starting point for novel NLP research for the Kyrgyz language.

**Acknowledgments.** This work was supported by the Russian Science Foundation grant # 23-11-00358. We also thank the anonymous reviewers whose comments have allowed us to improve the paper.

# References

1. Alekseev, A., et al.: Medical crossing: a cross-lingual evaluation of clinical entity linking. In: Proceedings of the Thirteenth Language Resources and Evaluation Conference, pp. 4212–4220 (2022)
2. An, B.: Prompt-based for low-resource Tibetan text classification. ACM Trans. Asian Low-Resour. Lang. Inf. Process. (2023, just accepted)
3. Apishev, M., Koltsov, S., Koltsova, O., Nikolenko, S.I., Vorontsov, K.: Mining ethnic content online with additively regularized topic models. Computación y Sistemas **20**(3), 387–403 (2016)
4. Arikoglu, E.: Dictionary project of the modern Kyrgyz language. In: Society, Language and Culture in the 21st Century, pp. 85–91 (2021). (in Russian)
5. Baisa, V., Suchomel, V.: Turkic language support in sketch engine. In: Proceedings of the International conference "Turkic Languages Processing" TurkLang-2015, pp. 214–223 (2015)
6. Bakasova, P.S., Israilova, N.A.: Algorithm for the formation of word forms to automate the procedure for updating the dictionary database. Proceedings of the Kyrgyz State Technical University named after I. Razzakov (2), 23–27 (2016). (in Russian)
7. Benli, I.: Ud_kyrgyz-ktmu: Ud for kyrgyz (2023). https://github.com/UniversalDependencies/UD_Kyrgyz-KTMU/
8. Bird, S., Klein, E., Loper, E.: Natural language processing with Python: analyzing text with the natural language toolkit. "O'Reilly Media, Inc." (2009)
9. Boizou, L., Mambetkazieva, D.: From Kyrgyz internet texts to an xml full-form annotated lexicon: a simple semi-automatic pipeline. In: TurkLang 2017: The Fifth International Conference on Computer Processing of Turkic Languages: Proceedings V 1. Kazan: Tatarstan Academy of Sciences Publishing House (2017)
10. Buraya, K., Farseev, A., Filchenkov, A., Chua, T.S.: Towards user personality profiling from multiple social networks. In: Proceedings of the AAAI Conference on Artificial Intelligence, vol. 31 (2017)

11. Cetin, M.A., Ismailova, R.: Assisting tool for essay grading for Turkish language instructors. MANAS J. Eng. **7**(2), 141–146 (2019)
12. Conneau, A., et al.: Unsupervised cross-lingual representation learning at scale. In: Proceedings of the 58th Annual Meeting of the Association for Computational Linguistics, pp. 8440–8451 (2020)
13. Cruz, J.C.B., Cheng, C.: Establishing baselines for text classification in low-resource languages. CoRR abs/2005.02068 (2020). https://arxiv.org/abs/2005.02068
14. Devlin, J., Chang, M.W., Lee, K., Toutanova, K.: BERT: pre-training of deep bidirectional transformers for language understanding. In: Proceedings of the 2019 Conference of the North American Chapter of the Association for Computational Linguistics: Human Language Technologies (Volume 1: Long and Short Papers), pp. 4171–4186 (2019)
15. Dhingra, B., Zhou, Z., Fitzpatrick, D., Muehl, M., Cohen, W.: Tweet2Vec: character-based distributed representations for social media. In: Proceedings of the 54th Annual Meeting of the Association for Computational Linguistics, Berlin, Germany (Volume 2: Short Papers), pp. 269–274. Association for Computational Linguistics (2016)
16. Ein-Dor, L., et al.: Active learning for BERT: an empirical study. In: Proceedings of the 2020 Conference on Empirical Methods in Natural Language Processing (EMNLP), pp. 7949–7962. Association for Computational Linguistics, Online (2020)
17. Eleftherios Spyromitros, Grigorios Tsoumakas, I.V.: An empirical study of lazy multilabel classification algorithms. In: Proceedings of the 5th Hellenic Conference on Artificial Intelligence (SETN 2008) (2008)
18. Fesseha, A., Emiru, E., Diallo, M., Dahou, A.: Text classification based on convolutional neural networks and word embedding for low-resource languages: Tigrinya. Information **12**, 52 (2021)
19. Forcada, M.L., et al.: Apertium: a free/open-source platform for rule-based machine translation. Mach. Transl. **25**, 127–144 (2011)
20. Greene, D., Cunningham, P.: Practical solutions to the problem of diagonal dominance in kernel document clustering. In: Proceedings of the 23rd International Conference on Machine Learning, ICML 2006, pp. 377–384. Association for Computing Machinery, New York (2006)
21. Homskiy, D., Maloyan, N.: DN at semeval-2023 task 12: low-resource language text classification via multilingual pretrained language model fine-tuning (2023)
22. Israilova, N.A.: Algorithm for morphological analysis and synthesis in a translator. Mod. Probl. Mech. **28**, 11–19 (2017). (in Russian)
23. Israilova, N.A., Bakasova, P.S.: Morphological analyzer of the Kyrgyz language. In: Proceedings of the V International Conference on Computer Processing of Turkic Languages Turklang, vol. 2, pp. 100–116 (2017)
24. Israilova, N.A., Bakasova, P.S.: Ontological models of morphological rules of the Kyrgyz language. In: Proceedings of the VII International Conference on Computer Processing of Turkic Languages, TurkLang 2019, Simferopol, Crimea, 3–5 October 2019 (2019). (in Russian)
25. Joshi, P., Santy, S., Budhiraja, A., Bali, K., Choudhury, M.: The state and fate of linguistic diversity and inclusion in the NLP world. In: Proceedings of the 58th Annual Meeting of the Association for Computational Linguistics, pp. 6282–6293 (2020)
26. Karabaeva, S.D.: Implementing grammatical rules in prolog. Bishkek Hum. Univ. Bull. **2**, 231–233 (2011). (in Russian)

27. Karabaeva, S., Dolmatova, P., Imanalieva, A.: Computer-mathematical modeling of national specificity of spatial models in Kyrgyz language. In: Proceedings of the International conference "Turkic Languages Processing" TurkLang-2015, pp. 416–422 (2015)

28. Kasieva, A.A., Kadyrbekova, A.K.: Corpus annotation tools: Kyrgyz language corpus (using Turkic lexicon Apertium and Penn treebank tools). In: Society, Language and Culture in the 21st Century, pp. 207–214 (2021)

29. Kasieva, A.A., Satybekova, A.T.: Parts-of-speech annotation of the newly created Kyrgyz corpus. Herald KRSU **20**(6), 67–72 (2020)

30. Kasieva, A., Knappen, J., Fischer, S., Teich, E.: A new Kyrgyz corpus: sampling, compilation, annotation. In: Jahrestagung der Deutschen Gesellschaft für Sprachwissenschaft (poster session) (2020). http://hdl.handle.net/21.11119/0000-0004-B62D-D

31. Kochkonbaeva, B.O.: Development of the algorithm for the machine analysis of natural language texts, from Russian to Kyrgyz. Proceedings of the Kyrgyz State Technical University named after I. Razzakov (2), 52–54 (2016). (in Kyrgyz)

32. Kochkonbaeva, B.O., Egemberdieva, Dzh.S.: Modeling of morphological analysis and synthesis of word forms of the natural language. Bull. Sci. Pract. **6**(9), 435–439 (2020)

33. Koltsova, O., Koltsov, S., Nikolenko, S.I.: Communities of co-commenting in the Russian LiveJournal and their topical coherence. Internet Res. **26**(3), 710–732 (2016)

34. Lang, K.: NewsWeeder: learning to filter netnews. In: Prieditis, A., Russell, S. (eds.) Machine Learning Proceedings 1995, pp. 331–339. Morgan Kaufmann, San Francisco (1995)

35. Lazaridou, A., et al.: Mind the gap: assessing temporal generalization in neural language models. In: Neural Information Processing Systems (2021)

36. Lewis, D.D., Yang, Y., Rose, T.G., Li, F.: RCV1: a new benchmark collection for text categorization research. J. Mach. Learn. Res. **5**, 361–397 (2004)

37. Li, X., Li, Z., Sheng, J., Slamu, W.: Low-resource text classification via cross-lingual language model fine-tuning. In: Proceedings of the 19th Chinese National Conference on Computational Linguistics, Haikou, China, pp. 994–1005. Chinese Information Processing Society of China (2020)

38. Liu, J., Chang, W.C., Wu, Y., Yang, Y.: Deep learning for extreme multi-label text classification. In: Proceedings of the 40th International ACM SIGIR Conference on Research and Development in Information Retrieval, SIGIR 2017, pp. 115–124. ACM, New York (2017)

39. Loshchilov, I., Hutter, F.: Decoupled weight decay regularization. In: 7th International Conference on Learning Representations, ICLR 2019, New Orleans, LA, USA, 6–9 May 2019 (2019)

40. Madjarov, G., Kocev, D., Gjorgjevikj, D., Džeroski, S.: An extensive experimental comparison of methods for multi-label learning. Pattern Recogn. **45**(9), 3084–3104 (2012). Best Papers of Iberian Conference on Pattern Recognition and Image Analysis (IbPRIA'2011)

41. Manning, C.D., Raghavan, P., Schütze, H.: Introduction to Information Retrieval. Cambridge University Press, Cambridge (2008)

42. Mirzakhalov, J.: Turkic interlingua: a case study of machine translation in low-resource languages. Ph.D. thesis, University of South Florida (2021)

43. Mirzakhalov, J., et al.: A large-scale study of machine translation in Turkic languages. In: Proceedings of the 2021 Conference on Empirical Methods in Natural Language Processing, pp. 5876–5890 (2021)

44. Mirzakhalov, J., et al.: Evaluating multiway multilingual NMT in the Turkic languages. In: Proceedings of the Sixth Conference on Machine Translation, pp. 518–530 (2021)

45. Momunaliev, K.Z.: Parsing and annotation of the Turkish-Kyrgyz dictionary. Proceedings of the Kyrgyz State Technical University named after I. Razzakov (2), 68–81 (2016). (in Russian)

46. Moskvichev, A., Dubova, M., Menshov, S., Filchenkov, A.: Using linguistic activity in social networks to predict and interpret dark psychological traits. In: Filchenkov, A., Pivovarova, L., Žižka, J. (eds.) AINL 2017. CCIS, vol. 789, pp. 16–26. Springer, Cham (2018). https://doi.org/10.1007/978-3-319-71746-3_2

47. Musaev, S.Dzh., Karabaeva, S.Dzh., Imanalieva, A.I.: Problems and prospects for the development of computational linguistics in Kyrgyzstan. In: Proceedings of the International Conference on Computer processing of Turkic Languages (TurkLang-2013), pp. 34–37 (2013). (in Russian)

48. Nikolenko, S.I.: Topic quality metrics based on distributed word representations. In: Proceedings of the 39th International ACM SIGIR Conference on Research and Development in Information Retrieval, pp. 1029–1032 (2016)

49. Nikolenko, S.I., Koltsova, O., Koltsov, S.: Topic modelling for qualitative studies. J. Inf. Sci. **43**(1), 88–102 (2017)

50. Nivre, J., Zeman, D., Ginter, F., Tyers, F.: Universal dependencies. In: Proceedings of the 15th Conference of the European Chapter of the Association for Computational Linguistics: Tutorial Abstracts, Valencia, Spain. Association for Computational Linguistics (2017)

51. Oleynik, M., Kugic, A., Kasáč, Z., Kreuzthaler, M.: Evaluating shallow and deep learning strategies for the 2018 n2c2 shared task on clinical text classification. J. Am. Med. Inform. Assoc. **26**(11), 1247–1254 (2019)

52. OpenAI: GPT-4 technical report (2023)

53. Pedregosa, F., et al.: Scikit-learn: machine learning in Python. J. Mach. Learn. Res. **12**, 2825–2830 (2011)

54. Polat, Y., Zakirov, A., Bajak, S., Mamatzhanova, Z., Bishkek, K.: Machine translation for Kyrgyz proverbs-google translate vs. Yandex translate-from Kyrgyz into English and Turkish. In: Proceedings of the 6th International Conference on Computer Processing of Turkic Languages «TurkLang-2018», Tashkenk, Uzbekistan, 18–20 October 2018 (2018)

55. Read, J., Pfahringer, B., Holmes, G., Frank, E.: Classifier chains for multi-label classification. Mach. Learn. **85**, 333–359 (2011)

56. Reddi, S.J., Kale, S., Kumar, S.: On the convergence of Adam and beyond. In: International Conference on Learning Representations (2018)

57. Reimers, N., Gurevych, I.: Sentence-BERT: sentence embeddings using Siamese BERT-networks. In: Proceedings of the 2019 Conference on Empirical Methods in Natural Language Processing. Association for Computational Linguistics (2019)

58. Rusnachenko, N., Loukachevitch, N., Tutubalina, E.: Distant supervision for sentiment attitude extraction. In: Proceedings of the International Conference on Recent Advances in Natural Language Processing (RANLP 2019), pp. 1022–1030. INCOMA Ltd., Varna, Bulgaria (2019). https://doi.org/10.26615/978-954-452-056-4_118. https://aclanthology.org/R19-1118

59. Sadykov, T., Kochkonbayeva, B.: Model of morphological analysis of the Kyrgyz language. In: Proceedings of the V International Conference on Computer Processing of Turkic Languages Turklang, vol. 2, pp. 135–154 (2017)

60. Sadykov, T., Kochkonbaeva, B.: On optimization of the morphological analysis algorithm. In: Proceedings of the 6th International Conference on Computer Processing of Turkic Languages «TurkLang-2018», Tashkenk, Uzbekistan, 18–20 October 2018 (2018). (in Russian)
61. Sadykov, T., Sharshembaev, B.: On the creation of a national corpus of the epic of Manas. In: Computer Processing of Turkic Languages. The First International Conference Proceedings, no. 6, pp. 148–154 L. N. Gumilev Eurasian National University, Astana (2013). (in Kyrgyz)
62. Savchenko, A., Alekseev, A., Kwon, S., Tutubalina, E., Myasnikov, E., Nikolenko, S.: Ad lingua: text classification improves symbolism prediction in image advertisements. In: Proceedings of the 28th International Conference on Computational Linguistics, pp. 1886–1892 (2020)
63. Sennrich, R., Haddow, B., Birch, A.: Neural machine translation of rare words with subword units. In: Proceedings of the 54th Annual Meeting of the Association for Computational Linguistics (Volume 1: Long Papers), pp. 1715–1725 (2016)
64. Sharipbai, A.A., et al.: Comparison of ontological models of nouns in the Kazakh and Kyrgyz languages. In: Proceedings of the 6th International Conference on Computer Processing of Turkic Languages «TurkLang-2018», Tashkenk, Uzbekistan, 18–20 October 2018 (2018). (in Russian)
65. Shen, Z., Zhang, S.: A novel deep-learning-based model for medical text classification. In: Proceedings of the 2020 9th International Conference on Computing and Pattern Recognition, ICCPR 2020, pp. 267–273. Association for Computing Machinery, New York (2021)
66. Song, K., Tan, X., Qin, T., Lu, J., Liu, T.Y.: MPNet: masked and permuted pre-training for language understanding. arXiv preprint arXiv:2004.09297 (2020)
67. Szymański, P., Kajdanowicz, T.: A scikit-based Python environment for performing multi-label classification. arXiv e-prints (2017)
68. Tang, P., Jiang, M., Xia, B.N., Pitera, J.W., Welser, J., Chawla, N.V.: Multi-label patent categorization with non-local attention-based graph convolutional network. In: Proceedings of the AAAI Conference on Artificial Intelligence, vol. 34, no. 05, pp. 9024–9031 (2020)
69. Toleush, A., Israilova, N., Tukeyev, U.: Development of morphological segmentation for the Kyrgyz language on complete set of endings. In: Nguyen, N.T., Chittayasothorn, S., Niyato, D., Trawiński, B. (eds.) ACIIDS 2021. LNCS (LNAI), vol. 12672, pp. 327–339. Springer, Cham (2021). https://doi.org/10.1007/978-3-030-73280-6_26
70. Tukeyev, U., Karibayeva, A., Zhumanov, Z.: Morphological segmentation method for Turkic language neural machine translation. Cogent Eng. 7(1), 1856500 (2020)
71. Tutubalina, E., Nikolenko, S.: Inferring sentiment-based priors in topic models. In: Lagunas, O.P., Alcántara, O.H., Figueroa, G.A. (eds.) MICAI 2015. LNCS (LNAI), vol. 9414, pp. 92–104. Springer, Cham (2015). https://doi.org/10.1007/978-3-319-27101-9_7
72. Tutubalina, E., Nikolenko, S.I.: Constructing aspect-based sentiment lexicons with topic modeling. In: Proceedings of the 5th International Conference on Analysis of Images, Social Networks, and Texts, pp. 208–220 (2016)
73. Tutubalina, E., Nikolenko, S.I.: Exploring convolutional neural networks and topic models for user profiling from drug reviews. Multimed. Tools Appl. 77(4), 4791–4809 (2018)
74. Vu, H.T., Nguyen, M.T., Nguyen, V.C., Pham, M.H., Nguyen, V.Q., Nguyen, V.H.: Label-representative graph convolutional network for multi-label text classification. Appl. Intell. 53(12), 14759–14774 (2022)

75. Ward, J.H., Jr.: Hierarchical grouping to optimize an objective function. J. Am. Stat. Assoc. **58**(301), 236–244 (1963)
76. Washington, J.N., Salimzianov, I., Tyers, F.M., Gökırmak, M., Ivanova, S., Kuyrukçu, O.: Free/open-source technologies for Turkic languages developed in the Apertium project. In: Proceedings of the International Conference on Turkic Language Processing (TURKLANG 2019) (2019)
77. Washington, J.N., Ipasov, M., Tyers, F.M.: A finite-state morphological transducer for Kyrgyz. In: LREC, pp. 934–940 (2012)
78. Yao, L., Mao, C., Luo, Y.: Clinical text classification with rule-based features and knowledge-guided convolutional neural networks. CoRR abs/1807.07425 (2018). http://arxiv.org/abs/1807.07425
79. Yin, W., Hay, J., Roth, D.: Benchmarking zero-shot text classification: datasets, evaluation and entailment approach. In: Proceedings of the 2019 Conference on Empirical Methods in Natural Language Processing and the 9th International Joint Conference on Natural Language Processing (EMNLP-IJCNLP), pp. 3914–3923 (2019)
80. Yiner, Z., Kurt, A., Kulamshaev, K., Zafer, H.R.: Kyrgyz orthography and morphotactics with implementation in NUVE. In: Proceedings of International Conference on Engineering and Natural Sciences, pp. 1650–1658 (2016)
81. Zhang, M.L., Zhou, Z.H.: Multilabel neural networks with applications to functional genomics and text categorization. IEEE Trans. Knowl. Data Eng. **18**(10), 1338–1351 (2006). https://doi.org/10.1109/TKDE.2006.162
82. Zhang, M.L., Zhou, Z.H.: ML-KNN: a lazy learning approach to multi-label learning. Pattern Recogn. **40**(7), 2038–2048 (2007)
83. Zhang, Y., Surendran, A.C., Platt, J.C., Narasimhan, M.: Learning from multi-topic web documents for contextual advertisement. In: Proceedings of the 14th ACM SIGKDD International Conference on Knowledge Discovery and Data Mining, pp. 1051–1059 (2008)
84. Zhu, H., Lei, L.: The research trends of text classification studies (2000–2020): a bibliometric analysis. SAGE Open **12**(2), 21582440221089963 (2022)

# Transformers Compression: A Study of Matrix Decomposition Methods Using Fisher Information

Sergey Pletenev[1,2], Daniil Moskovskiy[1(✉)], Viktoriia Chekalina[1],
Mikhail Seleznyov[1], Sergey Zagoruyko[3], and Alexander Panchenko[1,4]

[1] Skolkovo Institute of Science and Technology, Moscow, Russia
{s.pletenev,d.moskovskiy,v.chekalina,a.panchenko}@skol.tech
[2] HSE University, Moscow, Russia
[3] MTS AI, Moscow, Russia
[4] Artificial Intelligence Research Institute, Moscow, Russia

**Abstract.** Transformer models have been a breakthrough in Natural Language Processing. However, the performance of these models comes with their enormous size, limiting options for their deployment. Facing this issue, in this paper, we compare different compression techniques, such as low-rank matrix and tensor factorization, for compressing these heavy layers. We focus on Singular Value Decomposition (SVD) and Tensor Train Matrix Decomposition (TTM) and extend previous work [10] by incorporating Fisher information into the TTM, introducing a novel approach which we call FWTTM.

In this work, we provide a comprehensive analysis of the performance of the compressed models on different setups and compression levels. We observe a performance increase when using FWTTM compared to other methods on low ranks (high compression rates) for both encoder-only and encoder-decoder models.

**Keywords:** Transformer · Compression · Low Rank Approximation · Natural Language Processing · SVD · Tensor Train · Fisher information

## 1 Introduction

The field of Natural Language Processing has made significant progress with the development of large language models based on Transformer architecture. Nevertheless, these models share a common challenge of expanding scale, presenting a formidable obstacle to model training. This scalability issue poses a bottleneck for scientific progress, impacting not only large-scale industries but also smaller research teams lacking comparable training resources. By optimising the training process, we use matrix (SVD) and tensor (TTM) decomposition techniques to represent some layers into the pre-trained Transformer models. To

---

Sergey, Daniil and Viktoriia contributed equally to this work.

© The Author(s), under exclusive license to Springer Nature Switzerland AG 2024
D. I. Ignatov et al. (Eds.): AIST 2023, LNCS 14486, pp. 36–48, 2024.
https://doi.org/10.1007/978-3-031-54534-4_3

avoid drawback in the model performance, we align the compression objective and the task objective by injecting Fisher Information into the decomposition algorithms. We considered time and memory required for forward-backward signal propagation within the decomposed layer, as well as the quality obtained by the compressed model on downstream tasks. Measurements were carried out at various compression levels as well as different form of TTM decomposition.

Natural Language Processing has had a major advance with the recent introduction of large pre-trained models. Although these models set performance benchmarks for many tasks, their massive size and computational requirements pose problems for deployment in resource-constrained environments. As a result, there's a growing trend to develop methods to optimize the size of these models without losing or with acceptable loss of quality [10, 13, 23, 24].

Compressed models typically end up having lower capacity, which might hurt quality. To counteract this, models are often tuned after compression until they meet a desired performance level. However, even for compressed models, this fine-tuning can be demanding in terms of computational resources, and there is a lack of compression methods that do not require additional fine-tuning while compressing the model to a desired degree.

A popular technique for reducing the amount of parameters in Transformers is to focus on its most voluminous component - the fully connected layers (see Table 1). A simple and widely used method is to use SVD to trim parameters without having a noticeable loss in performance. Although the SVD application may affect the capacity of the matrix [31], complementary strategies ensure an acceptable quality of the derived model. Authors of [10] presented the Fisher-Weighted SVD (FWSVD) method, which adjusts the compression process considering the importance of each parameter for the overall performance based on gradient values.

Another matrix decomposition method, which is, in our opinion, underestimated, is Tensor-train matrix decomposition (TTM) [20]. Here, a weight matrix transforms into a multidimensional tensor, which is subsequently represented through products of lower-dimensional entities. In our research, we extend the Fisher-Weighted SVD (FWSVD) approach to TTM, resulting in a FWTTM method.

Based on our numerical experiments with BERT [3] and BART [15] models, compression with FWTTM is on par with FWSVD, being slightly better at low ranks (high compression rates).

The key contributions of this work are the following:

- We extend the previous work by [10] and incorporate weighting based on Fisher information inside the TTM decomposition (we denote this approach as FWTTM).
- We also present an extensive and comprehensive experimental analysis of the proposed (FWTTM) and competing (SVD, FWSVD, TTM) approaches on various ranks (compression rates) applied to the BERT and BART models. Following previous work, we evaluate BERT on the GLUE benchmark [26] and BART on a sequence-to-sequence task of text detoxification [17].

**Table 1.** Number of parameters for different layers in various Transformer architectures.

| Layer/Model | BERT | | BART | |
|---|---|---|---|---|
| Full model | 109 M | 100% | 140 M | 100% |
| Fully connected layers | 57 M | 52% | 84 M | 60% |
| Embeddings | 24 M | 22% | 38 M | 27% |
| Attention | 28 M | 26% | 23 M | 16% |

## 2  Related Work

This section reviews methods related to model size reduction. It contains Knowledge Distillation, quantization and pruning methods, and low-rank approximation techniques.

The first approach, Knowledge Distillation (KD), learns a student model of smaller size (fewer layers, attention heads, etc.) guided by a larger teacher model. These methods can also transfer knowledge from a large teacher model to a smaller student model [8]. KD can improve the generalization performance of the student model and reduce its size and computational cost [12]. We use DistilBERT [23] – a distilled version of the BERT model as one of the strong baselines in our work.

Pruning is another powerful technique to reduce the number of parameters in deep neural networks. The goal of neural network pruning is to identify and remove unimportant neurons to reduce the model size without affecting network accuracy. Block pruning involves removing entire blocks of unimportant layers rather than individual neurons. This can result in a more structured and efficient network architecture. One example of Block Pruning is filter pruning, where entire filters in a convolutional neural network are removed [16]. Another example is channel pruning, where entire channels are removed from the network [7]. As opposed to movement pruning, this approach encourages pruning that can be optimized on dense hardware.

The quantization approach enables the reduction of the model size without compromising the parameter count, achieved by reducing the number of bits allocated to each parameter. The concept of quantization-aware training, which involves training the model with the reduced weights, came from general deep learning [6] to transformer-based encoders [27].

Low-rank approximation techniques provide an alternative way to achieve model compression. One such technique is SVD, which has been successfully applied to compress various components of neural networks, such as word embeddings [14], attention matrices [18], and transformer layers [11]. Another approximation technique is TTM, which decomposes high-order tensors into a sequence of low-order tensors [20]. TTM has been used for compressing word embeddings [9], CNNs [5], and even vision transformers [19].

# 3   Low-Rank Compression Methods

In this section, we describe four low-rank approximation methods used in our computational study to compress feedforward layers of Transformers: SVD, TTM, FWSVD, and FWTTM, with the last one being a novel approach.

## 3.1   Singular Value Decomposition (SVD)

We compress the initial model by replacing fully connected layers with their SVD low-rank representations.

Assuming that $W$ is a layer weight matrix, we define SVD as follows: $W = U \Sigma V^T$. Then we use truncated products of it: $U_r = U[:,: r], \Sigma_r = \Sigma[: r,: r], V_r = V[:,: r]$ to define weights for two sequential linear layers, with which we will replace the current:

$$W_2 = U_r \sqrt{\Sigma_r}, \tag{1}$$

$$W_1 = \sqrt{\Sigma_r} V_r^T. \tag{2}$$

As a result, we get an approximation of linear matrix $W \approx W_2 W_1$ and an approximation of the initial layer $Y \approx X W_1^T W_2^T + b$.

If $W$ has $n_{in}, n_{out}$ shape, the number of parameters in the layer before compression is $n_{in} \times n_{out}$; after representation by truncated SVD, it is $r \times (n_{in} + n_{out})$.

## 3.2   Tensor Train Matrix Decomposition (TTM)

In the TTM decomposition the matrix $\mathcal{M} \in \mathbb{R}^{I \times J}$ is represented in $2D$-order tensor $\mathcal{T} \in \mathbb{R}^{I_1 \times J_1 \times \cdots \times I_D \times J_D}$ in TTM format, where $I = \prod_{k=1}^{D} I_k, J = \prod_{k=1}^{D} J_k$. In other words, each element of $\mathcal{T}$ is computed as

$$\mathcal{T}_{i_1,j_1,\ldots,i_D,j_D} = \sum_{r_1,\ldots,r_{D-1}} \mathcal{G}^1_{r_0,i_1,j_1,r_1} \cdots \mathcal{G}^D_{r_{D-1},i_D,j_D,r_D}, \tag{3}$$

where $\mathcal{G}^d \in \mathbb{R}^{R_{d-1} \times I_d \times J_d \times R_d}$, $d = \overline{1, D-1}$ are *core tensors (cores)* of TTM decomposition, vector $(R_0, \ldots, R_D)$ is called TTM ranks. Note that $R_0 = R_D = 1$ and $R = max(R_0, \ldots, R_D)$.

The compression rate in a TTM layer, with respect to the number of parameters is defined as follows

$$\text{c\_rate} = \frac{R(I_1 J_1 + I_D J_D) + R^2 \sum_{d=2}^{D-1} I_d J_d}{\prod_{k=1}^{D} I_k J_k} \tag{4}$$

We use two implementations of this type of layer based on TNTorch[1] and custom implementation of algorithm [20].

---

[1] https://github.com/rballester/tntorch.

### 3.3  Fisher Weighted SVD (FWSVD)

Authors of [10] inject Fisher information into decomposition algorithms to minimize the gap between decomposition and task-oriented objectives. Fisher information determines the importance of parameters for a specific task [1]. We follow the approach introduced by [10] and approximate the Fisher matrix using dataset $D = \{d_1, \dots, d_{|D|}\}$, for each weight matrix $W \in \mathbb{R}^{I \times J}$:

$$I_W = \mathbb{E}\left[\left(\frac{\partial}{\partial W} \log p(D|W)\right)^2\right] \approx \frac{1}{|D|} \sum_{i=1}^{|D|} \left(\frac{\partial}{\partial W} L(d_i; W)\right)^2. \qquad (5)$$

Having this, ideally, we would want to solve weighted low-rank approximation:

$$\|\sqrt{I_W} * (W - \hat{W})\|^2 \to \min_{\text{rank } \hat{W}=r}. \qquad (6)$$

Unfortunately, this problem does not have a closed-form solution. Therefore, [10] propose to sum Fisher matrix by rows and solve low-rank approximation with row-wise weighting, which can be done using SVD:

$$\tilde{I}_W = \text{diag}\left(I_W \cdot \mathbf{1}\right), \hat{W} = \tilde{I}_W W = USV^T \qquad (7)$$

where $\mathbf{1} = (1, \dots, 1) \in \mathbb{R}^{J \times 1}$, $diag$ - diagonal matrix with size $I \times I$.

The resulted weighted factors for initial matrix $W \approx \hat{U}\hat{S}\hat{V}^T$ are computed as follows:

$$\hat{U} = \tilde{I}_W^{-1}U, \hat{S} = S, \hat{V} = V. \qquad (8)$$

As a result, we get low-rank approximations, which account for parameter importances for the target task.

### 3.4  Fisher Weighted TTM (FWTTM)

We present Algorithm 1, a novel way of integrating Fisher information into the TTM decomposition. The procedure of this algorithm unfolds as follows:

1. Start by calculating the Fisher matrix $I_W$ using the original layer weight matrix $W$.
2. Transform $I_W$ similarly as done with $W$ to derive the "Fisher tensor" represented as $\hat{I}_W$.
3. In every SVD phase within the TTM process, the Fisher matrix is employed exactly like in the FWSVD method. The initial low-rank term undergoes reshaping to form another core in the TTM. The subsequent term is processed in the upcoming iteration, and the Fisher matrix is then unfolded with $U$ ensuring it maintains its shape corresponding to the second term.

---

**Algorithm 1.** Fisher-Weighted TTM decomposition.

---

**Require:** Matrix of layer weights $\mathcal{W}$, matrix of Fisher weights $\mathcal{I}_\mathcal{W}$, shapes $I_1, J_1, \ldots, I_d, J_d$, ranks $r_0, \ldots, r_d$
**Ensure:** Cores $\mathcal{G}^k$, $k = 1 \ldots d$ of the TTM decomposition
1: $\mathcal{B} = \mathcal{W}$.reshape $(I_1, J_1, \ldots, I_d, J_d)$
2: $\mathcal{B}_\mathcal{I} = \mathcal{I}_\mathcal{W}$.reshape $(I_1, J_1, \ldots, I_d, J_d)$
3: $\mathcal{C} = \mathcal{B}$.permute $(1, d+1, 2, d+2, \ldots, d, 2d)$
4: $\mathcal{C}_\mathcal{I} = \mathcal{B}_\mathcal{I}$.permute $(1, d+1, 2, d+2, \ldots, d, 2d)$    $N_r = I_1 J_1 \ldots I_d J_d$
5: **for** $k$ in $\{1, \ldots, d-1\}$ **do**
6:      $N_k = I_k J_k$
7:      $N_r = \frac{N_r}{N_k}$
8:      $r = r_k$
9:      Unfolding $M = \mathcal{C}$.reshape $(N_k, r N_r)$,
10:      Unfolding $M_I = \mathcal{C}_\mathcal{I}$.reshape $(N_k, r N_r)$
11:      $\tilde{M}_I = \mathrm{diag}\,(M_I)$
12:      $\tilde{M}_I M = U S V^T$ truncated to $r_k$
13:      $\tilde{U} = \tilde{M}_I^{-1} U$
14:      $M = S V^T$
15:      $M_I = U^T M_I$
16:      $G_k = \tilde{U}$.reshape $(r_k, n_k, r_{k+1})$
17:·      $G_k = G_k$.permute $(2, 1, 3)$
18: **end for**

---

# 4  Transformer Compression Setup

This section describes our setup for compressing Transformer models using low-rank approximation approaches. We focus on two methods: TTM and SVD, *with* and *without* using Fisher information. We aim to reduce the number of parameters in the model while maintaining its performance. Furthermore, we assume we can access the task-oriented model-tuning process. We use the information obtained within this process to improve the quality of the compression and thus speed up the tuning by the desired values.

## 4.1  Baselines

We compare our compressed model to the model obtained by KD [12], Block Pruning [24] and inference of the original model with floating-point precision equal to 16. Note that distillation and Block Pruning are train-aware methods. For mixed precision training and evaluation, we use the FP16 library, which is built-in in PyTorch [21]. We set the optimization level to 01 and patched all torch functions and tensor methods, except those that benefit from FP32 precision (softmax, etc.) For the two analyzed models, we obtained a compression up to 52% for the BERT model and 54% for the BART model. However, since the tables show compression of the models relative to the number of their parameters, but FP16 quantization keeps the number of parameters the same, we indicate the actual number of parameters with a dagger (†).

**Table 2.** Ranks for different compression approaches.

| BERT | | | BART | | |
|---|---|---|---|---|---|
| C. Rate | SVD | TTM | C. Rate | SVD | TTM |
| 48% (53 M) | 6 | 10 | 60% (83 M) | 10 | 10 |
| 63% (69 M) | 183 | 60 | 74% (102 M) | 210 | 64 |
| 95% (102 M) | 534 | 110 | 90% (125 M) | 460 | 96 |

### 4.2  Experimental Setup

In this study, we evaluate the performance of the aforementioned methods applied to the BERT [3] and BART [15] models. We apply three different compression ratios to these models and present the resulting ranks in Table 2. Following previous work on this topic [10,24], we test the performance of encoder-only model BERT on the GLUE benchmark [26] and encoder-decoder model BART on a sequence-to-sequence task. More precisely, we assess the performance of the compressed BERT model on nine natural language understanding tasks from GLUE benchmark [26], including language acceptability [28], sentiment analysis [25], paraphrasing [2,4], and natural language inference [22,29]. The compressed BART model is evaluated on detoxification tasks [17].

Moreover, for compressing and evaluating models on the GLUE, ParaDetox [17] datasets, we fine-tune a model for each task, compress it and measure performance, and fine-tune the compressed model again on the same task.

### 4.3  Selection of Hyperparameters

The proposed layer structure assumes two sets of hyperparameters - TTM *cores shapes* and *ranks* for both TTM and SVD.

For the maximum compression rate in TTM, cores' non-rank shapes should be as close to each other as possible. We choose $I_k \cdot J_k$ so that they are equal or approximately equal to $(I \cdot J)^{1/D}$. Shapes selection is implemented with a custom algorithm which will be presented in the appendix and source code. As a cores, we take objects with sizes $[1 \times 32 \times 12 \times R]$, $[10 \times 3 \times 2 \times R]$, $[R \times 2 \times 2 \times R]$, $[R \times 16 \times 16 \times 1]$.

Rank $r$ for truncation in SVD and set of $R_1 \ldots R_{M-1}$ is selected based on the desired compression level.

## 5  Experiments with Encoder Transformers

In this section, we evaluate encoder-based Transformers using the BERT model as the base model. Precisely, we use `bert-base-uncased`[2] checkpoint from the Hugging Face [30] model hub.

---

[2] https://huggingface.co/bert-base-uncased.

**Table 3.** Comparative results of BERT compression techniques during task-specific fine-tuning. The top performance for each model size is in **bold**, while the best overall scores are underlined.

| Method | C. Rate | AVG | STSB | CoLA | MNLI | MRCP | QNLI | QQP | RTE | SST2 | WNLI |
|---|---|---|---|---|---|---|---|---|---|---|---|
| Full (109 mln.) | 100% | 0.79 | 0.88 | 0.57 | 0.84 | 0.90 | 0.91 | 0.87 | 0.67 | 0.92 | 0.54 |
| DistilBERT | 61% | 0.76 | 0.87 | 0.51 | 0.82 | 0.87 | 0.89 | 0.88 | 0.59 | 0.91 | 0.48 |
| FP16 eval | 100%† | 0.78 | 0.88 | 0.55 | 0.83 | 0.88 | 0.90 | 0.88 | 0.67 | 0.91 | 0.48 |
| Block Pruning (75%) | 61% | 0.72 | 0.85 | 0.24 | 0.83 | 0.83 | 0.86 | 0.87 | 0.52 | 0.88 | 0.56 |
| SVD | | 0.68 | **0.83** | 0.00 | **0.79** | 0.79 | **0.85** | **0.87** | 0.59 | **0.87** | 0.49 |
| FWSVD | 49% | 0.68 | 0.82 | 0.04 | **0.79** | 0.79 | **0.85** | **0.87** | 0.56 | 0.86 | **0.54** |
| TTM | | **0.69** | **0.83** | **0.15** | 0.78 | **0.81** | 0.84 | **0.87** | **0.60** | 0.86 | 0.43 |
| FWTTM | | **0.69** | **0.83** | **0.15** | 0.78 | **0.81** | 0.84 | **0.87** | **0.60** | 0.86 | 0.49 |
| SVD | | 0.75 | 0.86 | 0.43 | **0.83** | 0.84 | **0.89** | **0.88** | 0.64 | **0.90** | 0.50 |
| FWSVD | 63% | **0.77** | **0.87** | **0.47** | **0.83** | **0.85** | **0.89** | **0.88** | **0.65** | **0.90** | **0.56** |
| TTM | | 0.70 | 0.85 | 0.10 | 0.81 | 0.81 | 0.86 | **0.88** | 0.61 | 0.88 | 0.49 |
| FWTTM | | 0.70 | 0.85 | 0.08 | 0.81 | 0.82 | 0.86 | 0.87 | 0.61 | 0.88 | 0.53 |
| SVD | | 0.78 | **0.89** | 0.56 | **0.84** | 0.88 | **0.91** | **0.89** | 0.68 | **0.91** | 0.44 |
| FWSVD | 95% | **0.79** | **0.89** | 0.56 | **0.84** | 0.88 | 0.90 | **0.89** | **0.69** | **0.91** | 0.51 |
| TTM | | 0.77 | 0.88 | 0.52 | 0.83 | 0.83 | 0.89 | 0.88 | 0.68 | 0.90 | 0.51 |
| FWTTM | | 0.78 | 0.88 | 0.52 | 0.83 | 0.87 | 0.90 | 0.88 | 0.68 | 0.90 | **0.54** |

## 5.1 Experimental Settings

We perform experiments on the GLUE benchmark using the evaluation script and metrics provided by Hugging Face library[3]. Additionally, we run our experiments with five different random seeds and report the average performance across runs to ensure the robustness of our results.

## 5.2 Results

We report the evaluation results of the BERT model on the GLUE benchmark using different compression methods in Table 3. Our results show that TTM decomposition outperforms SVD at low ranks (i.e., high compression levels), while SVD performs better at higher ranks. Incorporating Fisher information consistently improves SVD and slightly improves TTM at high ranks without degrading its performance at other ranks. TTM performs poorly on some tasks, such as CoLA, but better on others, such as STSB. Low-rank compression methods, especially FWSVD, outperform fine-tuned baseline models of approximately the same size at medium compression rates.

Finally, we observe that FWSVD (at 63% compression rate) performs comparably or better than distillation. We also note that pruning baselines and could be combined with FP-16 quantization.

---

[3] https://github.com/huggingface/transformers/tree/main/examples/pytorch/text-cl assification.

**Table 4.** Comparative results of BART compression techniques in the task of text detoxification. The top performance for each model size is in **bold**, while the best overall scores are underlined. *Italic* results represent senseless model outputs.

| Method | C. Rate | STA | SIM | FL | J |
|---|---|---|---|---|---|
| bart-base | 100% | <u>0.89</u> | 0.60 | <u>0.82</u> | <u>0.44</u> |
| FP16 eval | 100%† | 0.89 | 0.60 | <u>0.82</u> | <u>0.44</u> |
| Block Pruning (95%) | 63% | *0.92* | *0.34* | *0.30* | *0.12* |
| Block Pruning (65%) | 74% | 0.82 | 0.60 | 0.73 | 0.36 |
| SVD | | 0.75 | **0.59** | 0.65 | 0.28 |
| FWSVD | 60% | **0.78** | **0.59** | **0.68** | **0.30** |
| TTM | | 0.74 | 0.58 | 0.64 | 0.27 |
| FWTTM | | 0.74 | 0.58 | 0.65 | 0.27 |
| SVD | | 0.82 | 0.60 | 0.77 | 0.38 |
| FWSVD | 74% | **0.87** | <u>**0.61**</u> | **0.80** | **0.42** |
| TTM | | 0.82 | <u>**0.61**</u> | 0.75 | 0.37 |
| FWTTM | | 0.84 | 0.60 | 0.75 | 0.38 |
| SVD | | 0.85 | <u>**0.61**</u> | 0.81 | 0.43 |
| FWSVD | 90% | **0.87** | <u>**0.61**</u> | 0.81 | 0.43 |
| TTM | | 0.86 | <u>**0.61**</u> | 0.80 | 0.41 |
| FWTTM | | 0.85 | <u>**0.61**</u> | 0.80 | 0.41 |

# 6    Experiments with Encoder-Decoder Transformers

In this section, performance of the models with sequence-to-sequence Transformers is tested. We test different layer compression methods on the encoder-decoder model BART [15] on a sequence-to-sequence task, a substask of textual style transfer - text detoxification. In our experiments, we use bart-base checkpoint[4] from the Hugging Face [30] model hub.

## 6.1    Experimental Setup

Text detoxification is the task of transforming a sentence with a toxic tone into one that is neutrally phrased, all while maintaining its original meaning and fluency.

For our experiments, we use the ParaDetox dataset [17], which consists of pairs of sentences in both toxic and neutral style, allowing the training of text detoxification models similar to the methods used in neural machine translation.

In evaluation, we follow the automatic detoxification evaluation framework proposed by [17], which is based on three metrics: **STA** (indicates style transfer accuracy from toxic to neutral style), **SIM** (indicates the resemblance in meaning

---

[4] https://huggingface.co/facebook/bart-base.

between the original and generated sentences), and **FL** (depicts the fluency of the generated text). We also considered their sentence-level average, called the **Joint** metric.

## 6.2  Results

Results of our compression experiments are presented in Table 4. Text generation also preserves the trend shown in natural language understanding tasks.

Overall, the TTM and SVD approaches show the best results at low ranks in all tasks, especially when considering the SIM metric. At medium and high ranks, FWSVD breaks ahead. FWSVD demonstrates the most significant benefits at medium ranks, while the impact becomes nearly imperceptible at higher compression levels. Therefore, it is essential to highlight that the Fisher information acquired for the language modeling task also contributes to improvements in metrics that are not directly related to LM (such as **STA** and **FL**); moreover, they are obtained by the auxiliary neural network model.

It's worth noting that while SVD excels particularly with the SIM metric, the performance of the model in FP16 evaluation mode is almost similar to the performance of the original `bart-base-detox`, while compressing the model to approximately 50% in terms of GPU memory. We leave experiments with using both the compression method and FP16 precision for the future work.

## 7  Conclusion

In this study, we delve into the efficacy of Transformer compression methods like SVD, TTM, FWSVD, and FWTTM across varying ranks. We apply these approaches to the BERT model on the GLUE benchmark and the text detoxification task for a sequence-to-sequence model BART.

Our experimental results support the hypothesis that integrating Fisher information (seen in FW* models) consistently elevates the compression quality for SVD without compromising TTM's effectiveness. In natural language understanding and sequence-to-sequence generation tasks, both TTM and FWTTM stand out in their quality, especially at low ranks. Interestingly, FWTTM's performance is similar to FWSVD's performance at a 95% compression rate. SVD consistently appears to be the best among other methods in most evaluation metrics for generating non-toxic text. Yet, a marked disparity in STA and FL metrics sets it apart from the SIM metric. As we lessen the compression rate, the distinction in performance among the various techniques narrows down. Upon reaching a medium compression level, baselines like distillation, pruning, and quantization show either lower or similar performance to their decomposition-based competitors in each category.

**Acknowledgement.** The work was supported by the Center in the field of Artificial Intelligence in the direction of optimizing management decisions in order to reduce the carbon footprint on the basis of the Skolkovo Institute of Science and Technology under Contract No. 70-2021-00145/10841 dated 02.11.2021.

# References

1. Bishop, C.M., Nasrabadi, N.M.: Pattern recognition and machine learning. J. Electron. Imaging **16**(4), 049901 (2007)
2. Cer, D.M., Diab, M.T., Agirre, E., Lopez-Gazpio, I., Specia, L.: Semeval-2017 task 1: semantic textual similarity - multilingual and cross-lingual focused evaluation. CoRR abs/1708.00055 (2017). https://arxiv.org/abs/1708.00055
3. Devlin, J., Chang, M., Lee, K., Toutanova, K.: BERT: pre-training of deep bidirectional transformers for language understanding. In: Burstein, J., Doran, C., Solorio, T. (eds.) Proceedings of the 2019 Conference of the North American Chapter of the Association for Computational Linguistics: Human Language Technologies, NAACL-HLT 2019, Minneapolis, MN, USA, 2–7 June 2019, Volume 1 (Long and Short Papers), pp. 4171–4186. Association for Computational Linguistics (2019). https://doi.org/10.18653/v1/n19-1423
4. Dolan, W.B., Brockett, C.: Automatically constructing a corpus of sentential paraphrases. In: Proceedings of the Third International Workshop on Paraphrasing, IWP@IJCNLP 2005, Jeju Island, Korea, October 2005, 2005. Asian Federation of Natural Language Processing (2005). https://aclanthology.org/I05-5002/
5. Garipov, T., Podoprikhin, D., Novikov, A., Vetrov, D.P.: Ultimate tensorization: compressing convolutional and FC layers alike. CoRR abs/1611.03214 (2016). https://arxiv.org/abs/1611.03214
6. Hawks, B., Duarte, J.M., Fraser, N.J., Pappalardo, A., Tran, N., Umuroglu, Y.: PS and QS: quantization-aware pruning for efficient low latency neural network inference. Front. Artif. Intell. **4**, 676564 (2021)
7. He, Y., Zhang, X., Sun, J.: Channel pruning for accelerating very deep neural networks. In: IEEE International Conference on Computer Vision, ICCV 2017, Venice, Italy, 22–29 October 2017, pp. 1398–1406. IEEE Computer Society (2017). https://doi.org/10.1109/ICCV.2017.155
8. Hinton, G.E., Vinyals, O., Dean, J.: Distilling the knowledge in a neural network. CoRR abs/1503.02531 (2015). https://arxiv.org/abs/1503.02531
9. Hrinchuk, O., Khrulkov, V., Mirvakhabova, L., Orlova, E.D., Oseledets, I.V.: Tensorized embedding layers. In: Cohn, T., He, Y., Liu, Y. (eds.) Findings of the Association for Computational Linguistics: EMNLP 2020, Online Event, 16–20 November 2020. Findings of ACL, vol. EMNLP 2020, pp. 4847–4860. Association for Computational Linguistics (2020). https://doi.org/10.18653/v1/2020.findings-emnlp.436
10. Hsu, Y., Hua, T., Chang, S., Lou, Q., Shen, Y., Jin, H.: Language model compression with weighted low-rank factorization (2022). https://openreview.net/forum?id=uPv9Y3gmAI5
11. Hu, P., Peng, X., Zhu, H., Aly, M.M.S., Lin, J.: OPQ: compressing deep neural networks with one-shot pruning-quantization. In: Thirty-Fifth AAAI Conference on Artificial Intelligence, AAAI 2021, Thirty-Third Conference on Innovative Applications of Artificial Intelligence, IAAI 2021, The Eleventh Symposium on Educational Advances in Artificial Intelligence, EAAI 2021, Virtual Event, 2–9 February 2021, pp. 7780–7788. AAAI Press (2021). https://ojs.aaai.org/index.php/AAAI/article/view/16950
12. Jiao, X., et al.: Tinybert: distilling BERT for natural language understanding. In: Cohn, T., He, Y., Liu, Y. (eds.) Findings of the Association for Computational Linguistics: EMNLP 2020, Online Event, 16–20 November 2020. Findings of ACL, vol. EMNLP 2020, pp. 4163–4174. Association for Computational Linguistics (2020). https://doi.org/10.18653/v1/2020.findings-emnlp.372

13. Lagunas, F., Charlaix, E., Sanh, V., Rush, A.M.: Block pruning for faster transformers. In: Moens, M., Huang, X., Specia, L., Yih, S.W. (eds.) Proceedings of the 2021 Conference on Empirical Methods in Natural Language Processing, EMNLP 2021, Virtual Event/Punta Cana, Dominican Republic, 7–11 November 2021, pp. 10619–10629. Association for Computational Linguistics (2021). https://doi.org/10.18653/v1/2021.emnlp-main.829

14. Lan, Z., Chen, M., Goodman, S., Gimpel, K., Sharma, P., Soricut, R.: ALBERT: a lite BERT for self-supervised learning of language representations. In: 8th International Conference on Learning Representations, ICLR 2020, Addis Ababa, Ethiopia, 26–30 April 2020. OpenReview.net (2020). https://openreview.net/forum?id=H1eA7AEtvS

15. Lewis, M., et al.: BART: denoising sequence-to-sequence pre-training for natural language generation, translation, and comprehension. In: Jurafsky, D., Chai, J., Schluter, N., Tetreault, J.R. (eds.) Proceedings of the 58th Annual Meeting of the Association for Computational Linguistics, ACL 2020, Online, 5–10 July 2020, pp. 7871–7880. Association for Computational Linguistics (2020). https://doi.org/10.18653/v1/2020.acl-main.703

16. Li, H., Kadav, A., Durdanovic, I., Samet, H., Graf, H.P.: Pruning filters for efficient convnets. In: 5th International Conference on Learning Representations, ICLR 2017, Toulon, France, 24–26 April 2017, Conference Track Proceedings. OpenReview.net (2017). https://openreview.net/forum?id=rJqFGTslg

17. Logacheva, V., et al.: Paradetox: detoxification with parallel data. In: Muresan, S., Nakov, P., Villavicencio, A. (eds.) Proceedings of the 60th Annual Meeting of the Association for Computational Linguistics (Volume 1: Long Papers), ACL 2022, Dublin, Ireland, 22–27 May 2022, pp. 6804–6818. Association for Computational Linguistics (2022). https://doi.org/10.18653/v1/2022.acl-long.469

18. Michel, P., Levy, O., Neubig, G.: Are sixteen heads really better than one? In: Wallach, H.M., Larochelle, H., Beygelzimer, A., d'Alché-Buc, F., Fox, E.B., Garnett, R. (eds.) Advances in Neural Information Processing Systems 32: Annual Conference on Neural Information Processing Systems 2019, NeurIPS 2019, 8–14 December 2019, Vancouver, BC, Canada, pp. 14014–14024 (2019). https://proceedings.neurips.cc/paper/2019/hash/2c601ad9d2ff9bc8b282670cdd54f69f-Abstract.html

19. Minh, H.P., Xuan, N.N., Son, T.T.: TT-ViT: vision transformer compression using tensor-train decomposition. In: Nguyen, N.T., Manolopoulos, Y., Chbeir, R., Kozierkiewicz, A., Trawinski, B. (eds.) ICCCI 2022. LNCS, vol. 13501, pp. 755–767. Springer, Cham (2022). https://doi.org/10.1007/978-3-031-16014-1_59

20. Oseledets, I.V.: Tensor-train decomposition. SIAM J. Sci. Comput. **33**, 2295–2317 (2011)

21. Paszke, A., et al.: Pytorch: an imperative style, high-performance deep learning library. In: Wallach, H.M., Larochelle, H., Beygelzimer, A., d'Alché-Buc, F., Fox, E.B., Garnett, R. (eds.) Advances in Neural Information Processing Systems 32: Annual Conference on Neural Information Processing Systems 2019, NeurIPS 2019, 8–14 December 2019, Vancouver, BC, Canada, pp. 8024–8035 (2019). https://proceedings.neurips.cc/paper/2019/hash/bdbca288fee7f92f2bfa9f7012727740-Abstract.html

22. Rahman, A., Ng, V.: Resolving complex cases of definite pronouns: the winograd schema challenge. In: Tsujii, J., Henderson, J., Pasca, M. (eds.) Proceedings of the 2012 Joint Conference on Empirical Methods in Natural Language Processing and Computational Natural Language Learning, EMNLP-CoNLL 2012, 12–14 July

2012, Jeju Island, Korea, pp. 777–789. ACL (2012). https://aclanthology.org/D12-1071/

23. Sanh, V., Debut, L., Chaumond, J., Wolf, T.: Distilbert, a distilled version of BERT: smaller, faster, cheaper and lighter. CoRR abs/1910.01108 (2019). https://arxiv.org/abs/1910.01108

24. Sanh, V., Wolf, T., Rush, A.M.: Movement pruning: adaptive sparsity by fine-tuning. In: Larochelle, H., Ranzato, M., Hadsell, R., Balcan, M., Lin, H. (eds.) Advances in Neural Information Processing Systems 33: Annual Conference on Neural Information Processing Systems 2020, NeurIPS 2020, 6–12 December 2020, virtual (2020). https://proceedings.neurips.cc/paper/2020/hash/eae15aabaa768ae4a5993a8a4f4fa6e4-Abstract.html

25. Socher, R., et al.: Recursive deep models for semantic compositionality over a sentiment treebank. In: Proceedings of the 2013 Conference on Empirical Methods in Natural Language Processing, EMNLP 2013, 18–21 October 2013, Grand Hyatt Seattle, Seattle, Washington, USA, A meeting of SIGDAT, a Special Interest Group of the ACL, pp. 1631–1642. ACL (2013). https://aclanthology.org/D13-1170/

26. Wang, A., Singh, A., Michael, J., Hill, F., Levy, O., Bowman, S.R.: GLUE: a multi-task benchmark and analysis platform for natural language understanding. In: Linzen, T., Chrupala, G., Alishahi, A. (eds.) Proceedings of the Workshop: Analyzing and Interpreting Neural Networks for NLP, BlackboxNLP@EMNLP 2018, Brussels, Belgium, 1 November 2018, pp. 353–355. Association for Computational Linguistics (2018). https://doi.org/10.18653/v1/w18-5446

27. Wang, Z., Li, J.B., Qu, S., Metze, F., Strubell, E.: Squat: sharpness- and quantization-aware training for BERT. CoRR abs/2210.07171 (2022). https://doi.org/10.48550/arXiv.2210.07171

28. Warstadt, A., Singh, A., Bowman, S.R.: Neural network acceptability judgments. Trans. Assoc. Comput. Linguist. **7**, 625–641 (2019)

29. Williams, A., Nangia, N., Bowman, S.R.: A broad-coverage challenge corpus for sentence understanding through inference. In: Walker, M.A., Ji, H., Stent, A. (eds.) Proceedings of the 2018 Conference of the North American Chapter of the Association for Computational Linguistics: Human Language Technologies, NAACL-HLT 2018, New Orleans, Louisiana, USA, 1–6 June 2018, Volume 1 (Long Papers), pp. 1112–1122. Association for Computational Linguistics (2018). https://doi.org/10.18653/v1/n18-1101

30. Wolf, T., et al.: Huggingface's transformers: state-of-the-art natural language processing. CoRR abs/1910.03771 (2019). https://arxiv.org/abs/1910.03771

31. Yang, Z., Dai, Z., Salakhutdinov, R., Cohen, W.W.: Breaking the softmax bottleneck: a high-rank RNN language model. In: 6th International Conference on Learning Representations, ICLR 2018, Vancouver, BC, Canada, 30 April–3 May 2018, Conference Track Proceedings. OpenReview.net (2018). https://openreview.net/forum?id=HkwZSG-CZ

# Leveraging Taxonomic Information from Large Language Models for Hyponymy Prediction

Polina Chernomorchenko[1], Alexander Panchenko[2,3], and Irina Nikishina[4(✉)]

[1] HSE University, Moscow, Russia
pvchernomorchenko@edu.hse.ru
[2] Skolkovo Institute of Science and Technology, Moscow, Russia
[3] Artificial Intelligence Research Institute, Moscow, Russia
a.panchenko@skol.tech
[4] Universität Hamburg, Hamburg, Germany
irina.nikishina@uni-hamburg.de

**Abstract.** Pre-trained language models contain a vast amount of linguistic information as well as knowledge about the structure of the world. Both of these attributes are extremely beneficial for automatic enrichment of semantic graphs, such as knowledge bases and lexical-semantic databases. In this article, we employ generative language models to predict descendants of existing nodes in lexical data structures based on IS-A relations, such as WordNet. To accomplish this, we conduct experiments utilizing diverse formats of artificial text input containing information from lexical taxonomy for the English and Russian languages. Our findings demonstrate that the incorporation of data from the knowledge graph into a text input significantly affects the quality of hyponym prediction.

**Keywords:** taxonomy enrichment · IS-A relations · generative transformers · hyponym prediction

## 1 Introduction

Large pre-trained language models such as LLama-2 [24], Flan-T5 [4], Instruct-GPT [21] show impressive results in solving a wide range of tasks. However, the results of even such advanced models strongly depend on the input data [2,28]. In this paper, we assume that the prompting approach can be also extrapolated to lexical semantic tasks, such as IS-A relationship prediction. There are several studies exploring the ability of transformers to predict IS-A relationships through the use of natural language prompts [10,12]. However, they do not exploit sufficient information about taxonomy structure and the particular meaning of the lexeme which leads to the word sense disambiguation problem.

On the other hand, lexical taxonomies like WordNet [11] store a lot of important linguistic data. First of all, nodes (synsets) in such taxonomy graphs may accumulate several surface forms that represent the same meaning. Furthermore,

D. I. Ignatov et al. (Eds.): AIST 2023, LNCS 14486, pp. 49–63, 2024.
https://doi.org/10.1007/978-3-031-54534-4_4

they contain not only information about IS-A relations, but also words's synonyms, definitions and sense numbers specifying meaning of the particular word.

Taxonomy is a specific type of a knowledge graph that represents the relationships between the real-world entities and linguistic features. Taxonomies play a key role in a wide range of Natural Language Pricessing (NLP) tasks [9,15,27] and numerous studies are focused now on automatic enrichment of such structures [1,6]. Thus, this study aims to find out what information from the taxonomy can be useful for the accurate prediction of hyponyms using prompting. We focus on the hyponym prediction task, which aims at predicting new descendants for an existing node of the taxonomic graph. The formulation of the task is as follows: given taxonomy $G = (V, E)$; $G_{global} = (V_{global}, E_{global})$, $G \in G_{global}$. For given $v \in V$ find all $w \notin V : (v, w) \in E_{global}$ and $(v, w) \notin E$.

The main contributions of the paper are as follows: (i) we introduce new datasets for hyponym prediction for the Russian language; (ii) we explore the transformer-based generative architectures, specifically decoder and encoder-decoder models for hyponym prediction; (ii) we conduct experiments with various formats of artificial input data.

## 2    Related Work

Pre-trained language models demonstrate exceptional ability in encoding and understanding semantic information. For instance, Wiedemann et al. [26] demonstrate that homonyms can be differentiated using the k-NN search algorithm based on BERT embeddings [8]. Furthermore, BERT embeddings outperforms static embeddings in predicting lexical relationships [25].

In [10,12] authors exploring BERT's hypernymy knowledge. While BERT shows a decent level of acquisition of IS-A relations, there remain some limitations to prompting in natural language. Specifically, Ettinger notes in [10] that model predictions highly depend on the particular input, while in [12] is noted that universal prompts marking IS-A relations do not provide enough information for the model to distinguish homonyms.

These limitations can be addressed through the use of manually created prompts, although this process can be labor-intensive. Lexical taxonomies already provide structured information about lexemes and their relationships. Nevertheless, it is possible that incorporating information from graphs into language models can lead to more accurate hyponymy and hypernymy predictions.

Recent investigations have explored the incorporation of graph embeddings into language models [3,13,20]. In [3], for instance, vector representations of graphs were concatenated with text embeddings to provide model with knowledge of medical domain. In [20], the authors projected graph embeddings into the BERT space to enrich lexical taxonomies. Drawing inspiration from the significant achievements of language models in understanding and solving NLP tasks from bare text input, as exemplified by Brown et al. [2], we propose providing models with information about graph structure in textual format to improve their ability in comprehending IS-A relations.

# 3   Datasets

In this section, we present the datasets for both English and Russian languages. For each language, we perform our experiments on two different types of dataset: randomly and manually curated. We assume that the results in [20] for the English dataset collected automatically might be too low because the dataset comprises very uncommon and specific words from different domains (e.g. biological "protoctist family"), which significantly affects the results. Therefore, we also test on a smaller version of English dataset, where manually selected nodes are located at least 5 hops away from the root node and have 1–4 hop to descendants. We try to select similar words in both languages when creating similar datasets for Russian and present them within their features in Appendix A. We also collect two training datasets consisting of nonterminal nodes that are not included in any of the test datasets. Training datasets contain 15,000 and 10,000 synsets for English and Russian, respectively. The contents of the manually collected datasets as well as data on statistical parameters of the datasets can be found in Appendix A in Tables 6.

## 3.1   English Datasets

We utilize the CHSP dataset [20], consisting of 1000 preterminal nodes randomly selected from English WordNet [11], as an automatically generated dataset for English. One of the advantages of the CHSP dataset is that it closely resembles real data for the taxonomy enrichment task. However, the dataset contains highly specific and uncommon concepts, making it challenging to evaluate how well the models assimilate hyponymic relations. Based on the literature review on the acquisition of hyponymy with transformer-based models, we find out that past studies often use semantically simple datasets, as typified by the Battig dataset employed by Hanna and Mareček [12]. To provide an approximate understanding of the efficacy of proposed approach, we posit that a less intricate dataset is required. Therefore, to more accurately assess the models' hyponymy acquisition, we also test on a smaller dataset featuring 22 frequently used concepts from a common domain. While collecting the smaller dataset the formal criterion of a distance of at least five hops from the root while allowing nonterminal nodes is maintained. For the synset meanings, simple generic concepts that have at least 4 hyponyms including indirect ones are selected.

## 3.2   Russian Dataset

We generate same-sized random dataset for the Russian language based on formal criteria that match the CHSP dataset. When creating a manual dataset of common knowledge concepts, we try to find corresponding nodes in the Russian WordNet (RuWordNet)[1] for English ones. However, due to the different structures of the English and Russian taxonomies, some of the corresponding synsets

---

[1] https://ruwordnet.ru/en.

do not meet the formal criteria. For instance, the synsets with the meanings "room", "furniture", "monetary unit" and "board game" do not satisfy the condition for the distance from the root. Additionally, some synsets are replaced due to semantic inconsistency of concepts and specific hyponyms as a consequence. For example, the Russian synset with the meaning "color" has such hyponyms as *mimicry of organisms* and *animal's color* on the same taxonomy level as usual color names like *red* or *green*.

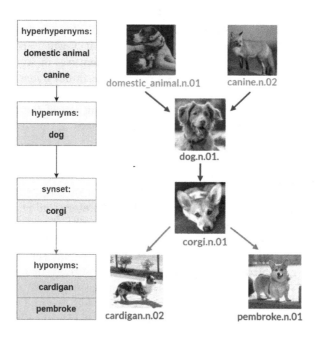

**Fig. 1.** Transformation of a graph data structure into a linear text representation.

## 4   Methodology

The current section presents a methodology for hyponymy prediction task. We compare artificial input formats to find out what information from the taxonomy the model needs for correct predictions and then use best input format to fine-tune models for both languages.

### 4.1   Artificial Prompt Selection for Fine-Tuning

There exist at least two studies exploring transformer-based models' IS-A relations acquisition through prompting [10,12]. While this approach offers a way to get knowledge from transformers with minimal computational costs, it also has some drawbacks. One major issue is the need to disambiguate polysemous words when enriching taxonomies. Universal prompts in natural language, without additional context, do not allow for this kind of disambiguation. For example, it is possible to end sentence *Bat is ...* with both *an animal* and *a wooden club*. Another challenge is that certain text patterns may only work for a small subset of concepts. For example, the hyponymy-related text pattern "My favorite X is a Y" cannot be applied to negative concepts, as using the word "favorite" would be inappropriate (e.g. "My favorite retinopathy is diabetic retinopathy" would not make sense). In addition, many languages specify particular names for subtypes of different entities. Thus, a subclass of a dog is a *breed*, and a subclass of a plant is a *variety* and none of these words can be replaced by a *type* or *kind*. To overcome this obstacle, we narrow the attention to artificial text patterns providing model with information from taxonomy structure.

The basic format of the artificial text pattern for English includes information about hypernym and hyperhypernym of the target node and looks as following:

*Example 1.* hyperhypernyms: $h_{1-n}$ | hypernyms: $h_{1-m}$ | synset: $s$ | hyponyms: $l_{1-k}$.

Here $s$ denotes the target vertex, $l_{1-k}$ is the list of correct hyponyms contained in the taxonomy, $h_{1-m}$ and $h_{1-n}$ are the parent and grandparent nodes relative to the target. Figure 1 shows the alignment between lexical taxonomy subgraph containing the target synset and proposed artificial input format.

In the English WordNet database [11], synsets contain not only the names of surface forms but also additional information that may be useful for predicting hyponyms. We take into account the following parameters: sense number, definitions and lemmas[2]. Based on these parameters, we create 8 artificial prefixes, the shortest of which contains only target synset and its hypernym, and the longest - all possible additional information. A description and examples of all types of artificial prefixes can be found in the Appendix A.

We fine-tune GPT-2 base [22] and T5-base [23] using constructed artificial patterns as input data. GPT-2 is fine-tuned for the language modeling task, while T5 is fine-tuned for seq2seq generation. Each experimental model is trained for three epochs with a batch size of 8 and evaluated with manually curated test data.

The structure of stored lexical information is the same in both the Russian and English WordNet, except for the absence of sense numbers in the former. Based on this fact we assume that outcomes of the experiments for English WordNet can be also extrapolated to the Russian data. Therefore, we restrict our comparison of artificial prefixes solely to the English WordNet.

## 4.2  Fine-Tuning Generative Transformers for Hyponymy Prediction Task

Based on the comparison of artificial prefixes, we select the best performing prompt for the decoder model. Then we conduct fine-tuning for three models

---

[2] In WordNet lemma represents a specific sense of a particular word. Each synset can contain multiple lemmas. (e.g., *color* and *colour* are two different lemmas, but have the same meaning).

in both languages: the decoder, the encoder-decoder and a third model which is instruction-based decoder of larger size. Thus, we use GPT2-large [22], T5-large [23] and Dolly-v1-6b [5] for English, and RuGPT2-large[3], RuT5-large[4] and Saiga-7b-LoRa[5] for Russian.

The justification for fine-tuning the third model is two-fold. Firstly, we assume that higher capacity of large language models to store information about the world and language will allow them to predict a wider variety of candidates from the texts it had seen during pre-training. Secondly, given the observed success of the instruction-scaled models [4], we expect that the model that has seen numerous instructions would finetune better and faster for yet another linguistic task of hyponym prediction.

For the first two models, we perform the full fine-tuning procedure and for the third - parameter efficient one using LoRa [14] and 8-bit version [7] of the model. For Saiga-7b fine-tuning we merge trained LoRa on Russian data adapter with the base LlaMA model [16] and than train new LoRa adapter via our data.

Models for English are trained during 3 epochs with batch size equal to 16. RuGPT2 and RuT5 are trained during two epochs with the same batch size, while Saiga is trained for one epoch with batch size equal to 4. All computations are performed on hardware of type NVIDIA RTX A6000-48GB.

## 5 Evaluation Setup

To compute evaluation metrics, we generate 50 sequences up to 15 tokens long (excluding prefix) for each input sample in the test dataset using top-$k$ sampling ($k = 20$). We believe that this experimental formulation provides a more accurate evaluation of the models' hyponymy acquisition than using greedy search.

Next, we split the output by comma (since the models learn the expected output format correctly) and sort the final list of n-grams by frequency of occurrence.

### 5.1 Evaluation Metrics

Predicted n-grams are compared with the actual hyponyms in the taxonomy using a set of metrics. We use the Precision@$k$ (P@$k$) metric to evaluate precision, which calculates the proportion of correct results achieved at a predetermined rank $k$. This helps to determine the number of accurate answers among the top-$k$ results. Additionally, we use the Mean Reciprocal Rank (MRR) metric, which evaluates the multiplicative inverse of the rank of the first correct answer. To assess the sum of correct answers and their rank in the candidate list, we also use the Mean Average Precision (MAP). We also use Recall to consider the coverage.

---

[3] https://huggingface.co/ai-forever/rugpt3large_based_on_gpt2.

[4] https://huggingface.co/ai-forever/ruT5-large.

[5] https://huggingface.co/IlyaGusev/saiga_7b_lora.

**Table 1.** Results of fine-tuning with different formats of artificial prefix

| Prefix format | MAP | MRR | P@1 | P@2 | P@5 | P@10 | R@a |
|---|---|---|---|---|---|---|---|
| GPT2-base | | | | | | | |
| Sense num | 0.05 | **0.64** | **0.50** | 0.43 | 0.30 | 0.27 | 0.17 |
| Default | 0.05 | 0.63 | 0.46 | **0.50** | **0.40** | **0.31** | 0.17 |
| Lemmas | 0.05 | 0.58 | 0.41 | 0.36 | 0.38 | 0.30 | 0.20 |
| Definition | 0.04 | 0.59 | 0.41 | 0.39 | 0.29 | 0.26 | 0.20 |
| Add. lemmas | 0.04 | 0.55 | 0.36 | 0.39 | 0.28 | 0.28 | 0.21 |
| Hyperdefinition | 0.04 | 0.59 | 0.46 | 0.36 | 0.31 | 0.26 | 0.21 |
| Hypernym only | 0.04 | 0.55 | 0.36 | 0.39 | 0.30 | 0.31 | 0.15 |
| All additions | 0.03 | 0.44 | 0.32 | 0.25 | 0.18 | 0.17 | 0.21 |
| T5-base | | | | | | | |
| Default | 0.07 | **0.67** | 0.50 | **0.57** | **0.46** | 0.32 | 0.13 |
| Lemmas | 0.07 | 0.65 | **0.55** | 0.53 | 0.41 | 0.32 | 0.12 |
| Sense num | 0.07 | 0.61 | 0.50 | 0.46 | 0.38 | 0.33 | 0.14 |
| Hyperdefinition | 0.07 | 0.59 | 0.46 | 0.48 | 0.41 | **0.36** | 0.14 |
| Hypernym only | 0.07 | 0.50 | 0.36 | 0.43 | 0.38 | 0.31 | 0.14 |
| Add. lemmas | 0.06 | 0.63 | 0.55 | 0.43 | 0.39 | 0.32 | 0.13 |
| Definition | 0.06 | 0.60 | 0.50 | 0.46 | 0.37 | 0.31 | 0.13 |
| All additions | 0.05 | 0.50 | 0.32 | 0.39 | 0.37 | 0.30 | 0.14 |

## 6  Results

In this section we discuss the results of comparing artificial inputs as well as the results of fine tuning.

### 6.1  Artificial Prefixes Comparing

The results, presented in Table 1, demonstrate that both GPT2 and T5 models achieve the best results by using prefixes with either no or very few number of additions, such as "default", "sense numbers", and "lemmas". This suggests that we can specify the meaning of the target synset by pointing to higher levels of the taxonomic structure. Surprisingly, the most informative input data format in both cases result in the lowest scores. "All additions" prefix variation contains sense numbers and also provides additional lemmas and definitions for each synset specified in the input. On the other hand, a minimalist input data format, which only indicates the parent vertex of the target synset, produced higher scores. We assume that such results may be related to the fact that long definitions combined with lemmas violate the formal data structure and make it harder for the model to understand the information. Additionally, prefix lengthening causes reduction of training examples due to input size constraints,

**Table 2.** Fine-tuning results for hyponymy prediction on automatically generated datasets for English and Russian

| Method | MAP | MRR | P@1 | P@2 | P@5 | P@10 | R@a |
|---|---|---|---|---|---|---|---|
| English | | | | | | | |
| GPT-2 | 0.039 | 0.172 | 0.110 | 0.104 | 0.091 | 0.074 | 0.243 |
| T5 | 0.048 | 0.189 | 0.123 | 0.115 | 0.096 | 0.076 | 0.127 |
| Dolly-6b 8-bit LoRa | **0.111** | **0.324** | **0.226** | **0.202** | **0.164** | **0.122** | 0.318 |
| Russian | | | | | | | |
| RuGPT-large | **0.082** | **0.244** | 0.142 | **0.142** | **0.117** | **0.092** | **0.275** |
| RuT5-large | 0.072 | 0.240 | **0.162** | **0.142** | 0.104 | 0.079 | 0.197 |
| Saiga-7b | 0.069 | 0.224 | 0.157 | 0.140 | 0.104 | 0.076 | 0.202 |

**Table 3.** Fine-tuning results for hyponymy prediction on manually collected datasets for English and Russian

| Method | MAP | MRR | P@1 | P@2 | P@5 | P@10 | R@a |
|---|---|---|---|---|---|---|---|
| English | | | | | | | |
| GPT2-large | 0.13 | 0.65 | 0.50 | 0.50 | 0.47 | 0.45 | **0.34** |
| T5-large | 0.15 | 0.85 | 0.77 | 0.66 | 0.64 | 0.54 | 0.24 |
| Dolly-6b 8bit LoRa | **0.18** | **0.93** | **0.86** | **0.84** | **0.70** | **0.57** | 0.29 |
| Russian | | | | | | | |
| RuGPT-large | **0.177** | 0.734 | 0.591 | 0.591 | **0.600** | **0.523** | 0.317 |
| RuT5-large | 0.156 | 0.664 | 0.500 | 0.523 | 0.500 | 0.486 | 0.256 |
| Saiga-7b 8bit LoRa | 0.129 | **0.851** | **0.818** | **0.682** | 0.555 | 0.450 | **0.269** |

while a minimalistic data format, on the contrary, allows more hyponyms in the input sequence (however, it's worth noting that this is only true for decoders, since in the seq2seq task formulation input and output are separated). Despite this fact, the prefix format under consideration exhibits the highest Recall scores for both models. This valuable characteristic makes the format particularly useful in the context of the final task of taxonomy enrichment. It can also be observed that GPT2 yields a greater extent of correct hyponym coverage in comparison to T5, as it generates a more diverse list of candidates. Higher MAP scores of T5 can therefore be attributed to this fact.

## 6.2  Hyponym Prediction for English and Russian

To fine-tune final models, we opt for the prefix containing sense numbers for English, as it demonstrated superior performance on the decoder model (since 2 out of 3 models for each language are decoders). On the other hand, for Russian, the default prefix is employed due to the absence of sense numbers in the Russian

WordNet that would indicate the ordinal number of given token's value in the current synset.

Below are examples of the prefixes selected for English and Russian for the synset *coat*:

*Example 2.* hyperhypernyms: garment.n.01 | hypernyms: overgarment.n.01 | synset: coat.n.01 | hyponyms:

*Example 3.* гипергиперонимы: одежда | гиперонимы: верхняя одежда | синсет: куртка | гипонимы:

We present the results for manual and automatic datasets in Tables 3 and 2, respectively. The results indicate that, in general, small datasets tend to yield significantly higher scores, suggesting that generative models are adept at assimilating hyponymic relations of frequently used words. However, automatic datasets, as previously mentioned, often comprise of rare and narrow concepts that lead to lower performance.

We can also observe the already mentioned trend that the Recall scores for GPT2 are significantly higher than those of T5 for both languages.

Regarding the architecture of the models, it is challenging to determine whether decoders or encoder-decoders perform better since the results are inconsistent between English and Russian. Among the smaller English models, the best scores in terms of MAP are achieved by T5, while for Russian, GPT2 fares better. Comparison of the results for Russian and English is not straightforward due to the usage of different prefixes and a varying number of training vertices - 15,000 for English and 10,000 for Russian.

Upon comparing the language-specific data, it becomes evident that the performance of smaller models is slightly better for Russian than to English according to the MAP scores. We connect this finding to the presence of a larger number of noun synsets for English, signifying a higher degree of fine-grained concepts in the taxonomy.

Regarding the large-scale instructional models, Dolly exhibits a significant lead over smaller models in relation to its performance on English-language data. This finding highlights the superior ability of larger models to assimilate hyponymic relations of both frequent and rare concepts. However, the same substantial increase in performance is not readily apparent for Saiga-7b. This outcome can be elucidated by the fact that the LlaMA model upon which Saiga-7b is based was primarily trained on English-language data. Moreover, the additional LoRa fine-tuning was confined to a relatively modest corpus of artificially generated Russian dialogue data, resulting in fewer instances of Russian lexical diversity being encountered during the training of the model.

## 7   Conclusion

In the presented study, we introduce a novel approach for constructing input data by incorporating information on the structure of the lexical taxonomy into large pre-trained language models. Our method shows that generative models provided with information about the graph in a well-perceived textual form can significantly improve the quality of hyponymy prediction. In this paper, we also provide a manually assembled dataset of general concepts for English, as well as the first datasets for evaluating the quality of predicting hyponyms in the context of enriching taxonomies for the Russian language.

Despite the fact that the prediction of hypernyms has a wider practical application than the prediction of hyponyms, the findings of the presented study are valuable for creating comprehensive common domain lexical taxonomies from scratch, benefiting low-resource languages.

In our research, we fine-tune relatively small-sized models, which demonstrate decent performance on both English and Russian data. However, the results from Dolly-v1-5b demonstrate that larger models can yield substantial improvements. As observed by Logan IV et al. [19] when comparing his findings to other studies [17,18], full fine-tuning is more effective for smaller models than prompt-tuning. Authors assumes that as model size increases, prompt-tuning yields better results. Thus, we can observe a promising path for future work to prompt-tune sizable generative models. We also anticipate that our approach can be extended both for other languages and other taxonomy enrichment tasks such as inserting the new node in the middle of the graph.

**Acknowledgements.** This work was supported by the DFG through the project "ACQuA: Answering Comparative Questions with Arguments" (grants BI 1544/7- 1 and HA 5851/2- 1) as part of the priority program "RATIO: Robust Argumentation Machines" (SPP 1999).

## A   Appendix

**Formats of Artificial Prefixes:**

   **Default:** part of speech (POS) tags and sense number are excluded from synset names.

*Example 4.* hyperhypernyms: undertaking | hypernyms: assignment | synset: school assignment | hyponyms: classroom project, classwork, homework, prep, preparation, lesson

   **Sense number:** the full name of the synset is used, including POS tag and sense number.

*Example 5.* hyperhypernyms: undertaking.n.01 | hypernyms: assignment.n.05 | synset: school assignment.n.01 | hyponyms: classroom project, classwork, homework, prep, preparation, lesson

**Lemmas:** a list of lemmas is used instead of the name of the synset.

*Example 6.* hyperhypernyms: undertaking, project, task, labor | hypernyms: assignment | synset: school assignment, schoolwork | hyponyms: classroom project, classwork, homework, prep, preparation, lesson

**Additional lemmas:** the lemmas included in the synset are listed after its full name.

*Example 7.* hyperhypernyms: undertaking.n.01 (undertaking, project, task, labor) | hypernyms: assignment.n.05 (assignment) | synset: school assignment.n.01 (school assignment, schoolwork) | hyponyms: classroom project, classwork, homework, prep, preparation, lesson

**Definitions:** definitions for target synsets are given in parentheses.

*Example 8.* hyperhypernyms: undertaking.n.01 | hypernyms: assignment.n.05 | synset: school assignment.n.01 (a school task performed by a student to satisfy the teacher) | hyponyms: classroom project, classwork, homework, prep, preparation, lesson

**Hyperdefinition:** definitions for hypernyms and hyperhypernyms are given in parentheses.

*Example 9.* hyperhypernyms: undertaking.n.01 (any piece of work that is undertaken or attempted) | hypernyms: assignment.n.05 (an undertaking that you have been assigned to do (as by an instructor)) | synset: school assignment.n.01 | hyponyms: classroom project, classwork, homework, prep, preparation, lesson

**Hypernym only:** only the hypernym is given.

*Example 10.* hypernyms: assignment.n.05 | synset: school assignment.n.01 | hyponyms: classroom project, classwork, homework, prep, preparation, lesson

**All additions:** all possible information is used: sense number, definitions and lemmas.

*Example 11.* hypernyms: assignment.n.05 (assignment) (an undertaking that you have been assigned to do (as by an instructor)) | synset: school assignment.n.01 (school assignment, schoolwork) (a school task performed by a student to satisfy the teacher) | hyponyms: classroom project, classwork, homework, prep, preparation, lesson (Tables 4 and 5).

**Table 4.** Content and statistic of manually curated dataset for Russian. Here $d_l$ denotes to leaf distance, $d_r$ to root distance and $n_h$ to the total number of hyponyms including indirect ones.

| Id | Title | $d_l$ | $d_r$ | $n_h$ |
|---|---|---|---|---|
| 6892-N | ПАЛЬТО | 2 | 7 | 4 |
| 108048-N | КУРТКА | 1 | 7 | 7 |
| 108194-N | ШТАНЫ, БРЮКИ | 2 | 6 | 10 |
| 109093-N | МАКАРОННЫЕ ИЗДЕЛИЯ | 3 | 6 | 6 |
| 3921-N | СЫР | 4 | 8 | 16 |
| 1225-N | МЯСО | 1 | 5 | 29 |
| 8367-N | ВИНО | 1 | 7 | 25 |
| 5239-N | КОНФЕТА | 2 | 6 | 8 |
| 107283-N | ПИРОГ | 1 | 6 | 9 |
| 549-N | НАПИТОК | 2 | 5 | 82 |
| 107842-N | ЯГОДА | 1 | 9 | 28 |
| 109620-N | ДЕТСКАЯ ИГРУШКА | 1 | 6 | 20 |
| 7992-N | УДАРНЫЙ МУЗЫКАЛЬНЫЙ ИНСТРУМЕНТ | 1 | 6 | 10 |
| 107996-N | СТРУННЫЙ МУЗЫКАЛЬНЫЙ ИНСТРУМЕНТ | 2 | 6 | 20 |
| 1045-N | ВРАЧ | 1 | 5 | 85 |
| 354-N | ФРУКТ | 1 | 9 | 37 |
| 348-N | ОВОЩ | 1 | 8 | 27 |
| 107795-N | ХИЩНОЕ МЛЕКОПИТАЮЩЕЕ | 1 | 7 | 56 |
| 4454-N | СОБАКА | 2 | 7 | 53 |
| 109170-N | КОСМЕТИЧЕСКОЕ СРЕДСТВО | 1 | 6 | 21 |
| 4318-N | КРУПА | 1 | 6 | 12 |
| 965-N | ЦВЕТКОВОЕ РАСТЕНИЕ | 2 | 5 | 51 |

**Table 5.** Content and statistic of manually curated dataset for English. Here $d_l$ denotes to leaf distance, $d_r$ to root distance and $n_h$ to the total number of hyponyms including indirect ones.

| Id | Title | $d_l$ | $d_r$ | $n_h$ |
|---|---|---|---|---|
| 3057021 | coat.n.01 | 2 | 9 | 53 |
| 3045337 | cloak.n.02 | 2 | 9 | 29 |
| 4489008 | trouser.n.01 | 2 | 8 | 26 |
| 7698915 | pasta.n.02 | 1 | 5 | 26 |
| 7850329 | cheese.n.01 | 3 | 5 | 37 |
| 7649854 | meat.n.01 | 1 | 5 | 197 |
| 7891726 | wine.n.01 | 3 | 7 | 68 |
| 7597365 | candy.n.01 | 1 | 8 | 62 |
| 7625493 | pie.n.01 | 1 | 7 | 25 |
| 7881800 | beverage.n.01 | 3 | 5 | 339 |
| 7742704 | berry.n.01 | 4 | 7 | 21 |
| 3219135 | doll.n.01 | 1 | 6 | 8 |
| 3249569 | drum.n.01 | 1 | 9 | 8 |
| 3467517 | guitar.n.01 | 1 | 9 | 6 |
| 502415 | board_game.n.01 | 1 | 8 | 18 |
| 4105893 | room.n.01 | 1 | 7 | 195 |
| 3405725 | furniture.n.01 | 1 | 7 | 196 |
| 13388245 | coin.n.01 | 1 | 8 | 41 |
| 3597469 | jewelry.n.01 | 1 | 7 | 39 |
| 3714235 | makeup.n.01 | 1 | 8 | 11 |
| 4959672 | chromatic_color.n.01 | 1 | 6 | 91 |
| 1699831 | dinosaur.n.01 | 1 | 12 | 50 |

**Table 6.** Number of synsets and hyponyms in test datasets.

| Dataset | Synsets | Hyponyms |
|---|---|---|
| CHSP | 1000 | 13617 |
| RuCHSP | 1000 | 5673 |
| EnManual | 22 | 1546 |
| RuManual | 22 | 616 |

# References

1. Aly, R., Acharya, S., Ossa, A., Köhn, A., Biemann, C., Panchenko, A.: Every child should have parents: a taxonomy refinement algorithm based on hyperbolic term embeddings. In: Proceedings of the 57th Annual Meeting of the Association for Computational Linguistics, Florence, Italy, pp. 4811–4817. Association for Computational Linguistics (2019). https://doi.org/10.18653/v1/P19-1474. https://aclanthology.org/P19-1474

2. Brown, T.B., et al.: Language models are few-shot learners. In: Larochelle, H., Ranzato, M., Hadsell, R., Balcan, M., Lin, H. (eds.) Advances in Neural Information Processing Systems 33: Annual Conference on Neural Information Processing Systems 2020, NeurIPS 2020, 6–12 December 2020, virtual (2020)

3. Chang, D., Lin, E., Brandt, C., Taylor, R.: Incorporating domain knowledge into language models using graph convolutional networks for clinical semantic textual similarity (preprint). JMIR Med. Inform. (2020). https://doi.org/10.2196/23101

4. Chung, H.W., et al.: Scaling instruction-finetuned language models (2022)

5. Conover, M., et al.: Free Dolly: Introducing the World's First Truly Open Instruction-Tuned LLM (2023). https://www.databricks.com/blog/2023/04/12/dolly-first-open-commercially-viable-instruction-tuned-llm

6. Dale, D.: A simple solution for the taxonomy enrichment task: discovering hypernyms using nearest neighbor search (2020)

7. Dettmers, T., Lewis, M., Belkada, Y., Zettlemoyer, L.: Llm.int8(): 8-bit matrix multiplication for transformers at scale (2022)

8. Devlin, J., Chang, M.W., Lee, K., Toutanova, K.: BERT: pre-training of deep bidirectional transformers for language understanding. In: Proceedings of the 2019 Conference of the North American Chapter of the Association for Computational Linguistics: Human Language Technologies, Minneapolis, Minnesota (Volume 1: Long and Short Papers), pp. 4171–4186. Association for Computational Linguistics (2019). https://doi.org/10.18653/v1/N19-1423. https://aclanthology.org/N19-1423

9. Dietz, L., Kotov, A., Meij, E.: Utilizing knowledge graphs in text-centric information retrieval, pp. 815–816 (2017). https://doi.org/10.1145/3018661.3022756

10. Ettinger, A.: What BERT is not: lessons from a new suite of psycholinguistic diagnostics for language models. Trans. Assoc. Comput. Linguist. 8, 34–48 (2020). https://doi.org/10.1162/tacl_a_00298. https://aclanthology.org/2020.tacl-1.3

11. Fellbaum, C. (ed.): WordNet: An Electronic Lexical Database. Language, Speech, and Communication. MIT Press, Cambridge (1998)

12. Hanna, M., Mareček, D.: Analyzing BERT's knowledge of hypernymy via prompting. In: Proceedings of the Fourth BlackboxNLP Workshop on Analyzing and Interpreting Neural Networks for NLP, Punta Cana, Dominican Republic, pp. 275–282. Association for Computational Linguistics (2021). https://doi.org/10.18653/v1/2021.blackboxnlp-1.20. https://aclanthology.org/2021.blackboxnlp-1.20

13. He, B., et al.: BERT-MK: integrating graph contextualized knowledge into pre-trained language models. In: Findings of the Association for Computational Linguistics: EMNLP 2020, pp. 2281–2290. Association for Computational Linguistics, Online (2020). https://doi.org/10.18653/v1/2020.findings-emnlp.207. https://aclanthology.org/2020.findings-emnlp.207

14. Hu, E.J., et al.: Lora: low-rank adaptation of large language models (2021)

15. Huang, X., Zhang, J., Li, D., Li, P.: Knowledge graph embedding based question answering. In: Proceedings of the Twelfth ACM International Conference on Web

Search and Data Mining, WSDM 2019, pp. 105–113. Association for Computing Machinery, New York (2019). https://doi.org/10.1145/3289600.3290956

16. Izacard, G., Grave, E., Pilehvar, M.T., Alzantot, M., Baroni, M.: LLaMA: open and efficient foundation language models (2023)

17. Lester, B., Al-Rfou, R., Constant, N.: The power of scale for parameter-efficient prompt tuning. In: Proceedings of the 2021 Conference on Empirical Methods in Natural Language Processing, pp. 3045–3059. Association for Computational Linguistics, Online and Punta Cana, Dominican Republic (2021). https://doi.org/10.18653/v1/2021.emnlp-main.243. https://aclanthology.org/2021.emnlp-main.243

18. Li, X.L., Liang, P.: Prefix-tuning: optimizing continuous prompts for generation. In: Proceedings of the 59th Annual Meeting of the Association for Computational Linguistics and the 11th International Joint Conference on Natural Language Processing (Volume 1: Long Papers), pp. 4582–4597. Association for Computational Linguistics, Online (2021). https://doi.org/10.18653/v1/2021.acl-long.353. https://aclanthology.org/2021.acl-long.353

19. Logan, R.L., IV., Balaževič, I., Wallace, E., Petroni, F., Singh, S., Riedel, S.: Cutting down on prompts and parameters: simple few-shot learning with language models. CoRR abs/2106.13353 (2021). https://arxiv.org/abs/2106.13353

20. Nikishina, I., Vakhitova, A., Tutubalina, E., Panchenko, A.: Cross-modal contextualized hidden state projection method for expanding of taxonomic graphs. In: Proceedings of TextGraphs-16: Graph-Based Methods for Natural Language Processing, Gyeongju, Republic of Korea, pp. 11–24. Association for Computational Linguistics (2022). https://aclanthology.org/2022.textgraphs-1.2

21. Ouyang, L., et al.: Training language models to follow instructions with human feedback (2022)

22. Radford, A., Wu, J., Child, R., Luan, D., Amodei, D., Sutskever, I.: Language models are unsupervised multitask learners (2018). https://d4mucfpksywv.cloudfront.net/better-language-models/language-models.pdf

23. Raffel, C., et al.: Exploring the limits of transfer learning with a unified text-to-text transformer. J. Mach. Learn. Res. **21**(140), 1–67 (2020). http://jmlr.org/papers/v21/20-074.html

24. Touvron, H., et al.: LLaMA 2: open foundation and fine-tuned chat models (2023)

25. Vulić, I., Ponti, E.M., Litschko, R., Glavaš, G., Korhonen, A.: Probing pre-trained language models for lexical semantics. In: Proceedings of the 2020 Conference on Empirical Methods in Natural Language Processing (EMNLP), pp. 7222–7240. Association for Computational Linguistics, Online (2020). https://doi.org/10.18653/v1/2020.emnlp-main.586. https://aclanthology.org/2020.emnlp-main.586

26. Wiedemann, G., Remus, S., Chawla, A., Biemann, C.: Does BERT make any sense? interpretable word sense disambiguation with contextualized embeddings (2019)

27. Zhou, R., et al.: WCL-BBCD: a contrastive learning and knowledge graph approach to named entity recognition (2022)

28. Zhou, Y., et al.: Large language models are human-level prompt engineers. In: The Eleventh International Conference on Learning Representations (2023). https://openreview.net/forum?id=92gvk82DE-

# Content Selection in Abstractive Summarization with Biased Encoder Mixtures

Daniil Chernyshev[1(✉)] and Boris Dobrov[1,2]

[1] Research Computing Center, Lomonosov Moscow State University, Moscow, Russia
chdanorbis@yandex.ru
[2] ISP RAS Research Center for Trusted Artificial Intelligence, Moscow, Russia

**Abstract.** Current abstractive summarization models consistently outperform extractive counterparts yet are unable to close the gap with Oracle extractive upper bound. Recent research suggests that the reason lies in the lack of planning and bad sentence-wise saliency intuition. Existing solutions to the problem either require new fine-tuning sessions to accommodate the architectural changes or disrupt the natural information flow limiting the utilization of accumulated global knowledge. Inspired by text-to-image result blending techniques we propose a plug-and-play alternative that preserves the integrity of the original model, Biased Encoder Mixture. Our approach utilizes attention masking and Siamese networks to reinforce the signal of salient tokens in encoder embeddings and guide the decoder to more relevant results. The evaluation on four datasets and their respective state-of-the-art abstractive summarization models demonstrate that Biased Encoder Mixture outperforms the attention-based plug-and-play alternatives even with static masking derived from sentence saliency positional distribution.

**Keywords:** Abstractive Summarization · Content Control · Attention Mechanism

## 1 Introduction

The quality of automatic summarization has improved substantially in the past decade thanks to advances in neural language modeling. The basis of the state-of-the-art approaches is pre-trained transformer language models that efficiently integrate global contextual knowledge through multi-head attention mechanism. The pre-trained models vary in both attention mechanism architecture and pre-training tasks, however, the baseline generative language models such as BART [1] and Pegasus [2] have the best average performance. Recent works [3–5] showed that with proper fine-tuning procedure, these generative models outperform methods with additional input [6] or complex post-generation result refinement [7,8]. At the same time, it was also shown that abstractive summarization approaches based solely on pre-trained language models lack the understanding of the extractive side of the task in terms of sentence-wise saliency [6,9].

D. I. Ignatov et al. (Eds.): AIST 2023, LNCS 14486, pp. 64–77, 2024.
https://doi.org/10.1007/978-3-031-54534-4_5

Improving extractive intuition can significantly boost the quality of generated summaries as well as provide better control over the generation process. However, the main issue with existing approaches is the need to alter either the architecture of the model [9–11] or the input format [6,12] which in turn requires a new fine-tuning session. This trait renders these approaches incompatible with existing state-of-the-art summarization models that were tailored to specific data representations and pre-trained language models [3–5].

Inspired by text-to-image result blending techniques [13], we propose Biased Encoder Mixture[1], an alternative solution to integrating the extractive summarization knowledge without interfering with the existing abstractive framework structure. Our method exploits the attention mechanism and Siamese networks to reinforce the signal of salient sentences in document embedding thus encouraging the decoder to better utilize the selected content. Unlike previous attention-based approaches [14–16] Biased Encoder Mixture retains the original information flow and doesn't introduce additional model parameters.

We test Biased Encoder Mixture on four diverse datasets of CNN/Daily Mail, Xsum, ArXiv, and SAMSum. The evaluation demonstrates that Biased Encoder Mixture consistently outperforms previous approaches even with static attention masking derived from Extractive Oracle sentence position distribution. Employing an auxiliary extractive summarization model for dynamic masking further boosts the quality of generated summaries bringing up to a 7% ROUGE-2 relative improvement to state-of-the-art abstractive summarization models.

## 2    Related Work

One of the first models to surpass extractive baselines was Pointer-generator network [17] which utilizes explicit copy-mechanism to bridge the gap between abstractive and extractive models. This model was further improved with an auxiliary content selection module [18] that limited the copy attention of Pointer generator to the most salient parts of the source document. Further development of the idea led to the introduction of EASE [10] joint extractive-abstractive training that optimizes content selector to maximize the quality of abstractive summarizer. The maximal effect of joint training was achieved with SEASON [9] model that integrated the content selector in Transformer encoder-decoder abstractive summarization model by training the encoder to provide sentence saliency weights for cross-attention.

A similar line of work augmented the content control signal through additional network components. Lebanoff et al. [12] proposed complementing word embeddings with highlight embeddings extracted from content selector to perform cascade summarization. Ji et al. [11] improved the approach by extending to inter-layer encoder embeddings and constructing highlight embeddings within the main model. For query-based content control Dou et al. [6] proposed GSum framework that encodes the guidance query in Siamese network fashion and passes it to an additional cross-attention block in the decoder layer. Cao

---

[1] https://github.com/dciresearch/BEM-ContentSelection.

et al. [14] adapted Bottom-up [18] attention modulation approach to Transformer architecture and proposed guiding individual cross-attention heads. Xiao et al. [15] combined attention modulation and GSum approaches by introducing relevance attention that biases cross-attention weights according to query relevance.

## 3 Motivation and Methodology

There exists evidence that humans perform abstractive summarization in a two-stage manner by initially selecting the salient fragments and then fusing and paraphrasing the fragments into sentences [19]. However, in practice, the neural abstractive summarization models lack advance content planning as they are trained in the same fashion as neural machine translation models, which is the estimation of the conditional probability distribution of the next generated token.

With the current abstractive summarization paradigm, the preliminary salient content selection can be done in two ways. Amid the rise of cloud LLMs (e.g. ChatGPT) the most popular is a two-step approach of reducing the document to the most salient fragments using the extractive summarization model and then summarizing these fragments with the abstractive model [20, 21]. While the approach is easy to implement it is let down by the suboptimality of extractive summarization systems [22].

A more reliable way to filter out the salient content is through input suppression. The idea is to assign weights to tokens with respect to relevance of their content and use the weights as extractive guidance in the generation process. This approach allows to retain the full information flow and at the same time ensures that the abstractive summarization model would follow the predefined content selection plan. Token weighting is commonly performed through either an attention mechanism [10, 14, 15, 18] or additional architectural components/input [6, 9, 11, 12]. In the era of pre-trained language models, the former plug-and-play approach has a higher preference as it doesn't interfere with knowledge accumulated during the pre-training stage.

### 3.1 Attention Distribution Discrepancy

Integrating the extractive knowledge has been repeatedly shown to significantly boost the quality of generated summaries [6, 14, 16]. However, none of the works investigated the reasons behind such positive changes. As most of the summarization datasets don't provide labels for extractive summarization it is common to estimate the reference with Extractive Oracle [23]:

$$\text{Oracle}(X, y) = \arg\max_{s_i \subset X} \text{ROUGE-2}(s_i, y) \tag{1}$$

where $s_i$ is a subset of sentences from document $X$, $y$ is the reference summary. While this solution serves as the performance upper bound of extractive systems, numerous works [6,10,23,24] showed that it proves to be challenging even for abstractive summarization approaches hinting at a bottleneck in extractive capabilities. To test whether it is the consequence of suboptimal content selection we must inspect the content extraction patterns of the abstractive summarization model. In Transformer models the pattern can be derived from cumulative attention distribution obtained by ALTI [25] method by aggregating total token-wise attention by sentences.

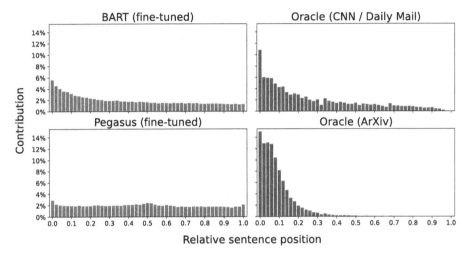

**Fig. 1.** Comparison of sentence-wise attention and Extractive Oracle positions.

We compared the attention and Oracle patterns of state-of-the-art abstractive summarization models on CNN Daily Mail and ArXiv (truncated) datasets (Fig. 1). Both datasets are known to exhibit strong leading sentence bias and the models are expected to reflect it in their attention patterns. CNN/Daily Mail focuses on news summarization and the news articles are written in accordance with the inverted pyramid scheme. This scheme dictates that the information in the text must be delivered in the order of its importance, so the reader skimming through the introductory sentences would get the same idea of the described event as the one who reads the full text. Similarly, scientific articles from ArXiv usually base their abstracts on the introduction section of the article and thus the respective sentences have the highest saliency. However, contrary to extractive distribution, BART is weakly biased and Pegasus possesses almost uniform attention distribution. Therefore, the benefits of extractive knowledge integration come from weak sentence-wise saliency intuition of pure abstractive summarization models.

## 3.2 Biasing Attention with Encoder Mixture

Correcting the attention distribution is the key to boosting the performance of fine-tuned models. A straightforward approach is to utilize the original binary attention masking mechanism to condition the model on the salient sentences. Given input mask $m = \{0,1\}^n$ the attention masking of head $h$ in encoder self-attention and decoder cross-attention is performed by nullifying the columns of stochastic matrix derived from query $q$ and key $K$ that is used to combine the components of value matrix $V$:

$$\text{Attention}(q, K, V) = \text{softmax}\left(\frac{qK^T}{\sqrt{d_h}} + (1-m)\cdot\infty\right) \tag{2}$$

where $d_h$ is the normalizing coefficient. Thus, binary masking of individual heads negates the effect of embeddings $v_i \in V$ corresponding to masked positions $m_i = 0$. The main issue is that the original implementation shares the mask $m$ across all layers and heads of the model, completely preventing the propagation of information from the masked positions. In the case of saliency-based masking this results in total loss of secondary information as if the input text was filtered before passing to the summarization model. The approach was tested in plug-and-play [11] and joint-training [10] scenarios and resulted in performance degradation even with additional pre-training.

Soft attention masking addresses the information loss by applying a variable penalty instead of $-\infty$ on selected positions to promote the remaining content, thus allowing an arbitrary information flow. This approach has been tested separately on encoder self-attention [16] and decoder cross-attention [11,14,16] with considerable success. However, unlike binary masking soft masking can't be used in zero-shot fashion as the optimal set of layer-wise and head-wise attention penalty coefficients differ from model to model and therefore must be tuned on validation set [14,16].

We hypothesize that a similar result may be achieved by altering embeddings directly. Previous works applied embedding modulation [11,12] to the encoder part of encoder-decoder models and achieved a significant improvement in the supervised setting. However, these methods can't be used without fine-tuning as well since they modify the input format and alter the intermediate embeddings of encoder layers.

To adapt embedding modulation to plug-and-play scenario the points of modification must be limited to model agnostic checkpoint, i.e., the final output of the encoder. This approach has been used in text-to-image DALLE-2 [13] model to merge the results of different prompts by interpolating the textual embeddings obtained by CLIP [26] encoder. Following this idea, we propose Biased Encoder Mixture.

Biased Encoder Mixture (BEM) alleviates the information gaps of binary attention masking by augmenting the original encoder-decoder abstractive summarization model with auxiliary attention biased input encoder (Fig. 2). Given input tokens $X = (x_1, x_2, ..., x_n)$ content selector produces the biased attention mask $\hat{m}$:

$$\hat{m}_i = \begin{cases} 1, & x_i \in \text{SalientSentences}(X) \\ 0, & \text{otherwise} \end{cases} \tag{3}$$

Salient sentence subset is determined by sampling the upper $p$-th percentile of saliency score distribution. The distribution can be estimated in two ways, stati-

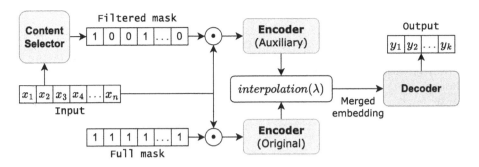

**Fig. 2.** Illustration of Biased Encoder Mixture

cally from Oracle sentence positional distribution (Fig. 1) and dynamically from sentence ranking predicted by the auxiliary extractive summarization model.

The biased mask $\hat{m}$ and input $X$ are then used to produce two document embeddings:

$$E = Encoder(X, m_0) \qquad (4)$$

$$\hat{E} = Encoder_{aux}(X, \hat{m}) \qquad (5)$$

where $Encoder_{aux}$ is an auxiliary encoder, $m_0 = (1, 1, ..., 1)$ is full attention mask. The original text-to-image approach [13] uses standard Siamese network ($Encoder_{aux} = Encoder$), however, we hypothesized that an auxiliary identity encoder that boosts the probability of copying the fragments from the source document could be beneficial.

The original and biased embeddings are merged by either of the two interpolation strategies:

$$\mathrm{lerp}(E, \hat{E}, \lambda) = (1 - \lambda)E + \lambda\hat{E} \qquad (6)$$

$$\mathrm{slerp}(E, \hat{E}, \lambda) = \frac{\sin\left((1 - \lambda)\Omega\right)}{\sin\Omega}E + \frac{\sin\left(\lambda\Omega\right)}{\sin\Omega}\hat{E} \qquad (7)$$

$$\Omega = \arccos\left(E \cdot \hat{E}\right) \qquad (8)$$

where $\lambda \in [0, 1]$ is an interpolation coefficient. The combined embedding is passed to decoder cross-attention with full attention mask $m_0$ to perform generation as if the information was provided by a sole encoder without any content filtering.

## 4   Experiments

In this section, we show the results of different attention modulation methods conditioned on extractive prior knowledge. To explain the effect on the summary generation process we provide an analysis of the methods in terms of content changes.

**Table 1.** Statistics of experimental datasets

| Dataset | # examples in Train/Val/Test | # words | | Extractiveness | | |
|---------|------------------------------|---------|--------|--------|----------|---------|
| | | Source | Target | Comp. | Coverage | Density |
| CNN/Daily Mail | 287K/13K/11K | 698.6 | 49.53 | 0.09 | 88% | 3.77 |
| Xsum | 203K/11K/11K | 383.17 | 21.74 | 0.1 | 64% | 1.06 |
| ArXiv | 203K/6K/6K | 5179.22 | 257.44 | 0.04 | 87% | 3.94 |
| SAMSum | 14K/818/819 | 97.23 | 21 | 0.25 | 68% | 1.46 |

### 4.1 Experimental Setting

**Datasets.** For the experiments, we chose four abstractive summarization datasets that were used in recent works [3–5]. CNN/Daily Mail [17] contains news articles and associated editor highlights acting as summaries from CNN and Daily Mail. Xsum [24] is a more abstractive counterpart with one-sentence summaries web scraped from BBC online news portal. ArXiv [27] is a collection of article-abstract pairs from ArXiv science article archive. SAMSum [28] is a human-annotated dialogue summarization dataset, which is created by asking linguists to create messenger-like conversations and then asking another group of linguists to write summaries. Dataset length statistics and extractive characteristics [29] are reported in Table 1.

**Baselines.** Besides backbone state-of-the-art abstractive summarization models, we compare the performance of our methods with common extractive baselines and popular attention-based content control methods. As for the abstractive backbone[2], we use BRIO (BART) [3] for CNN/Daily Mail, BRIO (Pegasus) [3] for Xsum, Pegasus [2] for ArXiv, BART [1] for SAMSum. For extractive we use Extractive Oracle [23] and BERTSumExt [23] model (denoted as BERTExt). For attention manipulation we consider three methods. Standard binary attention masking of the chosen subset of source sentences, Layer attention [14] which applies the binary mask to cross attention of selected decoder layer, and Relevant Attention [15] which modulates the cross attention with query-source similarity matrix (in our case query is concatenated sentence subset).

**Implementation Details.** Due to input length limitations of backbone abstractive summarization models, we truncate the input documents to the first 1024 tokens. To predict sentence saliency ranking, we use BERTExt model. The top-p sampling percentile for both static and dynamic distributions is determined by the 90th percentile of the reference-source sentence count ratio to ensure reference-source sentence alignment and accommodate for saliency ranking error. Identity encoders are obtained by freezing the decoders of abstractive summarization models and training the encoders on the corresponding dataset on auto-encoding task (generate the input document) for 1 epoch (to avoid

---

[2] We use model weights available at https://huggingface.co/models.

**Table 2.** Optimal parameters for attention correction methods

| Data | Distribution estimation | top-p | Layer attention | Relevance attention | BEM-id | BEM-original |
|------|------|------|------|------|------|------|
| CNN/Daily Mail | static | 0.25 | 5 | 0.05 | 0.1/slerp | 0.1/lerp |
| | dynamic | | 3 | 0.05 | 0.1/slerp | 0.2/slerp |
| Xsum | static | 0.2 | 2 | 0.01 | 0.1/lerp | 0.2/lerp |
| | dynamic | | 1 | 0.01 | 0.1/slerp | 0.1/slerp |
| ArXiv | static | 0.26 | 4 | 0.01 | 0.1/lerp | 0.1/lerp |
| | dynamic | | 3 | 0.03 | 0.1/slerp | 0.2/slerp |
| SAMSum | static | 0.32 | 5 | 0.03 | 0.1/slerp | 0.1/slerp |
| | dynamic | | 3 | 0.03 | 0.1/lerp | 0.1/slerp |

overfitting) with batch size 32, Adam optimizer with learning rate 3e-5 and linear scheduler with warmup ratio 0.1. BERTExt is trained on Extractive Oracle labels for 3 epochs with batch size 16, early stopping, and the same optimizer configuration. The interpolation method and $\lambda$ coefficient for Biased Encoder Mixture are determined for each dataset independently with grid search with grid size 0.1. Similarly, we use grid search to find the optimal masking layer and relevance modulation coefficient for Layer attention and Relevant attention. The exact values used in experiments can be found in Table 2 (BEM coefficients are explained in Sect. 4.2).

## 4.2   Results

The results of attention correction experiments are reported in Table 3. Each data subsection is split into Extractive Oracle and BERTExt saliency distribution column blocks, each showing the performance of *best subset* of top-p sampled sentences, the original abstractive summarization model, and the results of attention biasing methods.

As can be seen with static Oracle position distribution-based attention masking the improvements in the majority of cases are marginal. Binary masking being the strictest form of attention control only degrades the performance of the original model as it completely restricts the input information. A more precise approach such as Layer attention is more stable yet improves only certain ROUGE aspects. Relevance attention has comparable performance, however, is less efficient on more abstractive datasets such as XSum. Biased Encoder Mixture is more consistent with performance improvements. An identity auxiliary encoder mixture (BEM-id) is the most stable approach as it is the only one to show the improvements on SAMSum dataset. A simpler Siamese BEM variant (BEM-original) better utilizes the attention guidance on BRIO contrastive-tuned models suggesting a correlation with encoder discrimination power.

**Table 3.** Results of attention correction experiments

| Data | Model | Bias method | Sentence Saliency Evaluator | | | | | |
|---|---|---|---|---|---|---|---|---|
| | | | Extractive Oracle | | | BERTExt | | |
| | | | R1 | R2 | RL | R1 | R2 | RL |
| CNN/Daily Mail | Best subset | – | 53.88 | 32.73 | 49.97 | 42.21 | 20.25 | 38.87 |
| | BRIO (BART) | – | 47.14 | 23.75 | 44.33 | | | |
| | | Binary | 46.38 | 22.55 | 43.28 | 33.36 | 12.60 | 32.02 |
| | | Layer attention | **47.68** | 23.70 | **44.98** | **48.69** | **24.56** | **45.46** |
| | | Relevance attention | 47.22 | 23.81 | 44.31 | 47.50 | 24.06 | 45.02 |
| | | BEM-id | 47.44 | 24.36 | 44.46 | 47.40 | 24.45 | 44.89 |
| | | BEM-original | **47.63** | 24.21 | **44.77** | **49.08** | **25.57** | **46.18** |
| Xsum | Best subset | – | 27.52 | 7.59 | 21.46 | 21.55 | 4.90 | 14.41 |
| | BRIO (Pegasus) | – | 48.96 | 25.29 | 40.11 | | | |
| | | Binary | 47.51 | 23.59 | 38.23 | 35.99 | 14.60 | 28.05 |
| | | Layer attention | **49.22** | 25.30 | **40.32** | 49.14 | 25.44 | **40.41** |
| | | Relevance attention | 49.14 | **25.44** | 40.22 | **49.22** | **25.48** | 40.29 |
| | | BEM-id | 49.06 | 25.26 | 40.06 | 49.37 | **25.79** | 40.24 |
| | | BEM-original | **49.85** | **25.69** | **40.65** | **49.56** | 25.66 | **40.37** |
| ArXiv | Best subset | – | 53.24 | 23.31 | 32.42 | 41.25 | 15.63 | 29.88 |
| | Pegasus | – | 43.44 | 16.22 | 25.66 | | | |
| | | Binary | 42.54 | 14.85 | 24.90 | 42.64 | 14.95 | 24.64 |
| | | Layer attention | 43.44 | 16.34 | 25.67 | **43.70** | **16.41** | **25.66** |
| | | Relevance attention | 43.46 | **16.37** | 25.67 | 43.62 | 16.19 | 25.63 |
| | | BEM-id | **44.18** | **16.54** | 25.61 | **44.64** | 16.86 | 25.81 |
| | | BEM-original | 43.62 | 16.30 | 25.64 | 44.52 | **17.13** | **26.03** |
| SAMSum | Best subset | – | 45.21 | 17.58 | 39.55 | 34.07 | 12.48 | 30.69 |
| | BART | – | 53.39 | 28.32 | 44.12 | | | |
| | | Binary | 49.51 | 24.14 | 40.60 | 34.20 | 15.84 | 28.47 |
| | | Layer attention | 53.37 | 28.26 | 44.00 | **54.57** | **29.33** | **45.30** |
| | | Relevance attention | 53.11 | 27.99 | 43.65 | 53.65 | 28.21 | 44.03 |
| | | BEM-id | 53.20 | **28.76** | **44.45** | 52.69 | 27.95 | 43.70 |
| | | BEM-original | 53.29 | 28.15 | 44.18 | **53.93** | **28.61** | **44.52** |

With extractive summarization-based attention masking the quality improvements are much more noticeable. Layer attention stabilizes and shows quality gains on all ROUGE aspects. Relevance attention on the other hand struggles to utilize the improved guidance signal, retaining similar scores. In contrast, BEM ranking reverses as BEM-original demonstrates the best overall performance. The Siamese variant achieves almost 8% and 6% relative ROUGE-2 improvement over the original model on CNN/Daily Mail and ArXiv respectively. At the same time, the efficiency is much lower on datasets with higher abstractiveness of reference summaries (Table 1, extractive Density) as extractive strategies aren't optimal and thus the extractive summarizer can't properly estimate the alignment of source sentences with the reference summary.

**Table 4.** Examples of Content controlled generation

| |
|---|
| **Reference** |
| U.S. Navy is developing an unmanned drone ship to track enemy submarines to limit their tactical capacity for surprise. The vessel would be able to operate under with little supervisory control. Advances are necessary to maintain technological edge on Russia and China, admiral tells House panel. |
| **Original Prediction** |
| U.S. must rethink role of manned submarines, report says. China and other nations are rapidly expanding their submarine forces. The U.S., the U.K. and others are developing unmanned drone ships to track enemy submarines. |
| **Relevance attention** |
| U.S. must rethink role of manned submarines, report says. The U.S., China are rapidly expanding their submarine forces. The Pentagon is developing an unmanned drone ship to track enemy submarines. |
| **Layer attention** |
| U.S. must rethink role of manned submarines, report says. China is rapidly expanding its submarine forces. The U.S., officials say, is developing an unmanned drone ship to track enemy submarines. |
| **BEM-original (Ours)** |
| The U.S. Navy is developing an unmanned drone ship to independently track enemy submarines. The unmanned drone vessel will be able to track the enemy over thousands of miles. The U.S. Navy says China is rapidly expanding its submarine forces. |

Table 4 provides a generation samples from different attention control methods for CNN/Daily Mail. The underlined parts correspond to content differences with the original BRIO prediction. The effect of Relevance attention and Layer attention is marginal as only individual sentence fragments are altered. BEM on the other hand forces the model to completely revise the summary, introducing the missing facts about "independent tracking" and operation distance of "over thousands of miles", thus better reflection the intended reference.

To understand the impact of attention modulation we investigated summary alteration patterns on CNN/Daily Mail. First, we compared tokens and inspected the positions that don't overlap with the original generated summary (Fig. 3). In practice, the original generation can disagree with the reference on any token (Fig. 3, left plot), i.e., perfect attention modulation method would have a uniform token alteration pattern. However, Relevance attention tends to introduce the changes at the latter stages of generation, thus retaining the possible errors at the beginning of generated summary. While Layer attention shares this pattern it is more radical with corrections, taking effect as early as 40% of the original generated length and guaranteeing an alternative ending. The distinguishing trait of Biased Encoder Mixture is a more uniform alteration pattern with a much lower bias towards concluding tokens, meaning that the generation process may start diverging from the original even at the first token.

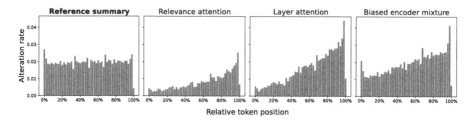

**Fig. 3.** Relative positions of the original generated summary altered by attention methods

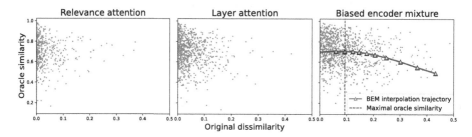

**Fig. 4.** Semantic distribution of summaries by attention correction methods

We additionally inspected the semantic characteristics of attention-biased summaries with Sentence Transformers[3] (Fig. 4). Beside later summary generation changes Relevant attention is also conservative about their semantic nature keeping the majority of new summaries within 0.9 of cosine similarity with the original. Following the previous trend, Layer attention has a higher variance of semantic changes and better utilizes the extractive signal to align with estimated Oracle yet still remains discreet. In contrast Biased Encoder Mixture seems to much less frequently preserve the semantic meaning of the summary diverging from the original by 0.5 in extreme cases. The additional slerp interpolation trajectory plot shows that Biased Encoder Mixture initially increases average similarity with Oracle sentences as the interpolation coefficient increases but past 0.2 value starts moving away from both Oracle and the original summaries suggesting a significant corruption of the information flow.

## 5   Conclusion

In this paper, we proposed Biased Encoder Mixture (BEM) a plug-and-play approach for content control in encoder-decoder abstractive summarization models based on Siamese networks and attention masking. We tested the approach on integration of extractive summarization knowledge in state-of-the-art pure abstractive summarization models. The evaluation on four popular datasets demonstrated that BEM is efficient even with static content selection strategies,

---

[3] We used 'all-mpnet-base-v2' model available at https://www.sbert.net.

outperforming the existing attention manipulation-based counterparts. Application of dynamic masking derived from extractive summarization results further boosts the quality of summaries bringing up to 7% ROUGE-2 relative improvements over the original model. Analysis of alteration patterns revealed that unlike other methods BEM produces much more diverse summaries having almost uniform positional editing distribution which aligns with the positional error rate between the original generated summary and the reference. Semantic analysis showed that BEM is more likely to diverge from the original content, however, the average magnitude of changes correlates with the interpolation coefficient and saliency changes are positive at lower values.

**Acknowledgements.** The work of Daniil Chernyshev (experiments, survey) was supported by Non-commercial Foundation for Support of Science and Education "INTELLECT". The work of Boris Dobrov (general concept, interpretation of results) was is supported by a grant for research centers in the field of artificial intelligence, provided by the Analytical Center for the Government of the Russian Federation in accordance with the subsidy agreement (agreement identifier 000000D730321P5Q0002) and the agreement with the Ivannikov Institute for System Programming of the Russian Academy of Sciences dated November 2, 2021 No. 70-2021-00142.

# References

1. Lewis, M., et al.: BART: denoising sequence-to-sequence pre-training for natural language generation, translation, and comprehension. In: Proceedings of the 58th Annual Meeting of the Association for Computational Linguistics, pp. 7871–7880 (2020)
2. Zhang, J., Zhao, Y., Saleh, M., Liu, P.J.: PEGASUS: pre-training with extracted gap-sentences for abstractive summarization. In: Proceedings of the 37th International Conference on Machine Learning, pp. 11328–11339 (2020)
3. Liu, Y., Liu, P., Radev, D., Neubig, G.: BRIO: bringing order to abstractive summarization. In: Proceedings of the 60th Annual Meeting of the Association for Computational Linguistics (Volume 1: Long Papers), pp. 2890–2903 (2022)
4. Zhang, X., et al.: Momentum calibration for text generation. arXiv:2212.04257 (2022)
5. Zhao, Y., Khalman, M., Joshi, R., Narayan, S., Saleh, M., Liu, P.J.: Calibrating sequence likelihood improves conditional language generation. arXiv:2210.00045 (2022)
6. Dou, Z.-Y., Liu, P., Hayashi, H., Jiang, Z., Neubig, G.: GSum: a general framework for guided neural abstractive summarization. In: Proceedings of the 2021 Conference of the North American Chapter of the Association for Computational Linguistics: Human Language Technologies, pp. 4830–4842 (2021)
7. Fabbri, A., Choubey, P.K., Vig, J., Wu, C.-S., Xiong, C.: Improving factual consistency in summarization with compression-based post-editing. In: Proceedings of the 2022 Conference on Empirical Methods in Natural Language Processing, pp. 9149–9156 (2022)
8. Ravaut, M., Joty, S., Chen, N.: SummaReranker: a multi-task mixture-of-experts re-ranking framework for abstractive summarization. In: Proceedings of the 60th Annual Meeting of the Association for Computational Linguistics (Volume 1: Long Papers), pp. 4504–4524 (2022)

9. Wang, F., et al.: Salience allocation as guidance for abstractive summarization. In: Proceedings of the 2022 Conference on Empirical Methods in Natural Language Processing, pp. 6094–6106 (2022)

10. Li, H., et al.: EASE: extractive-abstractive summarization end-to-end using the information bottleneck principle. In: Proceedings of the Third Workshop on New Frontiers in Summarization, pp. 85–95 (2021)

11. Ji, J., Kim, Y., Glass, J., He, T.: Controlling the focus of pretrained language generation models. In: Findings of the Association for Computational Linguistics: ACL 2022, pp. 3291–3306 (2022)

12. Lebanoff, L., Dernoncourt, F., Kim, D.S., Chang, W., Liu, F.: A cascade approach to neural abstractive summarization with content selection and fusion. In: Proceedings of the 1st Conference of the Asia-Pacific Chapter of the Association for Computational Linguistics and the 10th International Joint Conference on Natural Language Processing, pp. 529–535 (2020)

13. Ramesh, A., Dhariwal, P., Nichol, A., Chu, C., Chen, M.: Hierarchical text-conditional image generation with CLIP latents. arXiv:2204.06125 (2022)

14. Cao, S., Wang, L.: Attention head masking for inference time content selection in abstractive summarization. In: Proceedings of the 2021 Conference of the North American Chapter of the Association for Computational Linguistics: Human Language Technologies, pp. 5008–5016 (2021)

15. Xiao, W., Miculicich, L., Liu, Y., He, P., Carenini, G.: Attend to the right context: a plug-and-play module for content-controllable summarization. arXiv:2212.10819 (2022)

16. Cao, S., Wang, L.: HIBRIDS: attention with hierarchical biases for structure-aware long document summarization. In: Proceedings of the 60th Annual Meeting of the Association for Computational Linguistics (Volume 1: Long Papers), pp. 786–807 (2022)

17. See, A., Liu, P.J., Manning, C.D.: Get to the point: summarization with pointer-generator networks. In: Proceedings of the 55th Annual Meeting of the Association for Computational Linguistics, pp. 1073–1083 (2017)

18. Gehrmann, S., Deng, Y., Rush, A.: Bottom-up abstractive summarization. In: Proceedings of the 2018 Conference on Empirical Methods in Natural Language Processing, pp. 4098–4109 (2018)

19. Jing, H., McKeown, K.R.: The decomposition of human-written summary sentences. In: Proceedings of the 22nd Annual International ACM SIGIR Conference on Research and Development in Information Retrieval, pp. 129–136 (1999)

20. Suhara, Y., Wang, X., Angelidis, S., Tan, W.-C.: OpinionDigest: a simple framework for opinion summarization. In: Proceedings of the 58th Annual Meeting of the Association for Computational Linguistics, pp. 5789–5798 (2020)

21. Zhang, H., Liu, X., Zhang, J.: SummIt: iterative text summarization via ChatGPT. arXiv:2305.14835 (2023)

22. Zhang, S., Wan, D., Bansal, M.: Extractive is not faithful: an investigation of broad unfaithfulness problems in extractive summarization. arXiv:2209.03549 (2022)

23. Liu, Y., Lapata, M.: Text summarization with pretrained encoders. In: Proceedings of the 2019 Conference on Empirical Methods in Natural Language Processing and the 9th International Joint Conference on Natural Language Processing, pp. 3730–3740 (2019)

24. Narayan, S., Cohen, S.B., Lapata, M.: Don't give me the details, just the summary! topic-aware convolutional neural networks for extreme summarization. In: EMNLP 2018, pp. 1797–1807 (2018)

25. Ferrando, J., Gállego, G.I., Alastruey, B., Escolano, C., Costa-jussà, M.R.: Towards opening the black box of neural machine translation: source and target interpretations of the transformer. In: EMNLP 2022, pp. 8756–8769 (2022)

26. Radford, A., et al.: Learning transferable visual models from natural language supervision. arXiv:2103.00020 (2021)

27. Xu, J., Gan, Z., Cheng, Y., Liu, J.: Discourse-aware neural extractive model for text summarization. arXiv:1910.14142 (2019)

28. Gliwa, B., Mochol, I., Biesek, M., Wawer, A.: SAMSum corpus: a human-annotated dialogue dataset for abstractive summarization. In: Proceedings of the 2nd Workshop on New Frontiers in Summarization, pp. 70–79 (2019)

29. Grusky, M., Naaman, M., Artzi, Y.: Newsroom: a dataset of 1.3 million summaries with diverse extractive strategies. In: NAACL 2018: Human Language Technologies, vol. 1, pp. 708–719 (2018)

# RuCAM: Comparative Argumentative Machine for the Russian Language

Maria Maslova[1], Stefan Rebrikov[1], Anton Artsishevski[1], Sebastian Zaczek[2], Chris Biemann[2], and Irina Nikishina[2]([✉])

[1] HSE University, Moscow, Russia
{mdmaslova,sarebrikov}@edu.hse.ru
[2] Universität Hamburg, Hamburg, Germany
{sebastian.zaczek,chris.biemann,irina.nikishina}@uni-hamburg.de

**Abstract.** Comparative question answering is one of the question answering subtasks which requires not only to choose between two (or more) objects, but also to explain the choice and support it with arguments. ChatGPT-like models are able nowadays to generate a coherent answer in a natural language, however, they are not fully reliable as they are not publicly accessible and tend to hallucinate. Another solution is a Comparative Argument Machine (CAM), which however, has been developed for English only. In this paper, we describe the development of RuCAM—comparative argumentative machine for Russian, as well as the challenges of the system adaptation for another language. It is the first open-domain system to argumentatively compare objects in Russian with respect to information extracted from the OSCAR corpus. We also introduce several datasets for the RuCAM subtasks: comparative question classification, object and aspect identification, comparative sentences classification. We provide models for each subtask and compare them with the existing baselines.

**Keywords:** comparative question answering · Russian language · comparative question identification · question answering

## 1 Introduction

The problem of choice has always been topical for people. In everyday life one may choose between different options of food or drink, various types of clothing. Besides, sometimes it is important to choose the right university, convenient smartphone or suitable operating system. While comparing several objects, people mainly look for substantiated answers why one item may be better than the other on a certain aspect. Hence, a possible NLP solution of this problem lies concurrently in the field of question answering and argument mining.

In this paper, we narrow our research to the specific type of questions—comparative questions, where two objects are compared with each other, optionally, by some aspect. For example, in the sentence: *"Who is a better friend, a cat*

D. I. Ignatov et al. (Eds.): AIST 2023, LNCS 14486, pp. 78–91, 2024.
https://doi.org/10.1007/978-3-031-54534-4_6

or a dog?",—"cat" and "dog" are the objects being compared and "better friend" is the aspect by which both objects are compared. The comparative question answering task aims at answering this question not only by choosing the winning object, but also by explaining and supporting the decision with arguments. Comparative questions with more objects and superlative questions like "Who is the tallest man in the world" are out of scope of our research.

The complexity of comparative question answering task makes it hard to find a universal instrument meeting all users' requests. Specific product comparison systems, such as Compare.com[1], provide detailed information about a narrow range of goods (electric vehicles, certain hotels etc). Question answering platforms like Quora, or StackExchange[2] contain few comparative question answers regarding the total number of topics. Modern services based on Large Language Models (LLMs) are also able to provide a comparison of objects. However, they cannot be fully reliable as they are either not fully open (e.g. ChatGPT, Bing AI search) or they might hallucinate [7,10] or the provenance of the arguments in such models cannot be derived.

One of the most known and prominent research in the field is Comparative Argumentative Machine (CAM) [3,21] based on argumentative structures extracted from web-scale text resources. It allows to retrieve and rank textual argumentative structures relevant to a comparative user input of two objects. However, it is English-oriented, whereas there are no analogues for Russian.

Therefore, in this paper we present RuCAM—a system aimed at comparing two objects from general domain in Russian with argumentative explanations. We also provide the link to the web service[3] and to the GitHub repository[4] with code and data for all steps. As compared with its predecessor, CAM, it has the following differences: (i) it allows to work with comparative questions in natural language; (ii) it has the component for object and aspect identification from comparative questions; (iii) it uses an Elasticsearch index of Open Super-large Crawled Aggregated coRpus (OSCAR) [18]. The system is aimed to help users to speed up the process of comparative answers search. Moreover, a summary provided in the answer serves to support a decision-making process.

In this paper, we do not only describe a similar system and a pipeline from the engineering prospective. We also pose the following research questions: *(RQ1)* What the the main peculiarities of CAM that need to be taken into account when adapting it to other languages? *(RQ2)* What are the main challenges while adopting CAM specifically to the Russian language? In addition to the answers at those RQs, we also present the following contributions of the research:

1. We present RuCAM—a system for argumentative comparison of two objects based on information extracted from Common Crawl. As a successor of CAM, it is adapted for the peculiarities of the Russian language, has additional components that allow to work with natural questions and applies more recent NLP models.

---

[1] https://compare.com.

[2] https://quora.com, https://stackexchange.com.

[3] https://rucam.ltdemos.informatik.uni-hamburg.de.

[4] https://github.com/stefanrer/MCQA_RUS.

2. We provide several datasets for the RuCAM subtasks: comparative question classification, object and aspect identification in comparative questions, comparative sentence classification on general domains.
3. We provide baselines for each subtask in the pipeline and compare model performance with the existing approaches, if available.

## 2  Related Work

Comparative question answering derives from two tasks in Natural Language Processing (NLP): question answering and argument mining. There are multiple works that considered each problem separately, whereas very few studies combine both of them. In this section, we quickly overview each of the above mentioned tasks and then discuss the CAM system as the predecessor of RuCAM.

Recent studies on Comparative question answering [2] also deal with several problems in the field: comparative question classification, object and aspect identification and sentence classification. The identification of these elements will help to identify the comparative character of a question and its stance. These subtasks have been performed with the help of manually annotated question dataset and different neural networks, among which the fine-tuned RoBERTa has shown the best results. The extension of the above idea is described by Chekalina et al. [8]. Natural Language Understanding module is added to the argument search engine as well as Answer Generation module. The whole system is able to process a user's request, to wit, extract objects, aspects and predicates, find the pros and cons argumentation for the objects found, and generate a short answer which is a summary of arguments for the English language.

As for the Russian language, some research studies on question classification have been conducted in recent years. One of the deepest analysis of comparative questions for Russian as well as developing methods for their classification has been done by [3]. The authors introduce 10 subclasses of comparative questions and claim to collect 50,000 questions and analyse 6,250 questions in Russian, however, they are not allowed to disclose the data even for academical purposes. In this paper, we compare our model with the model developed by [3] and make the data public. Other papers devoted to the comparative questions in Russia, are [16,17], where the authors apply regular expressions, machine learning and neural networks methods to classify questions, including comparative ones, with respect to the predetermined typology.

Considering previous research on Argument Mining, there exist multiple publications on the topic for the Russian language. The most popular dataset and a variety of classification tasks are presented in [14], which describes different methods participated in the organised shared task. In [12], the authors present the first publicly available argument-annotated corpus of Russian based on Argumentative Microtext Corpus and experiment with feature-based machine learning approaches for argument identification. They also explore the possibility to classify argumentative discourse units (ADU) using traditional machine learning and deep learning methods [11]. More recent publications for the Russian language tackle the problem of argument generation given aspect [13] and end-to-end Argument Mining over varying rhetorical structures [9].

## 2.1  Comparative Argumentative Machine

As we stick to the concept of CAM search engine, it is necessary to point out its features that might be challenging to implement to other languages. Comparative Argumentative Machine [21] is aimed to help with answering comparative questions using argument mining techniques. It consists of 5 components: sentence retrieval, sentence classification, sentence ranking, aspect retrieval and result presentation.

First of all, CAM has no request processing step, whereas it is a significant part of RuCAM (Subsects. 3.1, 3.2). Objects and aspects are entered manually by user via interface. The first CAM step is the retrieval of relevant sentences from the CommonCrawl corpus. It searches for indexes corresponding entered objects and aspects using ElasticSearch. The RuCAM search is performed similarly (see Subsect. 3.3) and is based on a cleaned and a preprocessed version of CC—Open Super-large Crawled Aggregated coRpus (OSCAR) [18]. The main adaptation challenge at this step is the availability of tools for large-scale text preprocessing.

The second CAM step is the sentence classification. The detailed process for this subtask is described in [19]. In a series of experiments with classification models, XGBoost shows the best results. For this task, RuCAM applies a similar baseline, but uses the Transformer-based models for Russian. This step might be challenging in terms of collecting good-quality annotations for training.

The next steps are similar for CAM and RuCAM: sentence ranking using a combination of Elastic Search and classification scores, and additional aspect retrieval. The final step—displaying results is almost similar for both systems, however RuCAM allows users to choose between entering a comparative question in Russian or to enter objects and aspects manually (see Fig. 2 for details).

# 3  System Design

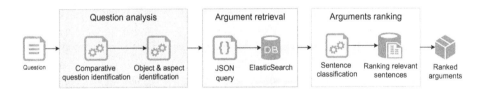

**Fig. 1.** Design of the RuCAM system.

As mentioned earlier, RuCAM is aimed at comparing two objects on the basis of argumentative structures extracted from web-scale text resources. This approach can be split in two consecutive steps: question analysis and argument retrieval. The first step is about identification of interrogative nature of sentence (is it comparative or not) as well as the sequence labelling subtask for object and aspect identification. The second step constitutes the search of argumentative structures relevant for objects extracted from the input question, their classification (*"Is the sentence in favour of one object or another?"*) and ranking (*"Which sentences are more relevant for comparison?"*).

The design of our RuCAM system is shown in Fig. 1. It consists of the following stages, which are described in more detail in the corresponding subsections: comparative question identification (Subsect. 3.1), object and aspect identification (Subsect. 3.2), argument retrieval (Subsect. 3.3), sentence classification (Subsect. 3.4), and sentence ranking (Subsect. 3.5).

## 3.1   Question Identification

The processing of the request starts with the identifying the question type (comparative or not). To do that, we first compile the dataset that satisfies our needs. We start with the Russian dataset of questions [16]: 146 items are used with positive tags (with classes "choice" and "comparison") and all other sentences (2,121) as negative. As the major Russian dataset from [3] with comparative questions cannot be disclosed for training, we only take its open part used for testing the fine-grained classifier. Additionally, we translate query questions from [4–6] for argument mining Touché competitions using the "EN_RU" translation model from [15]. We manually check and improve comparative sentences in terms of fluency and grammaticality, as their number is relatively small.

However, the number of positive entries in the compiled dataset might be still too small to work with, therefore, we automatically translate the existing English dataset for comparative question identification [2,3] with the model mentioned beforehand. The dataset used as well as their statistics and application is presented in Table 1. The example entries of the compiled Russian dataset could be seen in Table 2.

**Table 1.** Statistics and specifications of the datasets used for comparative question identification.

| Dataset | Non-Comparative | Comparative | Total |
|---|---|---|---|
| Nikolaev et al. [16] | 2,040 | 143 | 2,183 |
| Bondarenko et al. (fine-grained test) [3] | 0 | 1,240 | 1,240 |
| Touché (2020–2022) (translated) [4–6] | 0 | 100 | 100 |
| Webis-CompQuestions-20 (translated) [3] | 1,431 | 1,429 | 2,860 |
| Webis-CompQuestions-22 (translated) [2] | 2,529 | 3,088 | 5,617 |
| Total (selected) | 6,000 | 6,000 | 12,000 |

At this step, we implement and compare the following approaches for question identification: a rule-based baseline, a ru-tinyBERT model finetuned on the compiled dataset in Russian, and the fine-tuned BERT model from [3] (both trained on 3 epochs with batch 32, other parameters are default). We discuss the results and compare their performance in 'Evaluation' section. As a rule-based baseline we implement the idea of special patterns contained in questions as the simplest identification mechanism. Therefore, we created patterns which include comparative forms ("лучше" ("better"), "хуже" ("worse")), explicit mentions of comparison (*"по сравнению"* ("in comparison with")), similarity (*"похожий"* ("similar"), *"одинаковый"* ("same as")) or difference (*"отличается"* ("different from"), *"непохожий"* ("not like")), advantages (*"преимущество"* ("advantages"), *"выигрывает"* ("wins over")) or disadvantages (*"недостаток"* ("disadvantage"), *"проигрывает"* ("lags behind")), verbs expressing choice (*"выбрать"* ("choose"), *"предпочесть"* ("prefer")).

**Table 2.** Example entries from the Russian dataset for comparative question identification. 1 stands for the comparative sentence, 0 for non-comparative.

| Sentence | Label |
| --- | --- |
| **Каковы преимущества и недостатки PHP по сравнению с Python?** | 1 |
| *Kakovy preimushhestva i nedostatki PHP po sravneniju s Python?* | |
| What are the advantages and disadvantages of PHP compared to Python? | |
| **Когда мне поступить в университет?** | 0 |
| *Kogda mne postupit' v universitet?* | |
| When should I go to university? | |

### 3.2  Object and Aspect Identification

After identifying the question as comparative, we need to extract objects and (optionally) aspects to further provide them for the argument retrieval stage. In order to create a dataset for the task, we take 6000 sentences from the previous step than were labelled as comparative and manually annotate them. Three experts in computational linguistics and NLP are required to highlight *"object-1"*, *"object-2"*, *"aspect"*, *"common object"* in a text, analogously to the guidelines in [1]. *Common object* is the specific structure with a noun subordinating two adjectives in a construction like "черный или зеленый чай" ("green or black tea"). The level of annotator agreement (Fleiss's kappa) amounts to the following levels: *object-1 & object-2*—0.83, *aspect*—0.71, *common object*—0.54. When creating the final dataset for models fine-tuning (2,328 sentences with exactly two objects), we use the annotation versions supported by the majority of annotators.

At this step, we also implement several approaches of object identification: a rule-based algorithm, fine-tuned Transformer encoders (10 epochs, batch 32, other parameters are default) and a few-shot approach on generative Transformers. A rule-based algorithm is founded on the idea that all requests have a certain structure. By this, we mean the existence of two comparison objects and a connective between them. We consider the following cases: two nouns, two verbs, combinations of noun and adjective, combinations of noun and two subordinate adjectives. We also expect a connective from the list of conjunctions and synthetic words expressing comparison between the objects: "или" ("or"), "лучше" ("better"), "лучше чем" ("better than"), "лучше ADV чем SCONJ" ("better than"), "лучше A чем CONJ по PR сравнению NOUN" ("better than in comparison with")).

## 3.3 Sentence Retrieval

In order to retrieve arguments in favour of one or the other object, we use Open Super-large Crawled Aggregated coRpus (OSCAR) [18] which comprises 21.5M documents for the Russian language which we split into 21B sentences. We use OSCAR instead of the Common Crawl, while it is claimed to be its filtered version. We store and index this data with Elasticsearch[5]—a search engine based on the Lucene library that enables fast search using HTTP web interface and schema-free JSON. To ensure quick and stable responses from Elasticsearch we deploy it in parallel on 16 Ubintu Server 16.04 nodes with $2 \times 10$ Cores (+Hyperthreading) CPU, 256 GB of RAM and 4TB HDD each.

When indexing documents, we decide to create two indexes: the first one is used for storing document information (the number of sentences in the document, its metadata and the web-link) and the second one for storing sentences. Each indexed sentence includes its document ID, previous and next sentence IDs, number of words in the sentence and the text itself.

To retrieve sentences, we first do the Snowball stemming of the query objects and aspects and then apply wildcards to be able to find all word forms. Lemmatization could be a fair alternative at this step, however, we stick to stemming because of the time constraints. We send a boolean json query and require that the clause must appear in matching documents. We consider this step to be the most challenging in the whole CAM for implementation, as Russian language has a highly fusional morphology which makes it much more difficult to retrieve sentences than in English, as query words may occur in any form.

## 3.4 Sentence Classification

**Table 3.** Examples with tags from the sentence classification dataset. Objects for comparison are in **bold**.

| Sentence | Tag |
|---|---|
| В любом случае, **запекать** гораздо полезнее, чем **жарить** <br> *V ljubom sluchae,* **zapekat'** *gorazdo poleznee, chem* **zharit'** <br> In any case, **baking** is much healthier than **frying** | BETTER |
| Тащить **лыжи** на гору сложнее, чем один **сноуборд** <br> *Tashhit'* **lyzhy** *na goru slozhnee, chem odin* **snoubord** <br> Carrying **ski** up the mountain is harder than just one **snowboard** | WORSE |
| Здесь лучше всего использовать **спонж** или широкую **кисть** <br> *Zdes' luchshe vsego ispol'zovat'* **sponzh** *ili shirokuju* **kist'** <br> Here it is best to use a **sponge** or a wide **brush** | NONE |

After the candidate sentences with possible arguments are found, it is necessary to understand whether a sentence argues in favour of the first or the second

---

[5] https://www.elastic.co.

object. Analogously to CAM [21], we collect a dataset of 1208 sentences from 67 object pairs and annotate them using Yandex.Toloka system for data crowd-sourcing [20]. To do this, we select same or similar pairs from the same domains as in English (e.g., programming languages, car manufacturers, food and drinks) and make a query to Elasticsearch as it is described in Sect. 3.3 to extract all sentences matching the query. Then we create three tags: *"BETTER"* (the first item is better or "wins")/*"WORSE"* (the first item is worse or "looses") or *"NONE"* (the sentence does not contain a comparison of the target items). When displaying classified sentences, *"BETTER"*-sentences support the first compared object, *"WORSE"*-sentences are used as pro-argument for the second object. Unfortunately, the annotated dataset is highly imbalanced: 75% of texts belong to the *"NONE"* tag, 16%—to the *"BETTER"* tag and only 9%—to the *"WORSE"* tag. Table 3 demonstrates some excerpts from the dataset within the label.

At this step, we also implement a rule-based baseline, several large language model classifiers based on Transformer Encoders (3 epochs, batch 32, other parameters are default) and few-shot approaches with generative Transformers, which allow to address in issue of data imbalance using only 5 examples from each class. The rule-based approach requires two lists of keywords with adjectives and adverbs with the meaning of superiority or inferiority of the first object over the second. We also take into account negation cases when the sense of a sentence is reversed.

### 3.5  Sentence Ranking and Object Comparison

The processes of sentence ranking and object comparison is identical to the one in CAM [21]: we score comparative sentences by combining the classifier confidence and the Elasticsearch score[6]. When displaying the arguments in RuCAM on a certain object, we sum up not only BETTER-arguments, where the current object is the first item, but also WORSE-arguments, where the object is the second item in the sentence. For instance, both sentences "Python лучше, чем Java" (class >) and "Java хуже, чем Python" (class <) are used in favour of Python when comparing with Java.

In addition to user-specified comparison aspects, CAM generates up to ten supplementary aspects (even when no comparison aspect at all was provided by the user) to display it for better output presentation. To do the same for RuCAM, we use three different methods for aspect mining: (1) searching for comparative adjectives and adverbs; (2) searching for phrases with comparative adjectives/adverbs and a preposition like "для" ("for"), "чтобы" ("to"), etc. (e.g., "быстрее для написания кода" ("better for code writing")); (3) searching for specific hand-crafted patterns like "из-за более высокой скорости" ("due to higher speed"), or "причина этого кроется в цене" ("the reason for this lies in the price"). An extracted aspect is assigned to the object with the higher co-occurrence frequency.

---

[6] https://www.elastic.co/guide/en/elasticsearch/guide/current/scoring-theory.html.

# 4   Evaluation

This section is devoted to the evaluation of all of the developed models for each of the pipeline steps of RuCAM. We present comparison tables and select the best-performed model from each step to the final pipeline.

## 4.1   Question Identification

**Table 4.** Model comparison table for the results in the Comparative Question Classification task.

| Model name | Model Parameters | Precision | Recall | F1-score |
|---|---|---|---|---|
| rule-based | – | 0.89 | 0.88 | 0.87 |
| Bondarenko et al. [3] | 167.3M | **0.95** | **0.95** | **0.95** |
| ruBERT-tiny[a] | 29.4M | 0.91 | 0.90 | 0.90 |

[a]https://huggingface.co/cointegrated/rubert-tiny2

From the results, presented in Table 4, we can see comparative questions are indeed very specific kind of questions and that they can be easily identified even with rule-based approaches. The best results are achieved by the existing model from [3], however, the questions the model was trained on are from the same dataset we took the major part of our testing questions from. This partially explains higher results of [3], as they were training on the data from the same distribution as the test set. Nevertheless, even with no access to this data, but only to machine-translated datasets, it is possible to train a well-performing model. The finetuned ruBERT-tiny achieves decent results outperforming the baseline with smaller number of parameters.

## 4.2   Object and Aspect Identification

This subsection presents the results for the developed token classification models for identifying objects and aspects. We test a rule-based approach, and several Transformer approaches: finetuning of the standard sequence taggers as well as testing few-shot generative Transformers. Table 5 presents the results for each model. As this subtask is formulated for the first time for the Russian language, there are no additional models to compare with.

Even though the rule-based model demonstrates quite high results, it still lags behind most of the provided LLMs. However, the results of this approach are still higher than most of the models on *"CommonObject"* identification. Generally, the results show that generative models perform on par or even slightly better than the baseline and significantly lag behind Trasnformer encoders. Nevertheless, we need to specify that generative Transformers might have higher potential as they were shown 5 examples only and they might perform much better after

**Table 5.** Results on F1-score for the object and aspect identification experiments.

| Model name | Object-1 | Object-2 | Aspect | Common Object | Average |
|---|---|---|---|---|---|
| rule-based baseline | 0.71 | 0.81 | 0.17 | 0.23 | 0.66 |
| wikineural-multilingual-ner[a] | 0.77 | 0.75 | 0.50 | **0.66** | 0.71 |
| ruBERT-tiny[b] | 0.87 | 0.78 | 0.17 | 0.00 | 0.69 |
| ruBERT-large[c] | **0.97** | **0.88** | **0.57** | 0.00 | **0.80** |
| bactrian-x-llama-13b[d] (5-shot) | 0.75 | 0.72 | 0.22 | 0.00 | 0.62 |
| ruGPT3 (5-shot) | 0.66 | 0.65 | 0.04 | 0.00 | 0.54 |

[a] https://huggingface.co/Babelscape/wikineural-multilingual-ner
[b] https://huggingface.co/cointegrated/rubert-tiny
[c] https://huggingface.co/ai-forever/ruBert-large
[d] https://github.com/mbzuai-nlp/bactrian-x

proper finetuning. We leave this question out of scope of our research and present models as baselines only. Regarding lower and zero scores for the *"Aspect"* and *"CommonObject"* labels for many models, we assume that the reason of that is the inconsistency in annotations. Similarly low results we also shown in [1], which might also indicate the difficulty of the *"Aspect"* and *"CommonObject"* identification in general.

### 4.3 Sentence Classification

**Table 6.** The results on F1-score for the comparative sentence classification models.

| Model | BETTER | WORSE | NONE | Average |
|---|---|---|---|---|
| rule-based | 0.34 | 0.33 | 0.82 | 0.69 |
| ruBERT-tiny[a] | **0.57** | **0.38** | **0.91** | **0.82** |
| ruBERT-large[b] | 0.00 | 0.00 | 0.87 | 0.76 |
| bactrian-x-llama-13b (5 × 3-shot)[c] | 0.30 | 0.12 | 0.32 | 0.36 |
| ruGPT3 (5 × 3-shot) | 0.26 | 0.24 | 0.19 | 0.23 |

[a] https://huggingface.co/cointegrated/rubert-tiny
[b] https://huggingface.co/ai-forever/ruBert-large
[c] https://github.com/mbzuai-nlp/bactrian-x

As in the previous steps, we compare a rule-based and several Transformer-based approaches. According to Table 6, the results for comparative sentence classification are inconsistent for each class and relatively low for all of the presented models, due to the class imbalance problem. The "WORSE" class always achieves the lowest score among the classes. We can see that the best results are achieved with the ruBERT-tiny model, while BERT-large overfits on the dataset with prevalent *"NONE"* class. Rule-based approaches also produce average results for all classes while the lowest scores are achieved by LLMs (ruGPT3 and bactrian-x-llama) with generative setup.

## 5    Demonstration System and Current Work

The main outcome of our research is the final system where we integrate all the parts described above. Figure 2 depicts the interface of the whole system. We have decided to apply ruBERT-tiny at each step as the compromise between speed, memory space and efficiency. The evaluation of the system is currently work in progress. We plan to evaluate RuCAM analogously to CAM evaluation pipeline, by asking whether users are faster answering correctly when using the CAM system and ask some users to "play" to collect some user experience feedback. The research is to be based on the collection of topics (two objects + one aspect) available for CAM- and keyword-based search. A topic is suitable for the research if it has more support sentences than the established lower bound. Additional descriptions fore some topics will help to avoid potential ambiguities or subjectivities. In the first part of the task the participants should give an answer as quickly as possible using both experimental systems alternatingly. For that purpose the collection of topics is randomly split into two groups. The second part of the task allows users to test the functionality and convenience of the system without time limitations and make as much comparisons as they want.

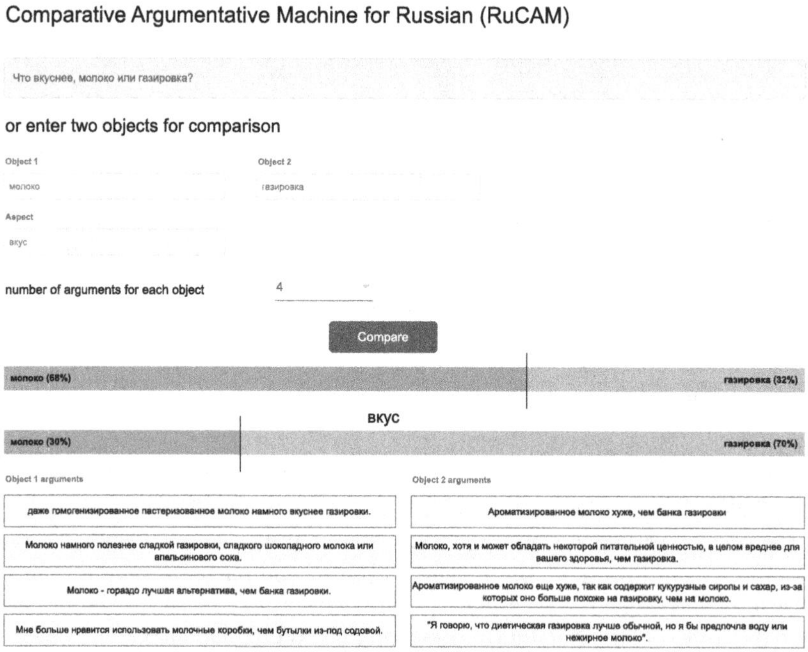

**Fig. 2.** Design interface of the RuCAM demonstration system.

# 6   Conclusion

In this article, we present RuCAM—the first instrument which helps to answer general-domain comparative questions in Russian. Inspired by the CAM system, we create a similar pipeline, adding new steps for comparative question classification, object and aspect identification, sentence classification. We also present several new datasets in Russian that might be further used for the fine-tuning of language models for each subtask. From the performed experiments, we can see that the rule-based approaches show decent results on all subtasks of comparative QA as well as few-shot generative Transformers (which need further investigation). As the answer on *(RQ1)*, we state that when transferring CAM to other languages, you should take the following peculiarities into account: (i) the difference in the notion of comparative sentences in different languages; (ii) the difference in the syntax and morphology of languages when re-implementing rule-based approaches; (iii) the existence of the relevant datasets and pre-trained Large Language Models for training and large text corpora containing comparative sentences for search in the target language. Nevertheless, as it has been shown for Russian, it might be quite smooth if at least some the required tools are available. As for *(RQ2)*, we can see that the main challenge is the complexity of Russian grammar for re-implementing rule-based approaches. Inflectional morphemes make it difficult to search for specific forms in the text at any step of our approach. Moreover, quite flexible word order which makes the process of matching regular expressions in the rule-based approaches more challenging. As future directions, we plan to incorporate a summarisation system that would be able to produce a coherent answer from two lists of arguments for each object. It will allow us to compare the results of various instruct-tuned models for Russian and ChatGPT with the RuCAM pipeline. As the instruct-tuned generative models are well-suited for such type of tasks, it would be a great study to understand in how many cases these models provide reasonable arguments, and how often they hallucinate in comparison to RuCAM. Another challenging direction is to apply on discourse analysis approaches to identify the argumentative sentences. Utilizing such methods may retrieve more coherent text spans.

**Acknowledgements.** This work was supported by the DFG through the project "ACQuA: Answering Comparative Questions with Arguments" (grants BI 1544/7- 1 and HA 5851/2- 1) as part of the priority program "RATIO: Robust Argumentation Machines" (SPP 1999).

# References

1. Beloucif, M., Yimam, S.M., Stahlhacke, S., Biemann, C.: Elvis vs. M. Jackson: who has more albums? Classification and identification of elements in comparative questions. In: Calzolari, N., et al. (eds.) Proceedings of the Thirteenth Language Resources and Evaluation Conference, LREC 2022, Marseille, France, 20–25 June 2022, pp. 3771–3779. European Language Resources Association (2022). https://aclanthology.org/2022.lrec-1.402

2. Bondarenko, A., Ajjour, Y., Dittmar, V., Homann, N., Braslavski, P., Hagen, M.: Towards understanding and answering comparative questions. In: Candan, K.S., Liu, H., Akoglu, L., Dong, X.L., Tang, J. (eds.) WSDM 2022: The Fifteenth ACM International Conference on Web Search and Data Mining, Virtual Event/Tempe, AZ, USA, 21–25 February 2022, pp. 66–74. ACM (2022). https://doi.org/10.1145/3488560.3498534

3. Bondarenko, A., et al.: Comparative web search questions. In: Caverlee, J., Hu, X.B., Lalmas, M., Wang, W. (eds.) WSDM 2020: The Thirteenth ACM International Conference on Web Search and Data Mining, Houston, TX, USA, 3–7 February 2020, pp. 52–60. ACM (2020). https://doi.org/10.1145/3336191.3371848

4. Bondarenko, A., et al.: Overview of Touché 2020: argument retrieval. In: Arampatzis, A., et al. (eds.) CLEF 2020. LNCS, vol. 12260, pp. 384–395. Springer, Cham (2020). https://doi.org/10.1007/978-3-030-58219-7_26

5. Bondarenko, A., et al.: Overview of touché 2022: argument retrieval. In: Faggioli, G., Ferro, N., Hanbury, A., Potthast, M. (eds.) Proceedings of the Working Notes of CLEF 2022 - Conference and Labs of the Evaluation Forum, Bologna, Italy, 5th–8th September 2022. CEUR Workshop Proceedings, vol. 3180, pp. 2867–2903. CEUR-WS.org (2022). https://ceur-ws.org/Vol-3180/paper-247.pdf

6. Bondarenko, A., et al.: Overview of Touché 2021: argument retrieval. In: Candan, K.S., et al. (eds.) CLEF 2021. LNCS, vol. 12880, pp. 450–467. Springer, Cham (2021). https://doi.org/10.1007/978-3-030-85251-1_28

7. Cao, M., Dong, Y., Cheung, J.: Hallucinated but factual! inspecting the factuality of hallucinations in abstractive summarization. In: Proceedings of the 60th Annual Meeting of the Association for Computational Linguistics, Dublin, Ireland (Volume 1: Long Papers), pp. 3340–3354. Association for Computational Linguistics (2022). https://doi.org/10.18653/v1/2022.acl-long.236

8. Chekalina, V., Bondarenko, A., Biemann, C., Beloucif, M., Logacheva, V., Panchenko, A.: Which is better for deep learning: Python or matlab? Answering comparative questions in natural language. In: Gkatzia, D., Seddah, D. (eds.) Proceedings of the 16th Conference of the European Chapter of the Association for Computational Linguistics: System Demonstrations, EACL 2021, Online, 19–23 April 2021, pp. 302–311. Association for Computational Linguistics (2021). https://doi.org/10.18653/v1/2021.eacl-demos.36

9. Chistova, E.: End-to-end argument mining over varying rhetorical structures. In: Findings of the Association for Computational Linguistics: ACL 2023, Toronto, Canada, pp. 3376–3391. Association for Computational Linguistics (2023). https://doi.org/10.18653/v1/2023.findings-acl.209. https://aclanthology.org/2023.findings-acl.209

10. Dale, D., Voita, E., Barrault, L., Costa-jussà, M.R.: Detecting and mitigating hallucinations in machine translation: model internal workings alone do well, sentence similarity Even better. In: Proceedings of the 61st Annual Meeting of the Association for Computational Linguistics, Toronto, Canada (Volume 1: Long Papers), pp. 36–50. Association for Computational Linguistics (2023). https://aclanthology.org/2023.acl-long.3

11. Fishcheva, I., Goloviznina, V., Kotelnikov, E.V.: Traditional machine learning and deep learning models for argumentation mining in Russian texts. CoRR abs/2106.14438 (2021). https://arxiv.org/abs/2106.14438

12. Fishcheva, I., Kotelnikov, E.: Cross-lingual argumentation mining for Russian texts. In: van der Aalst, W.M.P., et al. (eds.) AIST 2019. LNCS, vol. 11832, pp. 134–144. Springer, Cham (2019). https://doi.org/10.1007/978-3-030-37334-4_12

13. Goloviznina, V., Fishchev, I., Peskisheva, T., Kotelnikov, E.: Aspect-based argument generation in Russian. In: Computational Linguistics and Intellectual Technologies: Papers from the Annual Conference "Dialogue" (2023)
14. Kotelnikov, E.V., Loukachevitch, N.V., Nikishina, I., Panchenko, A.: RuArg-2022: argument mining evaluation. In: Computational Linguistics and Intellectual Technologies: Papers from the Annual Conference "Dialogue" (2022)
15. Ng, N., Yee, K., Baevski, A., Ott, M., Auli, M., Edunov, S.: Facebook FAIR's WMT19 news translation task submission. In: Proceedings of the Fourth Conference on Machine Translation, Florence, Italy (Volume 2: Shared Task Papers, Day 1), pp. 314–319. Association for Computational Linguistics (2019). https://doi.org/10.18653/v1/W19-5333. https://aclanthology.org/W19-5333
16. Nikolaev, K., Malafeev, A.: Russian-language question classification: a new typology and first results. In: van der Aalst, W.M.P., et al. (eds.) AIST 2017. LNCS, vol. 10716, pp. 72–81. Springer, Cham (2018). https://doi.org/10.1007/978-3-319-73013-4_7
17. Nikolaev, K., Malafeev, A.: Russian Q&A method study: from Naive Bayes to convolutional neural networks. In: van der Aalst, W.M.P., et al. (eds.) AIST 2018. LNCS, vol. 11179, pp. 121–126. Springer, Cham (2018). https://doi.org/10.1007/978-3-030-11027-7_12
18. Ortiz Suárez, P.J., Sagot, B., Romary, L.: Asynchronous pipelines for processing huge corpora on medium to low resource infrastructures. In: Proceedings of the Workshop on Challenges in the Management of Large Corpora (CMLC-7) 2019. Cardiff, 22nd July 2019, pp. 9–16. Leibniz-Institut für Deutsche Sprache, Mannheim (2019). https://doi.org/10.14618/ids-pub-9021. http://nbn-resolving.de/urn:nbn:de:bsz:mh39-90215
19. Panchenko, A., Bondarenko, A., Franzek, M., Hagen, M., Biemann, C.: Categorizing comparative sentences. In: Stein, B., Wachsmuth, H. (eds.) Proceedings of the 6th Workshop on Argument Mining, ArgMining@ACL 2019, Florence, Italy, 1 August 2019, pp. 136–145. Association for Computational Linguistics (2019). https://doi.org/10.18653/v1/w19-4516
20. Pavlichenko, N., Stelmakh, I., Ustalov, D.: CrowdSpeech and VoxDIY: benchmark dataset for crowdsourced audio transcription. In: Vanschoren, J., Yeung, S. (eds.) NeurIPS Datasets and Benchmarks 2021, December 2021, virtual (2021)
21. Schildwächter, M., Bondarenko, A., Zenker, J., Hagen, M., Biemann, C., Panchenko, A.: Answering comparative questions: better than ten-blue-links? In: Azzopardi, L., Halvey, M., Ruthven, I., Joho, H., Murdock, V., Qvarfordt, P. (eds.) Proceedings of the 2019 Conference on Human Information Interaction and Retrieval, CHIIR 2019, Glasgow, Scotland, UK, 10–14 March 2019, pp. 361–365. ACM (2019). https://doi.org/10.1145/3295750.3298916

# Paraphrasers and Classifiers: Controllable Text Generation for Text Style Transfer

Evgeny Orlov[1(✉)] and Murat Apishev[2]

[1] ITMO University, St. Petersburg, Russia
emorlov@niuitmo.ru
[2] MIPT, Moscow, Russia

**Abstract.** Text style transfer (TST) is an NLP task with a long history and a broad range of applications. Recently, it has seen success with the use of large pretrained language models (LMs). However, the size of contemporary LMs often makes fine-tuning for downstream tasks infeasible. For this reason, methods of controllable text generation (CTG) which do not aim at fine-tuning the original LM have received attention for solving TST tasks. In this work, we contribute to this line of research and adapt an existing CTG method, CAIF, for TST. The original CAIF is based on reweighting the logits of the generative LM according to a free-form style attribute classifier. To allow its use for TST, we replace the standard LM with a model capable of paraphrasing, making corresponding changes. We refer to the resulting unsupervised method as ParaCAIF. We illustrate its applicability by experimenting with detoxification, a relatively new yet practical TST subtask. We work with detoxification in two languages: Russian and English. For both languages, ParaCAIF significantly reduces the toxicity of the generated paraphrases as compared to plain paraphrasers. To the best of our knowledge, it is the first work that adapts a CTG method for Russian detoxification. For English, ParaCAIF outperforms an analogous adapted CTG method, ParaGeDi, in terms of style transfer accuracy. Although the overall performance of ParaCAIF remains lower than that of supervised approaches, it has a broader range of application, as it does not require parallel data which are not readily available for many TST tasks.

**Keywords:** text style transfer · controllable text generation · generative language models · paraphrasing · detoxification

## 1 Introduction

Text style transfer (TST) is the task of transforming a given input sentence from the source to the target style while preserving its semantic content. It has a wide range of applications which all serve the goal of making NLP products more user-centered. For example, it can be used to render a chat bot's persona more emotional and empathetic or serve in automatic writing assistants, etc. Along with many other NLP tasks, TST has seen great progress in recent years with the use of large pretrained language models (LMs). However, the size of contemporary LMs often makes fine-tuning for downstream tasks infeasible.

© The Author(s), under exclusive license to Springer Nature Switzerland AG 2024
D. I. Ignatov et al. (Eds.): AIST 2023, LNCS 14486, pp. 92–108, 2024.
https://doi.org/10.1007/978-3-031-54534-4_7

One of the solutions are methods of controllable text generation (CTG)—the task of generating text according to the given controlled element [26]. It has emerged in response to the need to control various aspects of the generated text in downstream applications and dangers of unreviewed web corpora used for large-scale pretraining [2]. One of broad groups of CTG approaches, post-processing, abandons fine-tuning, fixes the parameters of the original LM and aims at reweighting or reranking its outputs. Although CTG methods are mainly applied to standard prompt completion, they can also be used for sequence-to-sequence tasks, including TST. There is a nascent line of works which adapt post-processing CTG approaches for TST [5,12,23].

In this work, we adapt CAIF [33], an existing post-processing CTG method, for TST. CAIF differs from its analogues by its broader applicability, as it does not require training an additional LM for steering the generation and uses a free-form classifier as an attribute discriminator. We follow the idea of ParaGeDi [5] and replace the regular LM with a paraphraser, analogously referring to the resulting method as ParaCAIF. We illustrate the applicability of ParaCAIF by applying it to detoxification, a relatively new yet practical TST subtask. We conduct the experiments for two languages: English and Russian, which has seen much less work in detoxification. The contributions of this work can be summarized as follows:

– We adapt CAIF [33] for TST and illustrate the applicablity of the resulting method, ParaCAIF, applying it to detoxification.
– We conduct a series of experiments on Russian detoxification with ParaCAIF and a line of Transformer paraphraser models. To the best of our knowledge, this is the first application of a CTG method for Russian detoxification.
– We observe that for all models, ParaCAIF substantially reduces the toxicity of the generated paraphrases.
– We find that one of ParaCAIF models surpasses the supervised baseline in style transfer accuracy.
– We apply ParaCAIF to English detoxification and observe a multiple decrease in the toxicity of the generated paraphrases compared to the plain paraphraser baseline.
– We find that ParaCAIF outperforms ParaGeDi [5] in style transfer accuracy.

We make our code and data available in a Github repository[1].

## 2   Related Works

### 2.1   Controllable Text Generation

The common tasks for which CTG is used are attribute-based generation, dialogue generation, storytelling, data-to-text, data augmentation, debiasing, and format control. For comprehensive surveys of the state of the field, we refer to [36,41]. [41] distinguish three broad groups of CTG approaches: fine-tuning,

---

[1] https://github.com/BunnyNoBugs/ParaCAIF.

retraining or refactoring, and post-processing. In this research, we focus on post-processing methods which create a post-processing module to reweight or rerank the outputs of the original LM. [41] divide these methods into trainable strategies and guided strategies. The former require the post-processing module to be trained jointly with the model, while the model's weights are frozen [9]. In the latter, the post-processing module can be trained separately and guides the LM only in the inference stage. While PPLM [6] uses an arbitrary classifier to control the LM, a line of works train additional class-conditional LMs (CC-LMs) and use them as generative discriminators to shift the distribution of the original LM [19,23]. CC-LMs are trained to generate text while conditioning on a class variable $c$ which later serves as a "control code" describing, for example, sentiment or topic. Given a context $x_{<t}$ and a base language modeling distribution $P_{LM}(x_{1<t})$, the generative discriminator computes $P_{\theta}(c|x_t, x_{<t})$ for every possible next token $x_t$. Generation is then guided using a weighted decoding heuristic via

$$P_w(x_t|x_{<t}, c) \propto P_{LM}(x_t|x_{<t})P_{\theta}(c|x_t, x_{<t})^{\omega}, \quad (1)$$

where $\omega > 1$ to bias generation more strongly towards the correct class. CAIF [33] is similar to these methods but uses a free-form classifier instead of a CC-LM and is described in more detail in Sect. 3.

## 2.2   Text Style Transfer

In the variety of TST subtasks, not all of corresponding datasets provide parallel data. This is reflected in the range of existing methods for TST. While supervised approaches yield good results, unsupervised approaches receive more attention from researchers as capable of working with unparallel data and therefore having a broader range of application. For a comprehensive review of the state of the field, we refer to [16].

Most supervised approaches make use of the standard neural sequence-to-sequence (seq2seq) model with the encoder-encoder architecture. Unsupervised approaches can be divided into three broad groups: disentaglement, prototype editing, and pseudo-parallel corpus construction. We describe the first group in more detail as it is close to our research. Disentaglement is based on the following sequence of actions. The input text with the source attribute is encoded into a latent representation. The latent representation is manipulated to remove the source attribute and then decoded into an output text with the target attribute. The model that encodes and decodes the text can be implemented as an auto-encoder (AE; e.g., [32]), variational auto-encoder (VAE; e.g., [25]), or generative adversarial network (GAN; e.g., [42]). Further, there are different approaches to latent representation manipulation. [16] distinguish Latent Representation Editing [25], Attribute Code Control [32], and Latent Representation Splitting [17]. Disentaglement methods are the closest to ParaCAIF (and ParaGeDi), although the latter do not infuse the style into the model or a sentence representation, but impose it on the generator by another model.

Pseudo-parallel corpus construction group of approaches addresses the lack of parallel data by building pseudo-parallel corpora of style pairs. [20] propose

STRAP (Style Transfer via Paraphrasing) and create pseudo-parallel corpora by transferring styled sentences to neutral with a pretrained general-purpose paraphraser. [16] note that the tasks of paraphrasing and TST are close to each other and anticipate further research in this direction. They suggest that linguistic style transfer can be regarded as a subset of paraphrasing. STRAP [20], ParaGeDi [5], and our work fall into this direction of research.

### 2.3 Detoxification

Detoxification can tackle offensive or hateful speech in the Internet by rewriting it or prompting the user with a polite version of the text they just typed in. Existing works on this topic cast the task of detoxification as a form of TST, where the source style is toxic and the target style is neutral/non-toxic. The first work by [30] is an end-to-end seq2seq model trained on a non-parallel corpus with autoencoder loss, style classification loss and cycle-constistency loss.

There is a nascent line of works that adapt CTG methods for detoxification which our work falls into. [5] propose two unsupervised approaches, one of which, ParaGeDi, adapts GeDi [19] for detoxification, making several changes. First, the regular LM is substituted with an LM capable of paraphrasing. Second, the discriminator is conditioned only on the paraphrase, ignoring the source sentence. Third, an optional reranker is added which reweighs the hypotheses according to the target style. We implement the same ideas to turn CAIF into ParaCAIF, additionally using a similarity model to reweight the hypotheses. [5] also propose another approach, CondBERT, which follows a pointwise editing setup, first identifying tokens to mask in the input, then using a mask-filling model to replace them. It uses a lexicon-based approach to masking words by using weights from a whole-word toxic language logistic classifier. The approach proposed by [12] is inspired by DExperts [23] and consists of two steps: masking potentially toxic locations and replacing them in the context of the sentence. For the first step, a subsequence is considered potentially toxic and consequently masked according to a disagreement between non-toxic *expert* and toxic *anti-expert* LMs. At the second step, masked sequences are replaced by an auto-encoder LM which is steered by the same expert and anti-experts LMs.

However, all the above approaches have been proposed for English language. Detoxification for Russian language has not been addressed until recently. [7] is the first work on this topic which proposes a few simple baselines and two more sophisticated approaches. The first one is detoxGPT which comes in three variants: zero-shot, few-shot and fine-tuned. They differ by the size of the parallel dataset used to fine-tune the model. The second approach is condBERT which is further applied by the authors to English detoxification in [5] and has been described above. [8] organize RUSSE-2022 Detoxification competition (RUSSE Detox), the first such competition in Russian language with parallel data which at the time of publishing had no analogies in any languages. The organizers provided two model-based baselines for the task. The RuT5 baseline fine-tunes ruT5-base[2] model on the parallel data. RuPrompts baseline is based on the ruPrompts library[3] and the Continuous Prompt Training method [18] which consists in training embeddings corresponding to the prompts. The prompts are tuned for the ruGPT3-large model[4]. Analyzing the results of the competition, the organizers come to the conclusion that RuT5, a baseline of a large pretrained LM fine-tuned on parallel data, was hard to beat for the contestants.

## 3   Method

Unlike GeDi [19] and DExperts [23] which are designed to work with CC-LMs, CAIF is based on a free-form classifier. During generation, the classifier estimates the probability of possible continuations given the desired control attribute. The logits output by the classifier are combined with those of the LM, thus re-weighting the original token probability distribution. The authors propose several solutions to tackle the computational complexity of applying the classifier model to all candidate tokens at every generation step. First, they propose to re-weight only $j$ most probable tokens with the discriminator and then use them for top-$k$ sampling. The authors come up with an optimal value of 100 for $j$ during their experiments. Second, the authors suggest performing the re-weighting only at specific generation steps. They analyze two strategies: periodic and entropy criteria for CAIF. The periodic criterion consists of adjusting the logits at every $p$-th step. The entropy criterion is based on the entropy of token probabilities output by the model at every generation step. A threshold of $e$ is defined, and logit reweighting is applied only at steps when entropy is greater than $e$.

Another hyperparameter used for CAIF is the importance of the classifier for re-weighting, i.e., the style strength. The authors rely on the same solution of Bayesian inference, used, for example, in GeDi (Eq. 1), but denote the parameter of style strength as $\alpha$. Interestingly, they find that, while [19] only use $\alpha \geq 1$, it is possible to use any $\alpha \in \mathbb{R}$. With positive $\alpha$, LM steering is done with $\left(1 - P(c|x_{\leq i})\right)^{\alpha}$ (namely, inverse probability weighting). However, the authors observe that it is possible to perform weighting with $p(c|x_{\leq i})$ and $\alpha < 0$. Moreover, they find that negative $\alpha$ shows better results than inverse probability sampling in terms of attribute accuracy and perplexity.

We adapt CAIF for paraphrasing and TST, making several changes. First, we replace the regular LM with an LM capable of paraphrasing. Second, we add support for encoder-decoder models, as the original CAIF is applicable only to decoder models. On the one hand, it is important for paraphrasing, as encoder-decoder models like T5 [27] are initially designed to preserve content. On the other hand, it can be seen as an improvement of the original CAIF that broadens its applicability for CTG. Third, we add support for decoder-based paraphraser models. To this end, we limit the sequence considered by the classifier

---

[2] https://huggingface.co/sberbank-ai/ruT5-base.
[3] https://sberbank-ai.github.io/ru-prompts.
[4] https://github.com/sberbank-ai/ru-gpts.

to the output of the model after the input prompt. Fourth, we add posterior re-ranking of the output candidates. This practice is common in paraphrasing and is used, for example, in [11]. We generate $n$ candidates with the paraphraser model steered by CAIF and assess them with two metrics: style accuracy and similarity with the source sentence. However, with two metrics, we come to the problem of balancing them for the final ranking. We suggest the following sorting procedure:

1. Define a threshold $a$ for the accurcy score ·
2. Split the array of candidates into two parts: first with accuracy scores greater than $a$ and second with accuracy scores less than $a$
3. Sort the first part according to the similarity score and the second part according to the accuracy score
4. Return an array which is a concatenation of the first and the second part.

This procedure allows to control the balance between the scores by choosing different $a$. It prioritizes the similarity score, as our experiments have shown that it may be beneficial to lower the threshold for the accuracy score in order to raise the candidates more similar to the source sentence in the final ranking and obtain a greater joint score.

## 4    Experiments

### 4.1    Russian Detoxification

*Data.* We work with the data of RUSSE Detox. The competition dataset is parallel and contains toxic sentences with neutral rewritings. The source toxic sentences were taken from the Russian datasets of toxic messages from various social media: Odnoklassniki [1], Pikabu [31], and Twitter [29]. The target neutral sentences were obtained through manual rewriting by crowd workers who eliminated toxicity. Each toxic sentence has 1–3 variants of detoxification. The dataset is divided into three splits: train (6 948 toxic sentences), development (800 toxic sentences), and test (875 toxic sentences).

*Evaluation.* We employ the automatic evaluation setup used for RUSSE Detox. [8] follow the evaluation strategy of [20]. They evaluate the models' outputs on three parameters: style of text, content preservation and fluency of text. Style **(ACC)** is evaluated with the ruBERT model [21] fine-tuned for toxicity detection on the Odnoklassniki [1] and Pikabu [31] datasets. Content **(SIM)** is evaluated as the cosine similarity of the embeddings of the source and the transformed sentences generated by the LaBSE model [10]. Fluency **(FL)** is evaluated with an acceptability classifier trained on a synthetic Russian analogue of the CoLA dataset [35] for which sentences were corrupted by randomly replacing, deleting, or shuffling words. Following [20], the authors combine all three metrics at the sentence level by multiplying them, resulting into a Joint score **(J)**. In addition, chrF metric is calculated with the reference sentences.

*Models.* To be able to steer the paraphraser LM with CAIF, we need an attribute classifier. As virtually all available Russian datasets of toxicity were used to train the evaluation style accuracy classifier for RUSSE Detox, we train our classifier on the dataset released by the organizers. The model is trained on the train subset and evaluated on the development subset. Although the dataset provides 1–3 neutral rewritings for each toxic sentence, we use an equal number of toxic and neutral sentences for classifier training to preserve class balance. We note, however, that the training data for the attribute classifier need not to be parallel which significantly broadens the applicability of the system. We fine-tune RuBERT-tiny from [4] for 10 epochs with a learning rate of 2e−5, batch size of 32. The best model is then chosen with validation macro F1-score. It achieves a strong macro F1-score of 0.94 for both classes. We make this fine-tuned model available at Huggingface Hub[5].

For the generative paraphrasers, we explore two models from [11] and ruT5-paraphraser from [3]. The models presented in [11] are ruGPT3-large and mT5 [40]. They are fine-tuned for paraphrasing on news data from ParaPhraserPlus[6] and conversational data from subtitles and dialogues of users with chatbots. [3] train their ruT5-paraphraser model on Leipzig Russian web text corpus[7] paraphrased through back-translation.

All the above models are used as generators in ParaCAIF. As CAIF is designed to work with top-$k$ sampling, we perform the inference on all models with the following parameters: top-$k$ = 20, temperature = 1.0, number of candidates = 10. One exception is ruT5-paraphraser, which was initially presented to work with beam search. We test its plain version with the number of beams = 4. The top-$j$ parameter of CAIF is set to 100. The candidates are further scored with style accuracy classifier and similarity with the source sentence and reranked according to the scores. For the style accuracy classifier we use RuBERT-tiny described above. Following [11], we evaluate the semantic similarity with `paraphrase-xlm-r-multilingual-v1` model [28,34]. We calculate the cosine distance between the embeddings of source and output sentences generated by the model—strategy reported as the most stable by [11]. Applying these two models for reranking, we create some sort of proxies for the evaluation metrics.

We experiment with the hyperparameters available in our system. Following the experimental setup in [33], we choose from $\alpha \in [-5, -3, -2, -1, 0]$, where 0 corresponds to no CAIF sampling, and entropy threshold $e \in [0, 0.5, 1.5, 3.2, 5.0]$. We also explore the accuracy score threshold for candidates reranking $a \in [0.8, 0.9, 0.99]$.

### 4.2   English Detoxification

*Data.* We evaluate ParaCAIF on English toxicity data presented in [5], the work which proposed ParaGeDi. The overall dataset is a merge of the English parts of three datasets by Jigsaw [13–15], containing around 2 million examples. [5] split it into two nonoverlapping parts and fine-tune two RoBERTa [24] toxicity

---

[5] https://huggingface.co/BunnyNoBugs/rubert-tiny2-russe-toxicity.

[6] http://paraphraser.ru/download.

[7] https://wortschatz.uni-leipzig.de/en/download/Russian.

classifiers. One of them is used to rerank the candidates produced by ParaGeDi, and the other participates in the evaluation. The classifiers perform closely on the test set of the first Jigsaw competition, reaching the AUC-ROC of 0.98 and F1-score of 0.76. The data used for TST models comes from the first Jigsaw competition [13]. To prepare the toxic dataset, [5] divide the comments labelled as toxic into sentences and classify them with the evaluation classifier described above. The test set comprises 10 000 sentences with the highest toxicity score according to the classifier. For our evaluation, we select a random subset of 1 000 of the test set for computational complexity reasons.

*Evaluation.* We employ the automatic evaluation setup described in [5]. Along with [8], it follows [20] and is therefore similar to the one described in Sect. 4.1. Style accuracy (ACC) is measured with the evaluation classifier described above. Content preservation (SIM) is evaluated as the similarity of sentence-level embeddings of the original and transformed texts computed by the model of [37]. Fluency (FL) is measured with the classifier of linguistic acceptability trained on the CoLA dataset [35]. Joint score (J) is computed as the average of their sentence-level product.

*Models.* For the attribute classifier, we use one of the two classifiers described above which was used for reranking hypotheses in ParaGeDi[8]. For the paraphrasing model, we use the paraphraser which is employed in ParaGeDi as well as is featured as a standalone baseline in [5]. This model is a pre-trained T5-based [27] paraphraser, fine-tuned on a random subsample of the ParaNMT dataset [38]. The original paraphraser baseline used beam search with number of beams = 10, but CAIF was designed to work with top-$k$ sampling, so we perform the inference with the following parameters: top-$k$ = 20, temperature = 1.0, number of candidates = 10. CAIF parameters are: $\alpha = -5$, top-$j$ = 100, entropy threshold $e = 0$ (which corresponds to plain CAIF).

# 5   Results and Discussion

## 5.1   Russian Detoxification

Table 1 displays the results of our systems in detoxification evaluated on RUSSE Detox development subset. For all considered paraphraser models, we observe an approximately twofold increase in J score of ParaCAIF versions compared to plain sampling. None of them, however, reach the level of the supervised ruT5 baseline. The best J score of 0.33 is achieved by ParaCAIF ruGPT-3, but its SIM score is rather low. In terms of semantic similarity, the T5-based models perform quite better, which confirms the intuition that encoder-based models fare better in preserving content. On the other hand, the GPT model is consistently more fluent (FL) than T5 models.

In terms of style transfer accuracy (ACC), all ParaCAIF models perform closely to the supervised baseline, and ParaCAIF ruGPT-3 even outperforms it.

---

[8] https://huggingface.co/s-nlp/roberta_toxicity_classifier_v1.

**Table 1.** Detoxification results evaluated on RUSSE Detox development subset, sorted by J metric. The inference of all ParaCAIF models is done with $\alpha = -5$.

| Model | ACC | SIM | FL | J | chrF |
|---|---|---|---|---|---|
| ruT5-base (baseline) | 0.75 | 0.80 | 0.81 | 0.50 | 0.57 |
| ParaCAIF ruGPT-3 | 0.80 | 0.51 | 0.78 | 0.33 | 0.32 |
| ParaCAIF ruT5 | 0.68 | 0.65 | 0.69 | 0.30 | 0.32 |
| ParaCAIF mT5-large | 0.68 | 0.59 | 0.67 | 0.27 | 0.37 |
| ruT5 paraphraser (beam search) | 0.34 | 0.74 | 0.75 | 0.16 | 0.32 |
| ruT5 paraphraser (top-k) | 0.44 | 0.59 | 0.71 | 0.16 | 0.26 |
| mT5 paraphraser | 0.38 | 0.60 | 0.79 | 0.15 | 0.35 |
| ruGPT-3 paraphraser | 0.53 | 0.34 | 0.88 | 0.15 | 0.21 |

Comparing the decoding strategies for ruT5 paraphraser, we observe that top-k sampling allows to increase the ACC score but lowers the SIM score compared to beam search originally used in [3].

Overall, we observe that ParaCAIF models reach the level of the supervised baseline in style transfer accuracy, but insufficient content preservation hinders them from achieving high joint scores. Although formally the best J score is achieved by ParaCAIF ruGPT-3, we focus the following experiments on Para-CAIF ruT5 because its scores are more balanced across all metrics. On RUSSE Detox test subset, this model achieves the following scores: ACC 0.71, SIM 0.68, FL 0.72, J 0.35, chrF 0.33. Below, we conduct a more detailed analysis of Para-CAIF performance.

First, we study the influence of the CAIF $\alpha$ parameter. To begin with, we set it to 0 and are left with plain sampling and the reranking procedure described in Sect. 3. We compare the performance of plain ruGPT-3 and ruT5 paraphrasers with and without reranking and report the results in Table 2.

We can see that simply generating more candidates and appropriately reranking them significantly increases the ACC and J scores of the models. The SIM

**Table 2.** Detoxification results evaluated on RUSSE Detox development subset, sorted by J metric. The inference on all models is done with plain sampling.

| Model | ACC | SIM | FL | J | chrF |
|---|---|---|---|---|---|
| ruGPT-3 paraphraser + reranking (threshold 0.8) | 0.72 | 0.47 | 0.80 | 0.28 | 0.28 |
| ruGPT-3 paraphraser + reranking (threshold 0.9) | 0.73 | 0.46 | 0.80 | 0.28 | 0.28 |
| ruT5 paraphraser + reranking | 0.61 | 0.60 | 0.71 | 0.25 | 0.30 |
| ruGPT-3 paraphraser + reranking | 0.74 | 0.41 | 0.80 | 0.25 | 0.26 |
| ruGPT-3 paraphraser | 0.47 | 0.51 | 0.85 | 0.16 | 0.30 |
| ruT5 paraphraser | 0.44 | 0.59 | 0.71 | 0.16 | 0.26 |

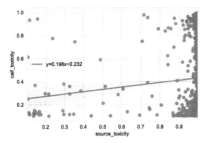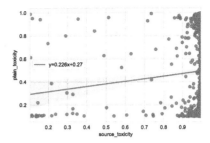

**Fig. 1.** Relative toxicity of ParaCAIF ruT5 (left) and ruT5 paraphraser (right) outputs depending on the toxicity of the source sentence, evaluated on RUSSE Detox development subset. The difference in regression coefficients is significant with $p < 0.01$, based on ANCOVA.

score of ruGPT-3 paraphraser after reranking is quite low, but lowering the threshold for the style transfer accuracy reranker allows to increase the SIM and consequently J score practically without a sacrifice in the ACC score. Moreover, the resulting J scores with reranking are close to those of the corresponding ParaCAIF models. However, a more in-depth comparison shows that ParaCAIF consistently produces less toxic paraphrases.

We average detoxification metrics of ruT5 paraphraser and its ParaCAIF version across 10 candidates for each source sentence. We observe that the paraphrases generated by the ParaCAIF models are by a large margin less toxic (precise scores are available in Appendix A). However, this comes with a sacrifice in semantic similarity and fluency. In addition, we calculate the average number of candidates considered toxic by the evaluation classifier with a threshold of 0.50. Plain ruT5 paraphraser generates 6.48 toxic paraphrases per 10 candidates on average, while its ParaCAIF version is toxic only in 2.81 candidates out of 10. This difference is statistically significant with $p < 0.01$, based on a two-sample t-test. Overall, these results imply that using ParaCAIF in practice is more computationally efficient, as we can find a non-toxic paraphrase in a significantly smaller amount of candidates.

Further, we compare the relative toxicity of the models' outputs, depending on the toxicity of the source sentence. Toxicity is scored with the style transfer accuracy classifier used in evaluation, where 0 is neutral and 1 is toxic. Figure 1 displays the results. We draw a regression line on both plots to find the correlation and compare the slope coefficients. The slope of ParaCAIF ruT5 is smaller (0.19 vs. 0.22), indicating that it is less dependent on the source toxicity and is more effective in detoxifying more toxic sentences than ruT5 parapharser. The intercept coefficient of ParaCAIF ruT5 is also smaller (0.23 vs. 0.27), implying that the overall toxicity level of its paraphrases is lower than that of ruT5 parapharser. The difference in regression coefficients is significant with $p < 0.01$, based on analysis of covariance (ANCOVA).

Finally, we explore the influence of different CAIF hyperparameters values on ParaCAIF ruT5 detoxification performance. The graph on the left in Fig. 2 dis-

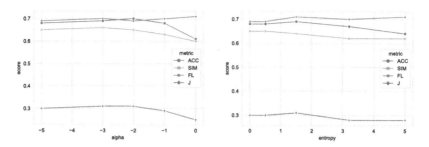

**Fig. 2.** Detoxification performance of ruT5 paraphraser depending on different $\alpha$ (left) and entropy (right) values, evaluated on RUSSE Detox development subset.

plays the performance depending on different $\alpha$ values. The dependence between the $\alpha$ value and the level of toxicity is less consistent than that reported by [33] in toxicity avoidance experiments. Decreasing the style strength from $-5$ up to $-3$ results in an increase of style transfer accuracy with no loss of semantic similarity and fluency. The peak ACC and J scores are achieved at $\alpha = -2$. This shows a discrepancy from the findings of previous studies [5,12,33] and, therefore, needs more thorough investigation. However, setting $\alpha$ to $-1$ and 0 which corresponds to weak and no style strength, respectively, results in low ACC and J scores, as expected. For examples of sentences detoxified by ParaCAIF with different $\alpha$ values, see Appendix B.

The graph on the right in Fig. 2 shows the performance of ParaCAIF depending on different entrpopy thresholds. Here, the peak J score is achieved with entropy threshold = 1.5. This is also beneficial for the practical applicability and computational effectiveness, as the inference with entropy threshold = 1.5 on RUSSE Detox test subset was around 1.4 times faster than with plain Para-CAIF. Analogously with $\alpha$, the fact that the ACC score does not consistently decrease with the increase in entropy threshold value goes against the results described by [33] who report that "entropy CAIF could perform with negligible performance loss compared to plain CAIF on the toxicity avoidance task". However, we are in line with [33], observing that the outputs of the model become more fluent with the increase in entropy threshold value.

### 5.2   English Detoxification

Table 3 reports the results of ParaCAIF in English detoxification evaluated on the test data from [5]. In addition to ParaGeDi and CondBERT, it also features two models considered in [5]: T5 paraphraser baseline, and the third-best model, Mask&Infill [39]. For each source sentence, T5 paraphraser baseline generates 10 candidates and the least toxic of them is then chosen by the toxicity classifier described in Sect. 4.2. However, even after reranking, only 15% of its outputs are classified as non-toxic. On the other hand, we observe that applying ParaCAIF to the T5 paraphraser baseline significantly increases its detoxifying ability, however, with a sacrifice in semantic similarity and fluency. As a result, ParaCAIF

**Table 3.** Detoxification results evaluated on the test data from [5], sorted by J metric. ParaCAIF models are evaluated on a random subset of 1k samples.

| Model | ACC | SIM | FL | J |
|---|---|---|---|---|
| ParaGeDi | 0.95 | 0.66 | 0.80 | 0.50 |
| CondBERT | 0.94 | 0.69 | 0.77 | 0.50 |
| ParaCAIF T5 | 0.99 | 0.65 | 0.56 | 0.38 |
| ParaCAIF T5 (threshold 0.9) | 0.99 | 0.65 | 0.57 | 0.38 |
| ParaCAIF T5 (threshold 0.8) | 0.99 | 0.65 | 0.57 | 0.38 |
| Mask&Infill [39] | 0.78 | 0.80 | 0.49 | 0.31 |
| T5 paraphraser (baseline) | 0.15 | 0.90 | 0.87 | 0.11 |

outperforms the second-best baseline from [5], Mask&Infill [39], by the J score. Furthermore, the ACC score of 0.99 outperforms ParaGeDi itself. Lowering the threshold for the style accuracy of the reranker marginally increases the fluency of the outputs.

Summarizing the results of detoxification experiments in Russian and English, we can claim that ParaCAIF is able to significantly increase the detoxifying ability of paraphraser models. ParaCAIF is potentially applicable to other TST tasks. However, the resulting style transfer accuracy may depend on the size of the training dataset used to fine-tune the attribute discriminator. While the best ParaCAIF model for Russian scores 0.80, ParaCAIF for English achieves a near-perfect ACC score of 0.99. The size of the training datasets for the respective attribute classifiers may provide an explanation. Although the toxicity classifier for Russian achieves a F1-score of 0.94 on the test set, it is trained on a relatively small dataset of 14 000 samples and, therefore, has a lower exposure to real-world toxicity. On the other hand, the toxicity classifier for English is trained on a dataset of around one million examples.

Although ParaCAIF can outperform supervised models in terms of style transfer accuracy, its overall performance is nonetheless inferior to supervised approaches. On the other hand, we note that ParaCAIF, being an unsupervised approach, has a broader range of application, as parallel data are not readily available for many TST tasks. The main reason behind the underperformance of ParaCAIF is insufficient content preservation and fluency of the outputs. It is thus subject to the problem of trade-off between fluency, style accuracy, and meaning preservation, attested by many TST works [5,20,22].

## 6    Conclusion and Further Work

In this work, we adapt CAIF, an existing method which implements a guided strategy to perform CTG, for TST. We follow the idea of ParaGeDi and replace the regular LM in the original method with a model capable of paraphrasing. We illustrate the applicability of the resulting unsupervised method, ParaCAIF, by addressing detoxification, a TST subtask with a high potential for practical application. We conduct our experiments in two languages. First, we evaluate our method on the data of RUSSE-2022 Detoxification competition, thus contributing to the relatively understudied area of Russian detoxification. To the best of our knowledge, this is the first work that adapts a CTG method for Russian detoxification. We experiment with a line of Transformer paraphraser models and observe that for all models, ParaCAIF substantially reduces the toxicity of the generated paraphrases. We find that the GPT-based ParaCAIF model surpasses the supervised baseline in style transfer accuracy.

Applying ParaCAIF to English detoxification, we observe a multiple decrease in the toxicity of the generated paraphrases compared to the plain paraphraser baseline. Moreover, we find that ParaCAIF outperforms ParaGeDi, another previously proposed CTG-based TST approach, in style transfer accuracy. However, the overall performance of ParaCAIF remains lower than that of supervised approaches, mainly because of insufficient content preservation and fluency of the outputs. This result falls in line with previous TST research, which attests a trade-off between fluency, style accuracy, and meaning preservation. On the other hand, ParaCAIF, being an unsupervised approach, has a broader range of application, as it does not require parallel data which are not readily available for many TST tasks.

Several future research directions open up on the basis of this work. First, it is important to assess the performance of ParaCAIF with human evaluation. TST work has noted that automatic evaluation can serve as a proxy, but cannot fully replace human evaluation (e.g., [8]). Second, it can be promising to adapt ParaCAIF for beam search. The results of our experiments with reranking suggest that plain paraphrasers have the potential for detoxification, it just takes more generation rounds to find a non-toxic candidate. On the other hand, an increase in the entropy threshold does not lead to a drastic fall of style transfer accuracy, while decreasing the inference time. Therefore, evaluating longer beam candidates with the attribute classifier at positions with high entropy may prove to be more computationally effective than sampling multiple candidates and further reranking them. Finally, it is of interest to expand CAIF to a multi-class setting, as suggested for GeDi in [19]. Using two or more classifiers would benefit both CAIF, allowing for a more fine-grained control of the desired attributes, and ParaCAIF, enabling the evaluation of additional parameters, like semantic similarity, during generation.

# A ParaCAIF vs. Reranking

Table 4 displays the comparison results of ParaCAIF model and plain paraphraser with added reranking procedure..

**Table 4.** Detoxification metrics evaluated on RUSSE Detox development subset, averaged across 10 candidates for each source sentence. The differences between models in all metrics are statistically significant with $p < 0.01$, based on a two-sample t-test.

| Model | ACC | SIM | FL |
|---|---|---|---|
| ParaCAIF ruT5 | 0.72 | 0.53 | 0.68 |
| ruT5 paraphraser + reranking | 0.42 | 0.63 | 0.72 |

# B ParaCAIF Russian Detoxification Examples

Tables 5 and 6 demonstrate examples of sentences from RUSSE Detox detoxified by ParaCAIF ruT5 with different $\alpha$ values. We observe that $\alpha = -5$ tends to delete all strong obscene words, while a less strict $\alpha$ of $-1$ allows some direct toxicity (e.g., *блин* 'damn it').

**Table 5.** Examples of sentences from RUSSE Detox detoxified by ParaCAIF ruT5 with $\alpha = -5$.

| source toxic sentence | ParaCAIF ruT5 ($\alpha = -5$) |
|---|---|
| а мы просим тебя иди на\*\*й (and we ask you to go f\*ck yourlself) | Мы просим тебя лучше сходить (We ask that you better go) |
| это не цирк это уже бл\*\*ь шоу. (it's not a circus, it's a f\*cking show) | Оно не цирк, а всякую бредятину шоу. (It not circus, but some nonsense show) |
| возле какого там 10 августа так что х\*\*\*\* несёшь ты где бл\*\*ь это прочитала дура е\*\*\*\*\* (around what August 10 so you're talking sh\*t where have you f\*cking read that you f\*cking idiot) | Да что черт возьми здесь 10 августа (What the h\*ll is here August 10) |
| подстрекатель родину гад не любиш ..... (you instigator you don't love your homeland you bastard .....) | Привет, не любишь родину и мать. (Hi, you don't love your homeland and your mother.) |
| это что за высер придурка обиженного (what kind of offended \*sshole's sh\*t is that) | Да что это за фигня (What the h\*ll is this) |

**Table 6.** Examples of sentences from RUSSE Detox detoxified by ParaCAIF ruT5 with $\alpha = -1$.

| source toxic sentence | ParaCAIF ruT5 ($\alpha = -1$) |
|---|---|
| сашок пзь**ц, позорище про**ал последний бой!!! ((((( (sasha's a f*cking disgrace, f*cked up last battle!!! ((((( ) | Саша блин, позорник проехал последний бой! :)(( (Sasha's a crappy disgrace, bucked up last battle! :)(( ) |
| это не цирк это уже бл**ь шоу. (it's not a circus, it's a f*cking show) | А это не цирк, это уже блин шоу. (Now that's not a circus, that's a show, damn it.) |
| это твари а не люди (they're creatures not people) | Все это не люди а звери (All these aren't people they are animals) |
| Что за лошак? Его мнение важно? (Who's the ass? Does his opinion matter?) | Что за лошадь? У нее важное мнение? (Who's the horse? Does she have an important opinion?) |
| кто сочиняет этот бред?!!! дибилы!!!!! (who makes up this nonsense?!!! morons!!!!!) | Кто сочиняет этот бред?!!!!!!!!! (Who makes up this nonsense?!!!!!!!!) |

# References

1. Belchikov, A.: Russian language toxic comments (2019). https://www.kaggle.com/datasets/blackmoon/russian-language-toxic-comments
2. Bender, E.M., Gebru, T., McMillan-Major, A., Shmitchell, S.: On the dangers of stochastic parrots: can language models be too big? In: FAccT 2021 (2021). https://doi.org/10.1145/3442188.3445922
3. Dale, D.: Перефразирование русских текстов: корпуса, модели, метрики (2021). https://habr.com/ru/post/564916/
4. Dale, D.: Маленький и быстрый BERT для русского языка (2021). https://habr.com/ru/post/562064/
5. Dale, D., et al.: Text Detoxification Using Large Pre-trained Neural Models (2021). https://aclanthology.org/2021.emnlp-main.629
6. Dathathri, S., et al.: Plug and Play Language Models: A Simple Approach to Controlled Text Generation (2020). https://openreview.net/forum?id=H1edEyBKDS
7. Dementieva, D., et al.: Methods for Detoxification of Texts for the Russian Language (2021). https://arxiv.org/abs/2105.09052
8. Dementieva, D., et al.: RUSSE-2022: findings of the first Russian detoxification task based on parallel corpora (2022)
9. Deng, Y., et al.: Residual Energy-Based Models for Text Generation (2020). https://openreview.net/forum?id=B1l4SgHKDH
10. Feng, F., Yang, Y., Cer, D., Arivazhagan, N., Wang, W.: Language-agnostic BERT sentence embedding (2020)
11. Fenogenova, A.: Russian Paraphrasers: Paraphrase with Transformers (2021). https://aclanthology.org/2021.bsnlp-1.2
12. Hallinan, S., Liu, A., Choi, Y., Sap, M.: Detoxifying Text with MaRCo: Controllable Revision with Experts and Anti-Experts (2022). https://arxiv.org/abs/2212.10543
13. Jigsaw: Toxic Comment Classification Challenge (2018). https://www.kaggle.com/c/jigsaw-toxic-comment-classification-challenge
14. Jigsaw: Jigsaw Unintended Bias in Toxicity Classification (2019). https://www.kaggle.com/c/jigsaw-unintended-bias-in-toxicity-classification

15. Jigsaw: Jigsaw Multilingual Toxic Comment Classification (2020). https://www. kaggle.com/c/jigsaw-multilingual-toxic-comment-classification
16. Jin, D., Jin, Z., Hu, Z., Vechtomova, O., Mihalcea, R.: Deep Learning for Text Style Transfer: A Survey (1) (2022). https://aclanthology.org/2022.cl-1.6
17. John, V., Mou, L., Bahuleyan, H., Vechtomova, O.: Disentangled Representation Learning for Non-Parallel Text Style Transfer (2019). https://www.aclweb.org/ anthology/P19-1041.pdf
18. Konodyuk, N., Tikhonova, M.: Continuous prompt tuning for Russian: how to learn prompts efficiently with RuGPT3? In: Burnaev, E., et al. (eds.) AIST 2021. CCIS, vol. 1573, pp. 30–40. Springer, Cham (2022). https://doi.org/10.1007/978-3-031-15168-2_3
19. Krause, B., et al.: GeDi: Generative Discriminator Guided Sequence Generation (2021). https://aclanthology.org/2021.findings-emnlp.424
20. Krishna, K., Wieting, J., Iyyer, M.: Reformulating Unsupervised Style Transfer as Paraphrase Generation (2020). https://aclanthology.org/2020.emnlp-main.55
21. Kuratov, Y., Arkhipov, M.: Adaptation of deep bidirectional multilingual transformers for Russian language (2019)
22. Laugier, L., Pavlopoulos, J., Sorensen, J., Dixon, L.: Civil Rephrases Of Toxic Texts With Self-Supervised Transformers (2021). https://doi.org/10.18653/v1/2021.eacl-main.124
23. Liu, A., et al.: DExperts: Decoding-Time Controlled Text Generation with Experts and Anti-Experts (2021). https://arxiv.org/abs/2105.03023
24. Liu, Y., et al.: RoBERTa: A Robustly Optimized BERT Pretraining Approach (2019). https://arxiv.org/abs/1907.11692
25. Mueller, J., Gifford, D.K., Jaakkola, T.S.: Sequence to better sequence: continuous revision of combinatorial structures. In: Proceedings of Machine Learning Research (2017). https://proceedings.mlr.press/v70/mueller17a.html
26. Prabhumoye, S., Black, A.W., Salakhutdinov, R.: Exploring Controllable Text Generation Techniques (2020). https://aclanthology.org/2020.coling-main.1
27. Raffel, C., et al.: Exploring the Limits of Transfer Learning with a Unified Text-to-Text Transformer (2019). https://arxiv.org/abs/1910.10683
28. Reimers, N., Gurevych, I.: Making monolingual sentence embeddings multilingual using knowledge distillation (2020)
29. Rubtsova, Y.: Avtomaticheskoye postroyeniye i analiz korpusa korotkikh tekstov (postov mikroblogov) dlya zadachi razrabotki i trenirovki tonovogo klassifikatora. inzheneriya znaniy i tekhnologii semanticheskogo veba (2012)
30. Nogueira dos Santos, C., Melnyk, I., Padhi, I.: Fighting Offensive Language on Social Media with Unsupervised Text Style Transfer (2018). https://aclanthology. org/P18-2031
31. Semiletov, A.: Toxic Russian comments (2020). https://www.kaggle.com/datasets/ alexandersemiletov/toxic-russian-comments
32. Shen, T., Lei, T., Barzilay, R., Jaakkola, T.: Style transfer from non-parallel text by cross-alignment (2017)
33. Sitdikov, A., Balagansky, N., Gavrilov, D., Markov, A.: Classifiers are Better Experts for Controllable Text Generation (2022). https://arxiv.org/abs/2205. 07276
34. Thakur, N., Reimers, N., Daxenberger, J., Gurevych, I.: Augmented SBERT: data augmentation method for improving bi-encoders for pairwise sentence scoring tasks (2020)
35. Warstadt, A., Singh, A., Bowman, S.R.: Neural Network Acceptability Judgments (2019). https://aclanthology.org/Q19-1040

36. Weng, L.: Controllable Neural Text Generation (2021). https://lilianweng.github.io/posts/2021-01-02-controllable-text-generation/
37. Wieting, J., Berg-Kirkpatrick, T., Gimpel, K., Neubig, G.: Beyond BLEU: Training Neural Machine Translation with Semantic Similarity (2019). https://aclanthology.org/P19-1427
38. Wieting, J., Gimpel, K.: ParaNMT-50M: Pushing the Limits of Paraphrastic Sentence Embeddings with Millions of Machine Translations (2018). https://aclanthology.org/P18-1042
39. Wu, X., Zhang, T., Zang, L., Han, J., Hu, S.: "Mask and Infill": Applying Masked Language Model to Sentiment Transfer (2019). https://arxiv.org/pdf/1908.08039
40. Xue, L., et al.: mT5: A Massively Multilingual Pre-trained Text-to-Text Transformer (2021). https://aclanthology.org/2021.naacl-main.41
41. Zhang, H., Song, H., Li, S., Zhou, M., Song, D.: A Survey of Controllable Text Generation using Transformer-based Pre-trained Language Models (2022). https://arxiv.org/abs/2201.05337
42. Zhao, J.J., Kim, Y., Zhang, K., Rush, A.M., LeCun, Y.: Adversarially regularized autoencoders. In: Proceedings of Machine Learning Research (2018). https://proceedings.mlr.press/v80/zhao18b.html

# Less than Necessary or More than Sufficient: Validating Probing Dataset Size

Evgeny Orlov[1]([⊠]) [iD] and Oleg Serikov[2] [iD]

[1] ITMO University, St. Petersburg, Russia
411996@edu.itmo.ru
[2] AIR Institute, Moscow, Russia

**Abstract.** The vast body of research is dedicated to interpreting language models, particularly probing them for linguistic properties. As in many other NLP fields, probing works tend to reuse existing datasets, resulting in more and more specialized findings. Introducing new datasets, although necessary for truly typologically diverse studies, requires labor-intensive data annotation. Meanwhile, models become heavier, probing methods inventory enriches, and the cost of probing experiments grows accordingly. To minimize the amount of work annotating new data, and reduce the computational cost of experimenting with the existing data, it will be beneficial to assess dataset size.

We propose *fractions probing*, a novel method of validating probing dataset size. It includes *data redundancy test* to review existing datasets and *data sufficiency test* to provide guidance when collecting new ones. We illustrate the method's applicability with SentEval probing suite, finding that it can be safely reduced. Our experiments are conducted for two models, BERT and RoBERTa, showing the latter to consistently require more data. Fractions probing can be used to analogously investigate other datasets and models.

**Keywords:** probing · dataset analysis · language models

## 1 Introduction

As black-box models acquired prevalence in the majority of the nowadays NLP tasks, the need arose to interpret the behavior of such models. Indeed, while there is no way to explain the reasoning of used models and guarantee their reliable behavior, applying these models to valuable processes is risky.

Probing, a popular approach to revealing underlying structures in language models, addresses this interest by trying to systematize models' behavior from the linguistic point of view. Inspired by the success of the SentEval probing suite, researchers tend to conduct multi-dimensional probing studies. A significant amount of probing techniques output graphs which are further interpreted, as in layer-wise and amnesic [13] probing; we refer to such methods with an umbrella term of *visual probing*. A typical structural probing study considers

© The Author(s), under exclusive license to Springer Nature Switzerland AG 2024
D. I. Ignatov et al. (Eds.): AIST 2023, LNCS 14486, pp. 109–125, 2024.
https://doi.org/10.1007/978-3-031-54534-4_8

several models, languages, or grammatical categories and somehow ranks models' layers wrt. the category in a particular language. To address languages and categories at a typological scale, one would need to use the diverse groups of datasets, which could be extremely small or huge. Both extremes are the subject of the optimal dataset size research we conduct in this paper.

[36] point out the necessity of a complete "recipe" for probing dataset collection. They note several recent papers which call for similar goal of creating reliable benchmarks. [14] criticize the current leaderboards for focusing on the models' performance rather than practically necessary compactness, fairness, and energy efficiency. [31] suggest upgrading existing leaderboards so they better reflect if and where the progress is made.

The question of probing dataset size is by all means practical. Larger datasets provide the probing classifiers with more evidence to reflect the linguistic capability of the model. Yet collecting substantial datasets requires labor-intensive data annotation. Moreover, running ever-growing language models on them results into high computational and time costs. Ecological factor should not be excluded as well. Thus, it could be beneficial firstly to revise existing datasets in terms of size and secondly to have a guideline when building new ones.

We describe the contributions of this paper below. We propose *fractions probing*, a method of determining optimal dataset size for visual probing. To this end, we come up with means of comparing probing graphs by computing similarity metrics between them. Our method is based on analysing the "learning curves" formed by these metrics on increasing fractions of data.

Fractions probing can be applied in two ways. First, existing datasets can be reviewed with *data redundancy test*. We illustrate it with SentEval, a popular probing suite for English, and find that it can be effectively reduced up to 0.19 of its overall size (at least for examined models BERT and RoBERTa). Second, *data sufficiency test* can provide guidelines when collecting a new dataset. We simulate this setup with SentEval and compare the results with those of the first setup. Fractions probing can be easily adopted to other datasets and probing methods.

## 2　Related Works

Analysis and interpretation of neural language models has grown its popularity these days. The works vary from scrupulous studies analyzing the representedness of individual properties in the models [7] to survey papers [2,32] which summarize the whole field and provide the community with further research milestones.

In [7], authors bring the diagnostic classification probing mechanism to study the linguistic capabilities of neural language models. They pick several classification datasets, each one representing a specific linguistic feature, and learn to perform the classification by using transformer models layers activations on text as this text features. In [32], authors summarize the known advances in Transformer language models probing interpretation studies, and provide a distant overview of transformer models linguistic capabilities.

The popularity of probing research motivated the emergence of probing frameworks (e.g. [9,30]), allowing for researchers to focus on interpretation, and not implementation of experiments. The general trend consists of using wider sets of languages, models and linguistic categories in probing studies, resulting in the growth of published research computational cost.

One trend is worth special mention, as it concerns the correctness of probing studies in general. Diagnostic classification probing has been criticized for the misleading nature of accuracy scores widely used in probing studies. Several approaches to deal with this weakness were introduced, including MDL Probing [34], claiming that probing can be approached as a compression task. MDL uses the parameters of the probed model to effectively compress the information about the probing dataset. Such encoding is accomplished by extending the classical probing classifier with a regularizer term.

*Probing Dataset Size.* Little research is published on the subject of probing dataset size. As put in a systematic survey [2], "the effect of the probing dataset—its size, composition, etc.—is not well studied". Here we review the few works which seek to compare different dataset sizes.

[12] analyse different combinations of probing study design parameters, namely probing classifier and dataset size. They use correlation analysis to find a 'region of stability' – the set of pairs of parameters that order the encoders in a similar way. The authors experiment with SentEval and end up with a recommendation of 10k training samples and logistic regression as a classifier. The major differences from our approach are the following. First, [12] average the scores across all SentEval tasks, while we examine each one separately. Second, they consider only single scores achieved by embeddings of the last BERT layer, while we compare probing curves formed by all layers. Finally, they compare increasing data fractions with each other, while we compare all lower fractions with the original dataset.

[36] offer a method to determine how many additional data is needed to reliably replicate the comparison effect of a pilot experiment with two probing setups. The authors employ learning theory to move from generalization bounds to required sample size given the desired difference in classifiers' performances. They use power analysis to evaluate the reliability of the given recommendations. Similarly with [12], the main difference from our approach is that this method does not support comparing different probing tasks. Furthermore, [36] do not explicitly compare probing curves graphs.

*Sample Size Determination (SSD).* Our method is highly inspired by sample size determination (SSD), the task of determining sample size for study design. There are a number of different SSD methods to meet researchers' specific data requirements and goals [1,21,25]. For statistical studies, it means finding the sample size required to achieve sufficient statistical power to reject a null hypothesis [1,4,6,21]. For classification problems, such methods suppose predicting the sample size required for a classifier to reach a particular accuracy [11,17,20]. One of generic approaches is to fit a classifier's learning curve created using empirical data to inverse power law models [3,19,27]. This approach is based on prior

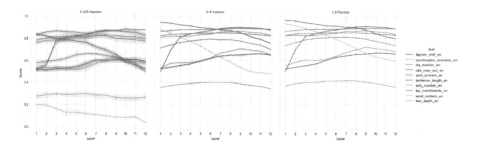

**Fig. 1.** All SentEval probing curves on selected fractions. Even with 40% of original amount of data, most of the probing curves are very similar to the original ones.

studies where it was shown that the classifier learning curves generally follow this law [8].

*Learning Curves.* Learning curves were initially introduced in educational and behavioral/cognitive psychology [28]. Applied to machine learning, the learning curve describes the performance of a given model for a problem as function of the training sample size [18].

As described, for example, in [15], the learning curves can typically be divided into three sections. In the first section, the values increase rapidly with an increase in the size of the training set. The second section is characterized by a turning point where the increase in the metric is less rapid. A final section begins where the model has reached its efficiency threshold, i.e. no (or only marginal) improvement in performance is observed with increasing training set size. In other words, the trend of diminishing returns is observed.

*Progressive Sampling.* The closest research area to our data sufficiency test is progressive sampling. Contrary to SSD, progressive sampling does not predict the required sample size beforehand, but gradually builds the model's learning curve. Together with active learning, it usually starts with a very small batch of instances and progressively increases the training data size until a termination criteria is met [5,22,23,29,35]. As progressive sampling seeks to minimize the amount of computation, the choice of data approximate the learning curve (called sampling schedule) becomes important. [29] compare different sampling schedules and come up with an effective geometric progression-based sampling schedule.

## 3   Our Method

### 3.1   Fractions Probing

To find a reasonable lower-bound for probing dataset size, we employ sample size determination (SSD) methods. Namely, we run layer-wise probing experiments on incrementally sized fractions of data (Fig. 1). The major difference from standard classification problems is that we do not have an interpretable quality metric right away to plot a learning curve. The only target parameter is the similarity of the probing curves graphs obtained with smaller fractions with that obtained with the largest fraction available. We try to find the smallest fraction that gives results sufficiently similar to the largest one.

An elaboration on how we perceive the similarity of probing graphs follows in Sect. 3.2. The proposed method can be used in two setups: *data redundancy test* (Sect. 3.3) for existing datasets and *data sufficiency test* for new ones (Sect. 3.4). Like when plotting a standard learning curve, we suggest repeating all the experiments an adequate number of times to reduce the effect of random sampling. Resembling the real-world setup, we change both the train and test set sizes.

## 3.2   Probing Curves Graphs Similarity

Having a sufficiently large visual probing dataset, we believe that the graphs obtained on it show a true picture of the analyzed model's language capability. When we aim to reduce the size of the dataset, we want the resulting graphs to be sufficiently similar to those reflecting the real knowledge of the model. This similarity is expressed in different features, and we analyze it both visually and computing numerical metrics.

**Visual Analysis.** When speaking of probing studies which involve graphs, the results are visual by themselves. For example, the probing curves' shape is used to draw conclusions about the distribution of linguistic knowledge between the layers of the model. Consequently, the visual component of the analysis cannot be underestimated.

Several parameters are significant for the interpretation of probing curves graphs and therefore will be important for the comparison. Of first priority is probing curves' *shape* which is crucial for any layer-wise probing study. When comparing probing curves, the practitioner should decide whether the shape can be interpreted the same way as the original one. Other parameters are *relative positioning* and *absolute numbers* of the probing curves. They may be important depending on the specifics of the probing study. In an ideal case these parameters of the fraction should be as close as possible to those of the original dataset. Furthermore, as a result of repeating the experiments, *"confidence intervals"* emerge, which in ideal case should be as narrow as possible.

Looking at the relevant features, one is trying to find the lowest suitable fraction. It is worth mentioning that this visual method is hard to formalize and, therefore, is somewhat subjective. To assist the researcher, we suggest computing numerical metrics which measure the difference between graphs. Moreover, numerical metrics allow us to proceed to plotting learning curves common for SSD studies.

**Metrics.** We compute several metrics between the fractions and the original dataset that aim to model the visual similarity described in the previous section.

*Pearson Correlation.* Following [12], we use Pearson correlation to compare results on different dataset sizes. However, in contrast to [12], we use correlation to compare probing curves, not single probing scores.

*Euclidean Distance.* We also compute Euclidean distance. We expect that Euclidean distance reflects the difference in absolute numbers of the curves.

*Fréchet Distance.* Fréchet distance is a measure of similarity between curves that takes into account the location and ordering of the points along the curves [16]. Its meaning is traditionally explained as the minimum length of a leash between a person and a dog who walk two separate curved paths. They can vary their speed and stop, but cannot walk backwards. We expect that it measures both the shape and absolute numbers of the curves.

We apply the metrics in two ways:

- task-wise, when we compute the metric between probing curves of different tasks represented as $l$-sized vectors (where $l$ is the number of layers in the model). Thus we compare the shape of the curves.
- layer-wise, when we compute the metric between layers represented as $t$-sized vectors (where $t$ is the number of tasks). This way we compare the relative positioning of the curves.

### 3.3   Data Redundancy Test

When analyzing an existing dataset, we subsequently compare the results on lower fractions with the results on the original dataset. During visual analysis, the researcher is supposed to find the lowest suitable fraction based on the parameters relevant for current study.

The changing values of the metrics allow us to plot some semblance of learning curves. These are not standard learning curves, as they are formed not by actual results of the model, but by posterior analysis. Their structure, however, adheres to the description of generic learning curves given in Sect. 2. This typical structure allows us to find the point before the final plateau of the learning curve, refining the results of visual analysis. For example, Fréchet distance for *Object number* in Fig. 2 plateaus after fraction 0.1.

### 3.4   Data Sufficiency Test

The major difference of this setup from data redundancy test is that we do not have the results on the "original" dataset to compare with. As the dataset is being developed and its size constantly increases, we cannot directly assess how well the results obtained on it reflect the model's linguistic capability. While visual comparison is unavailable, we suggest simulating the learning curve from data redundancy test. We compute the metrics between lower fractions and the full dataset available at the moment, as if it was the "original" dataset from the previous setup. As the dataset increases, the learning curve grows and its shape changes. The task is to determine whether the curve has reached its final plateau section. To do this, we consequently compute the first and second discrete difference of the learning curves available at each fraction. The first difference reflects the absolute change of the learning curve's values, while the second shows the rate of the change, irrespective of specific values. A fraction is considered suitable according to the differences when their graph starts to plateau. For example, Fig. 3 demonstrates the first difference of Fréchet distance with 0.5 fraction. At this moment *Bigram shift* clearly starts to plateau, which means that its recommended fraction according to first difference of Fréchet distance is 0.5.

As a result, data sufficiency test is effectively a form of progressive sampling (see Sect. 2).

## 4   Experiments

Here we describe the specific details of our experiments which illustrate the applicability of fractions probing. We note, however, that fractions probing is appplicable in any visual probing setup which outputs scores as $n$-sized vectors.

*Data.* We choose SentEval [7] for our experiments as a common ground for probing studies and apply our proposed method to it. SentEval is a probing suite for English sentence embeddings which covers different domains of language. It consists of 10 tasks which can be grouped by the type of linguistic information they probe for. All the tasks only require single sentence embeddings as input. See Appendix A for a detailed description.

*Models.* We use our method with a Transformer [33] model which has become an anchor point in probing studies, BERT [10]. We also extend our experiments to RoBERTa [24], an improved version of BERT, to prove that our method is model-inspecific.

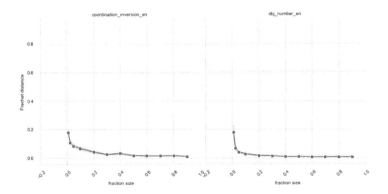

**Fig. 2.** Data redundancy test for *Bigram shift, Coordination inversion* and *Obj number* tasks, task-wise Fréchet distance.

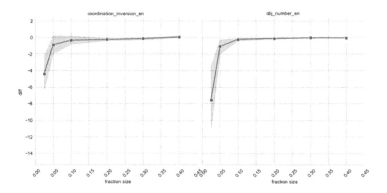

**Fig. 3.** Data sufficiency test for *Bigram shift*, *Coordination inversion* and *Obj number* tasks, first difference of Fréchet distance with 0.5 fraction.

*Embeddings.* As the original SentEval suite did not provide means to employ Transformer models, we rely on the code from RuSentEval [26] replacing Russian data with original English one. Thus, mean-pooled sentence embeddings are extracted from the models.

*Probing Classifier.* Following the default settings in RuSentEval, we use logisic regression as a probing classifier with the L2-regularization parameter $\in [0.1, ..., 1e^{-5}]$ tuned on the validation set.

### 4.1 Data Redundancy Test Experiments

We divide SentEval into fractions sized incrementally as follows: $[0.01, 0.025, 0.05, 0.1, 0.2, ..., 0.9]$. Using the terminology from [29], we choose an effective *sampling schedule* which is a combination of logarithmic and geometric schedules. We repeat the layer-wise probing experiments for 10 times on each fraction to minimize the effect of random sampling.

We then compare the results on the original 1.0 fraction with all lower fractions applying both visual analysis and metrics'[1] learning curves to find the lowest suitable fraction.

### 4.2 Data Sufficiency Test Experiments

We also use SentEval to simulate the setup of a new dataset that is being developed. To do this, we compute the metrics for fractions from 0.05 to 0.9 with all preceding fractions. We then compare the results of simulated data sufficiency test with data redundancy test on SentEval.

---

[1] We compute all the metrics in both ways, but interpret the layer-wise setup only for Pearson correlation. When computing layer-wise metrics, we omit *Word content*. Radical changes in absolute numbers of its learning curve override the signal from other tasks.

**Table 1.** Recommended fractions according to different methods for BERT. Data redundancy test: $r$ is Pearson correlation; $f$ is Fréchet distance. Data sufficiency test: $D1$ and $D2$ are first and second discrete differences. RoBERTa results are given in Appendix B.

| Model | Task | # classes | Score growth | visual method | correlation | | | euclidean | | | Frechet | | |
|---|---|---|---|---|---|---|---|---|---|---|---|---|---|
| | | | | | r | D1 | D2 | d | D1 | D2 | f | D1 | D2 |
| bert | bigram shift | 2 | no | 0.1 | 0.2 | 0.2 | 0.3 | 0.2 | 0.5 | 0.4 | 0.2 | 0.5 | 0.4 |
| bert | coordination | 2 | yes | 0.1 | 0.3 | 0.3 | 0.4 | 0.3 | 0.7 | 0.5 | 0.3 | 0.6 | 0.4 |
| bert | obj number | 2 | no | 0.2 | 0.2 | 0.4 | 0.4 | 0.2 | 0.5 | 0.4 | 0.1 | 0.3 | 0.4 |
| bert | odd man out | 2 | no | 0.1 | 0.2 | 0.3 | 0.4 | 0.2 | 0.4 | 0.4 | 0.1 | 0.4 | 0.4 |
| bert | past present | 2 | no | 0.1 | 0.2 | 0.4 | 0.4 | 0.1 | 0.4 | 0.4 | 0.1 | 0.3 | 0.4 |
| bert | sentence length | 6 | yes | 0.2 | 0.05 | 0.2 | 0.3 | 0.6 | 0.5 | 0.5 | 0.4 | 0.5 | 0.4 |
| bert | subj number | 2 | no | 0.1 | 0.2 | 0.3 | 0.4 | 0.2 | 0.5 | 0.4 | 0.2 | 0.3 | 0.4 |
| bert | top constituents | 20 | yes | 0.3 | 0.3 | 0.2 | 0.3 | 0.6 | 0.6 | 0.5 | 0.5 | 0.5 | 0.5 |
| bert | word content | 1000 | yes | 0.4 | 0.2 | 0.3 | 0.4 | 0.9 | 0.9 | 0.7 | 0.9 | 0.9 | 0.6 |
| bert | tree depth | 8 | yes | 0.1 | 0.2 | 0.4 | 0.4 | 0.4 | 0.6 | 0.5 | 0.3 | 0.5 | 0.4 |
| Mean fraction / Total dataset size | | | | 0.17 | 0.205 | 0.3 | 0.37 | 0.37 | 0.56 | 0.47 | 0.31 | 0.48 | 0.43 |
| Mean error | | | | | 0.105 | 0.17 | 0.2 | 0.2 | 0.39 | 0.3 | 0.16 | 0.31 | 0.26 |
| Corr (r) with visual method | | | | | 0.02 | -0.26 | -0.2 | 0.87 | 0.68 | 0.78 | 0.86 | 0.66 | 0.92 |
| Corr (r) of the diff with the original metric | | | | | | 0.1 | 0.22 | | 0.78 | 0.93 | | 0.9 | 0.91 |

## 5 Results Interpretation

### 5.1 Data Redundancy Test Results

Fig. 1 displays probing curves for all 10 SentEval tasks on selected fractions of the dataset: 0.05, 0.4 and 1.0[2]. We notice that the shapes of the curves on fraction 0.4 are already very close to the shapes on the original fraction of 1.0. Figure 4 shows probing curves from all fractions for selected tasks combined in one plot. A fine-grained description of the results follows.

**Visual Method.** Analyzing the parameters described above in Sect. 3.2, we end up with fractions for each task given in Table 1 (column *Visual method*). The following observations could be noted:

- It can be clearly seen that every SentEval task could be safely reduced in size to a certain degree.
- The tasks can be divided into two groups depending on the extent to which the absolute numbers of the curves change throughout the fractions (see column *Score growth* in Table 1). In the first group, the curves from different fractions are grouped in one region (e.g., *Obj number* in Fig. 4), while in the second they constantly move upwards until a certain fraction (e.g., *Sentence length* in Fig. 4).

## Metrics

*Task-Wise.* Table 1 displays the suitable fractions found by analyzing the learning curve of each metric. It also contains columns with absolute metric errors as compared with visual method. Additionally, we compute Pearson correlation of each metric with visual method.

---

[2] The reported graphs are for BERT model.

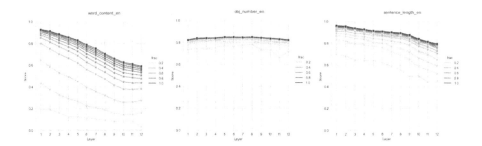

**Fig. 4.** *Word content*, *Obj number* and *Sentence length* tasks on different fractions.

Of the three metrics, Fréchet distance produces the lowest mean error. In terms of correlation with visual method, Euclidean distance has the highest score, although its error is higher than of Fréchet distance. Pearson correlation obtains negative correlation with visual method.

We find that Pearson correlation is less susceptible to the change of absolute values of probing curves than Euclidean and Fréchet distance. It can be seen in Fig. 5 that Euclidean and Fréchet distances tend to give higher fractions for tasks in the *Score growth* group, which is not noticed for Pearson correlation. A vivid example is *Sentence length*, whose learning curve continues to actively move upwards until fraction 0.4 (see Fig. 4), while the suitable fraction according to Pearson correlation is 0.05. As a result, Pearson correlation should not be used as a reference in probing experiments where curves' ordering is relevant.

*Layer-Wise.* The recommended optimal fractions that preserve the ordering of all 10 learning curves according to Pearson correlation are given in Table 2. To maintain correct curves' ordering across all layers simultaneously one should pick the maximum fraction (e.g. 0.4 for BERT). We can note that preserving the original curves' ordering requires more data on average than repeating single probing curves' shape.

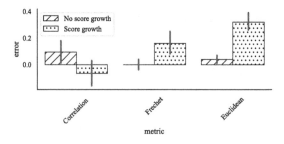

**Fig. 5.** Relationship of metrics' absolute errors and score growth of probing curves.

**Table 2.** Recommended fractions to preserve probing curves ordering according to Pearson correlation.

| Model | Layer | | | | | | | | | | | |
|---|---|---|---|---|---|---|---|---|---|---|---|---|
| | 1 | 2 | 3 | 4 | 5 | 6 | 7 | 8 | 9 | 10 | 11 | 12 |
| **BERT** | 0.3 | 0.3 | 0.3 | 0.3 | 0.3 | 0.3 | 0.3 | 0.3 | 0.3 | 0.4 | 0.4 | 0.4 |
| **RoBERTa** | 0.4 | 0.4 | 0.4 | 0.4 | 0.4 | 0.6 | 0.6 | 0.6 | 0.6 | 0.5 | 0.5 | 0.4 |

### 5.2  Data Sufficiency Test Results

In this section, we compare the results of the simulated data sufficiency test with the results of data redundancy test, which plays a role of oracle information in this context.

Although the discrete differences constantly recommend higher fractions than original metrics, their errors remain comparable with those of the original fractions (see Table 1). Moreover, the resulting fractions obtained with differences are highly correlated with those of the original metric (see line *correlation of the diff with the original metric* in Table 1). This adds evidence to the fact that the differences in data sufficiency test can be used reliably.

The second difference tends to produce less error than the first. This can be explained by the fact that the second difference is less strict since it reflects rate of the change, not the absolute increases or decreases. It can be clearly illustrated with the example of *Word content* task, which certainly belongs to the *Score growth* group (see Table 1). Consequently, it receives high fraction recommendations from the first differences of metrics which are sensible to absolute numbers (Fréchet and Euclidean distance). However, corresponding second differences recommend lower fractions.

## 6  Discussion

**SentEval "Distillation".** Our findings indicate that SentEval could be substantially (at least for BERT-like models) reduced without losing its probing explanatory power. The specific numbers depend on the experiment design the dataset is planned to be used in.

If one plans to use SentEval's tasks separately, the following numbers obtained with our visual method apply. Depending on the specific probing task, they can be reduced to 0.1–0.4 of their original size. Overall, SentEval could be effectively reduced up to 0.19 of its size, which results in saving of 966, 000 samples. Numerical metrics are more strict and testify that SentEval could be 3–5 times smaller (column *Mean fraction/Total dataset size* in Table 1).

If the dataset is used as a whole and probing curves' relative positions are important, one should address Pearson correlation computed in layer-wise manner. Our results suggest that 0.4–0.6 of the original SentEval can be safely used depending on the encoder.

Our results generally go in line with the previous work. [12] come up with a recommendation of stable 10k training examples and logistic regression for SentEval. [36] also experiment with SentEval. Although they do not provide a definitive recommendation for the whole dataset, they note that "most probing classification tasks contain more than enough data for, e.g., comparing BERT vs. InferSent". In general, our results together with the above-mentioned findings imply that many existing probing datasets could be revised in terms of size. This calls for a broader use of our method which can be easily applied to graph-related probing datasets.

**What Explains Specific Probing Task's Data Requirements?** The dependence between probing task specifics and the necessary amount of data remains an open question. Based on the numbers obtained with the visual method, all the tasks in SentEval can be divided into two equal groups:

- first, which can be reduced to the minimal fraction of 0.1 for both BERT and RoBERTa models,
- and second, which require more data on at least one of two models (*Obj number, Odd man out, Sentence length, Top constituents, Word content, Tree depth*).

Aside from linguistic specifics of the tasks, standard classification parameters like the number of classes remain relevant. Our results seem to give evidence for this claim, as all the tasks with the number of classes higher than 2 fall in the second group. This might explain the elasticity of *Word content* in terms of train data amount, as it simply has an extremely high number of classes. As for linguistic features, the above-mentioned grouping cannot be explained by belonging to one of the three SentEval "domains" (see Appendix A).

This poses a question of the extent to which linguistic and standard machine learning parameters contribute to the data amount requirements of specific probing tasks. We leave this analysis for further work.

**Models Comparison.** BERT and RoBERTa show some differences in our experiments. First, our visual method gives evidence that RoBERTa constantly requires more data to probe it than BERT[3]. This can be explained by the fact that RoBERTa outputs better-quality embeddings which contain more information to extract. This is not the only possible explanation, however, and this question is left open.

Second, as seen in Table 2, RoBERTa needs more data for the order of the curves to stabilize. It goes in line with the previous finding.

# 7   Conclusion and Future Work

In this paper, we propose *fractions probing*, a novel method to determine optimal dataset size for visual probing. It compares favourably with previous work as it

---

[3] The numerical metrics do not give such consistency. However, we suspect that they may be subject to biases like, for example, described in Sect. 5.1.

allows to compare whole probing curves and across different probing tasks. It can be employed in two setups: *data redundancy test* explores existing datasets and *data sufficiency test* provides guidance when collecting new ones. We show the proposed method's applicability with SentEval, a popular probing suite for English, in both setups. First, we examine it as an existing dataset and find that it could be substantially reduced while receiving comparable probing results. Then we simulate the setup of a new dataset with SentEval and compare the results with the previous experiment.

Several research directions open up based on the results of this paper. First, it is of interest if learning curves formed by distance metrics and correlation between probing curves graphs follow the inverse-power law. Consequently, it should be determined if it is meaningful to fit these learning curves in advance which could benefit data sufficiency test. Second, a promising and practical improvement of the proposed method is adding a numerical definition of the learning curves' plateau, e.g. a threshold. Finally, other visual probing datasets besides SentEval should be explored with fractions probing.

# A    Detailed SentEval Tasks Description

## A.1    Surface Information

These tasks test the extent to which sentence embeddings are preserving surface properties of the sentences they encode. **SentLen** is the task of predicting the length of sentences in terms of words. *Word content* (**WC**) is the task of predicting which word the sentence contains from a closed set of 1000.

## A.2    Syntactic Information

*Bigram shift* (**BShift** is the task of predicting whether the sentence has intact word order or contains two inverted random adjacent words. *Tree depth* (**TreeDepth**) is the task of determining the depth of the syntactic tree of the sentence. *Top constituent* (**TopConst**) is the task of determining the sequence of the top constituents immediately below the sentence (S) node.

## A.3    Semantic Information

**Tense** is the task of predicting the tens of the main-clause verb. *Subject number* (**SubjNum**) is the task of determining the number of the direct object of the main clause. Similarly, *object number* (**ObjNum**) tests for the number of the direct object of the main clause. *Semantic odd man out* (**SOMO**) is the task of predicting whether the sentence contains a replaced verb or noun which forms bigrams with the previous and following word of the same frequency as the original. *Coordination inversion* (**CoordInv**) is the task of determining whether the sentence has intact or inverted order of clauses.

Here we would like to note that the original SentEval domains partition should possibly be revised. For example, *Tense*, *Subj number* and *Obj number* were categorized as semantic information because the models did not have access to morphology. Nowadays, it is not the case with the models which use byte pair encoding and similar techniques.

## B    Recommended Fractions for BERT and RoBERTa

Table 3 displays the results of data redundancy and data sufficiency tests on SentEval for BERT and RoBERTa.

**Table 3.** Recommended fractions according to different methods for BERT and RoBERTa. Data redundancy test: $r$ is Pearson correlation; $f$ is Fréchet distance. Data sufficiency test: $D1$ and $D2$ are first and second discrete differences. *Mean fraction/Total dataset size* row simultaneously shows the mean fraction across the tasks and the resulting fraction of whole SentEval for each method.

| Model | Task | # classes | Score growth | visual method | correlation | | | euclidean | | | Frechet | | |
|---|---|---|---|---|---|---|---|---|---|---|---|---|---|
| | | | | | r | D1 | D2 | d | D1 | D2 | f | D1 | D2 |
| bert | bigram shift | 2 | no | 0.1 | 0.2 | 0.2 | 0.3 | 0.2 | 0.5 | 0.4 | 0.2 | 0.5 | 0.4 |
| bert | coordination | 2 | yes | 0.1 | 0.3 | 0.3 | 0.4 | 0.3 | 0.7 | 0.5 | 0.3 | 0.6 | 0.4 |
| bert | obj number | 2 | no | 0.2 | 0.2 | 0.4 | 0.4 | 0.2 | 0.5 | 0.4 | 0.1 | 0.3 | 0.4 |
| bert | odd man out | 2 | no | 0.1 | 0.2 | 0.3 | 0.4 | 0.2 | 0.4 | 0.4 | 0.1 | 0.4 | 0.4 |
| bert | past present | 2 | no | 0.1 | 0.2 | 0.4 | 0.4 | 0.1 | 0.4 | 0.4 | 0.1 | 0.3 | 0.4 |
| bert | sentence length | 6 | yes | 0.2 | 0.05 | 0.2 | 0.3 | 0.6 | 0.5 | 0.5 | 0.4 | 0.5 | 0.4 |
| bert | subj number | 2 | no | 0.1 | 0.2 | 0.3 | 0.4 | 0.2 | 0.5 | 0.4 | 0.2 | 0.3 | 0.4 |
| bert | top constituents | 20 | yes | 0.3 | 0.3 | 0.2 | 0.3 | 0.6 | 0.6 | 0.5 | 0.5 | 0.5 | 0.5 |
| bert | word content | 1000 | yes | 0.4 | 0.2 | 0.3 | 0.4 | 0.9 | 0.9 | 0.7 | 0.9 | 0.9 | 0.6 |
| bert | tree depth | 8 | yes | 0.1 | 0.2 | 0.4 | 0.4 | 0.4 | 0.6 | 0.5 | 0.3 | 0.5 | 0.4 |
| roberta | bigram shift | 2 | no | 0.1 | 0.05 | 0.2 | 0.3 | 0.2 | 0.4 | 0.4 | 0.1 | 0.5 | 0.5 |
| roberta | coordination inversion | 2 | yes | 0.1 | 0.2 | 0.4 | 0.4 | 0.3 | 0.6 | 0.5 | 0.1 | 0.6 | 0.5 |
| roberta | obj number | 2 | no | 0.2 | 0.5 | 0.6 | 0.5 | 0.2 | 0.5 | 0.4 | 0.2 | 0.4 | 0.4 |
| roberta | odd man out | 2 | no | 0.3 | 0.1 | 0.3 | 0.4 | 0.3 | 0.6 | 0.4 | 0.2 | 0.5 | 0.4 |
| roberta | past present | 2 | no | 0.1 | 0.4 | 0.7 | 0.5 | 0.1 | 0.4 | 0.4 | 0.1 | 0.4 | 0.4 |
| roberta | sentence length | 6 | yes | 0.3 | 0.05 | 0.2 | 0.3 | 0.6 | 0.6 | 0.5 | 0.3 | 0.7 | 0.5 |
| roberta | subj number | 2 | no | 0.1 | 0.3 | 0.5 | 0.4 | 0.1 | 0.4 | 0.4 | 0.1 | 0.4 | 0.4 |
| roberta | top constituents | 20 | yes | 0.3 | 0.05 | 0.3 | 0.4 | 0.6 | 0.6 | 0.5 | 0.4 | 0.6 | 0.5 |
| roberta | word content | 1000 | yes | 0.4 | 0.2 | 0.4 | 0.4 | 0.9 | 0.9 | 0.6 | 0.6 | 0.7 | 0.5 |
| roberta | tree depth | 8 | yes | 0.3 | 0.3 | 0.5 | 0.4 | 0.5 | 0.6 | 0.5 | 0.3 | 0.6 | 0.4 |
| **Mean fraction / Total dataset size** | | | | 0.195 | 0.21 | 0.355 | 0.385 | 0.375 | 0.56 | 0.465 | 0.275 | 0.51 | 0.44 |
| **Mean error** | | | | | 0.145 | 0.19 | 0.19 | 0.18 | 0.365 | 0.27 | 0.1 | 0.315 | 0.245 |
| **Corr (r) with visual method** | | | | | -0.18 | -0.15 | -0.09 | 0.85 | 0.74 | 0.69 | 0.78 | 0.67 | 0.59 |
| **Corr (r) of the diff with the original metric** | | | | | | 0.74 | 0.66 | | 0.84 | 0.91 | | 0.78 | 0.68 |

## C    Limitations

Our work has a number of limitations. First, our results with SentEval cannot be extrapolated to other datasets without rerunning experiments with our method on them. Second, our method does not allow to extrapolate the learning curves of the metric in data redundancy test. One should build them empirically by running experiments on bigger fractions of data.

Then, one should always keep in mind the known weaknesses of probing studies, such as the misleading nature of accuracy scores. Although fractions probing can be easily adapted for safer methods such as selectivity-based ones or MDL, in this work we have employed the vanilla classification technique. We encourage researchers to responsibly choose proper probing methods for their studies when using fractions probing.

# References

1. Adcock, C.J.: Sample size determination: a review. J. Roy. Stat. Soc.: Ser. D (Stat.) **46**(2), 261–283 (1997). https://doi.org/10.1111/1467-9884.00082, https://onlinelibrary.wiley.com/doi/abs/10.1111/1467-9884.00082, _eprint: https://online library.wiley.com/doi/pdf/10.1111/1467-9884.00082
2. Belinkov, Y.: Probing classifiers: promises, shortcomings, and advances. arXiv:2102.12452 [cs] (2021)
3. Boonyanunta, N., Zeephongsekul, P.: Predicting the relationship between the size of training sample and the predictive power of classifiers. In: Negoita, M.G., Howlett, R.J., Jain, L.C. (eds.) KES 2004. LNCS (LNAI), vol. 3215, pp. 529–535. Springer, Heidelberg (2004). https://doi.org/10.1007/978-3-540-30134-9_71
4. Briggs, A.H., Gray, A.M.: Power and sample size calculations for stochastic cost-effectiveness analysis. Med. Decis. Making: Int. J. Soc. Med. Decis. Making **18**(2 Suppl), S81-92 (1998). https://doi.org/10.1177/0272989X98018002S10
5. Brinker, K.: Incorporating diversity in active learning with support vector machines, pp. 59–66 (2003)
6. Carneiro, A.V.: Estimating sample size in clinical studies: basic methodological principles. Revista Portuguesa De Cardiologia: Orgao Oficial Da Sociedade Portuguesa De Cardiologia = Portuguese J. Cardiol.: Off. J. Portuguese Soc. Cardiol. **22**(12), 1513–1521 (2003)
7. Conneau, A., Kruszewski, G., Lample, G., Barrault, L., Baroni, M.: What you can cram into a single vector: probing sentence embeddings for linguistic properties. arXiv:1805.01070 [cs] (2018)
8. Cortes, C., Jackel, L., Solla, S., Vapnik, V., Denker, J.: Learning curves: asymptotic values and rate of convergence. In: NIPS (1993)
9. Dalvi, F., et al.: NeuroX: a toolkit for analyzing individual neurons in neural networks. In: AAAI Conference on Artificial Intelligence (AAAI) (2019). https://www.aaai.org/ojs/index.php/AAAI/article/view/5063
10. Devlin, J., Chang, M.W., Lee, K., Toutanova, K.: BERT: pre-training of deep bidirectional transformers for language understanding. arXiv:1810.04805 [cs] (2019)
11. Dobbin, K.K., Zhao, Y., Simon, R.M.: How large a training set is needed to develop a classifier for microarray data? Clin. Cancer Res.: Off. J. Am. Assoc. Cancer Res. **14**(1), 108–114 (2008). https://doi.org/10.1158/1078-0432.CCR-07-0443
12. Eger, S., Daxenberger, J., Gurevych, I.: How to probe sentence embeddings in low-resource languages: on structural design choices for probing task evaluation. In: Proceedings of the 24th Conference on Computational Natural Language Learning, pp. 108–118. Association for Computational Linguistics, Online (2020). https://doi.org/10.18653/v1/2020.conll-1.8, https://aclanthology.org/2020.conll-1.8

13. Elazar, Y., Ravfogel, S., Jacovi, A., Goldberg, Y.: Amnesic probing: behavioral explanation with amnesic counterfactuals. Trans. Assoc. Comput. Linguist. **9**, 160–175 (2021). https://doi.org/10.1162/tacl_a_00359, _eprint: https://direct.mit.edu/tacl/article-pdf/doi/10.1162/tacl_a_00359/1924189/tacl_a_00359.pdf

14. Ethayarajh, K., Jurafsky, D.: Utility is in the eye of the user: a critique of NLP leaderboards. In: Proceedings of the 2020 Conference on Empirical Methods in Natural Language Processing (EMNLP), pp. 4846–4853. Association for Computational Linguistics, Online (2020). https://doi.org/10.18653/v1/2020.emnlp-main.393, https://aclanthology.org/2020.emnlp-main.393

15. Figueroa, R.L., Zeng-Treitler, Q., Kandula, S., Ngo, L.H.: Predicting sample size required for classification performance. BMC Med. Inform. Decis. Making **12**(1), 8 (2012). https://doi.org/10.1186/1472-6947-12-8

16. Fréchet, M.: Sur quelques points du calcul fonctionnel. Rendiconti Circolo Mat. Palermo **22**, 1–72 (1884–1940)

17. Fukunaga, K., Hayes, R.: Effects of sample size in classifier design. IEEE Trans. Pattern Anal. Mach. Intell. **11**, 873–885 (1989). https://doi.org/10.1109/34.31448

18. Hastie, T., Tibshirani, R., Friedman, J.H., Friedman, J.H.: The Elements of Statistical Learning: Data Mining, Inference, and Prediction, vol. 2. Springer, Heidelberg (2009). https://doi.org/10.1007/978-0-387-84858-7

19. Hess, K.R., Wei, C.: Learning curves in classification with microarray data. Semin. Oncol. **37**(1), 65–68 (2010). https://doi.org/10.1053/j.seminoncol.2009.12.002

20. Kim, S.Y.: Effects of sample size on robustness and prediction accuracy of a prognostic gene signature. BMC Bioinform. **10**, 147 (2009). https://doi.org/10.1186/1471-2105-10-147

21. Lenth, R.: Some practical guidelines for effective sample-size determination. Am. Stat. **55** (2001). https://doi.org/10.1198/000313001317098149

22. Li, M., Sethi, I.: Confidence-based active learning. IEEE Trans. Pattern Anal. Mach. Intell. **28**, 1251–61 (2006). https://doi.org/10.1109/TPAMI.2006.156

23. Liu, Y.: Active learning with support vector machine applied to gene expression data for cancer classification. J. Chem. Inf. Comput. Sci. **44**(6), 1936–1941 (2004). https://doi.org/10.1021/ci049810a

24. Liu, Y., et al.: RoBERTa: a robustly optimized BERT pretraining approach (2019). https://doi.org/10.48550/arXiv.1907.11692, http://arxiv.org/abs/1907.11692

25. Maxwell, S.E., Kelley, K., Rausch, J.R.: Sample size planning for statistical power and accuracy in parameter estimation. Annu. Rev. Psychol. **59**, 537–563 (2008). https://doi.org/10.1146/annurev.psych.59.103006.093735

26. Mikhailov, V., Taktasheva, E., Sigdel, E., Artemova, E.: RuSentEval: linguistic source, encoder force! arXiv:2103.00573 [cs] (2021)

27. Mukherjee, S., et al.: Estimating dataset size requirements for classifying DNA microarray data. J. Computat. Biol.: J. Comput. Mol. Cell Biol. **10**(2), 119–142 (2003). https://doi.org/10.1089/106652703321825928

28. Perlich, C.: Learning curves in machine learning (2011). https://doi.org/10.1007/978-0-387-30164-8_452

29. Provost, F., Jensen, D., Oates, T.: Efficient progressive sampling. In: Proceedings of the fifth ACM SIGKDD International Conference on Knowledge Discovery and Data Mining, KDD 1999, pp. 23–32. Association for Computing Machinery, New York (1999). https://doi.org/10.1145/312129.312188

30. Ravishankar, V., Øvrelid, L., Velldal, E.: Probing multilingual sentence representations with x-probe. In: RepL4NLP@ACL (2019)

31. Rodriguez, P., Barrow, J., Hoyle, A.M., Lalor, J.P., Jia, R., Boyd-Graber, J.: Evaluation examples are not equally informative: how should that change NLP leaderboards? In: Proceedings of the 59th Annual Meeting of the Association for Computational Linguistics and the 11th International Joint Conference on Natural Language Processing (Volume 1: Long Papers), pp. 4486–4503. Association for Computational Linguistics, Online (2021). https://doi.org/10.18653/v1/2021.acl-long.346, https://aclanthology.org/2021.acl-long.346
32. Rogers, A., Kovaleva, O., Rumshisky, A.: A primer in BERTology: what we know about how BERT works. arXiv:2002.12327 [cs] (2020)
33. Vaswani, A., et al.: Attention is all you need. arXiv:1706.03762 [cs] (2017)
34. Voita, E., Titov, I.: Information-theoretic probing with minimum description length. In: Proceedings of the 2020 Conference on Empirical Methods in Natural Language Processing (EMNLP), pp. 183–196. Association for Computational Linguistics, Online (2020). https://doi.org/10.18653/v1/2020.emnlp-main.14, https://aclanthology.org/2020.emnlp-main.14
35. Warmuth, M.K., Liao, J., Rätsch, G., Mathieson, M., Putta, S., Lemmen, C.: Active learning with support vector machines in the drug discovery process. J. Chem. Inf. Comput. Sci. **43**(2), 667–673 (2003). https://doi.org/10.1021/ci025620t
36. Zhu, Z., Wang, J., Li, B., Rudzicz, F.: On the data requirements of probing. In: Findings of the Association for Computational Linguistics: ACL 2022, pp. 4132–4147. Association for Computational Linguistics, Dublin (2022). https://doi.org/10.18653/v1/2022.findings-acl.326, https://aclanthology.org/2022.findings-acl.326

# Unsupervised Ultra-Fine Entity Typing with Distributionally Induced Word Senses

Özge Sevgili[1(✉)], Steffen Remus[1], Abhik Jana[2], Alexander Panchenko[3,4], and Chris Biemann[1]

[1] Language Technology Group, Universität Hamburg, Hamburg, Germany
`oezge.sevgili.ergueven@studium.uni-hamburg.de`,
`{steffen.remus,chris.biemann}@uni-hamburg.de`
[2] School of Electrical Sciences, Indian Institute of Technology Bhubaneswar, Bhubaneswar, India
`abhikjana@iitbbs.ac.in`
[3] Skolkovo Institute of Science and Technology, Moscow, Russia
[4] Artificial Intelligence Research Institute, Moscow, Russia
`a.panchenko@skol.tech`

**Abstract.** The lack of annotated data is one of the challenging issues in an ultra-fine entity typing, which is the task to assign semantic types for a given entity mention. Hence, automatic type generation is receiving increased interest, typically to be used as distant supervision data. In this study, we investigate an unsupervised way based on distributionally induced word senses. The types or labels are obtained by selecting the appropriate sense cluster for a mention. Experimental results on an ultra-fine entity typing task demonstrate that combining our predictions with the predictions of an existing neural model leads to a slight improvement over the ultra-fine types for mentions that are not pronouns.

**Keywords:** entity typing · word senses · distributional semantics

## 1 Introduction

Ultra-fine entity typing (UFET) is the task of assigning semantic types to an entity mention in context [6]. There exist numerous diverse types, e.g., consider the sentence – "Olympic National Park came into the national park system in 1938 and has been a favorite destination for naturalists and tourists ever since." – the types for "Olympic National Park" are geographical_area, national_park, space, region, location, landmark, park, place. Ultra-fine types can be helpful for natural language understanding tasks, for example, Sui et al. [36] leverage ultra-fine entity types from entity descriptions in a zero-shot entity linking task. Yet, those large type sets lead to difficulties in annotating mentions for humans [7].

This causes a challenge of the scarcity of annotated data. There are more than 10K labels, in this task. Therefore, methods to create automatic annotations

© The Author(s), under exclusive license to Springer Nature Switzerland AG 2024
D. I. Ignatov et al. (Eds.): AIST 2023, LNCS 14486, pp. 126–140, 2024.
https://doi.org/10.1007/978-3-031-54534-4_9

have been proposed. We explore the leverage of distributionally induced word senses, since we believe the induced word senses can help to understand and disambiguate the given mention.

For this challenge, some studies propose to generate labels to be used as distant supervision with different strategies [6,7]. For example, Dai et al. [7] modify sentences by inserting [MASK] tokens with hypernym extraction patterns and obtain predictions from the pre-trained language model (PLM). In a similar vein, Ding et al. [9] and Li et al. [18] also leverage PLMs in different scenarios with different goals, i.e., to apply prompt learning or re-formulate the task, respectively. Qian et al. [32] and Liu et al. [20] generate labels under the setting without access to a knowledge base rather based on a large amount of data, and the underlying techniques of their solutions are quite similar to our investigation. However, their evaluations are on a fine-grained entity typing (FET) task. We rather focus on UFET task with richer type set described as free-form phrases. There are also some works for zero-shot [9,23,39], and unsupervised [12,13] way of solutions, mostly using either a labeled data or knowledge base, e.g. Wikipedia information. In our study, we do not make use of any such knowledge bases or labeled data.

In this work, to produce ultra-fine types, we leverage the API of the JoBimText framework [4,35], which provides sense clusters with hypernym labels (i.e., IS-As) for a queried term in an unsupervised and knowledge-free way based on a distributional thesaurus. The appropriate sense for a particular mention is selected based on the cosine similarity between vectorial representations of contextual information and each sense cluster information of the mention. The hypernym labels for the selected sense are our final prediction. Our goal in this work is to explore the potential of this approach in UFET task. Thus, the contribution of this paper is the investigation of the labels from the JoBimText in the UFET task in an unsupervised way. We experiment a combination of the neural approach predictions by Choi et al. [6] with JoBimText based predictions to explore their complementarity. We utilize predictions from Choi et al. [6] here, since they set the baselines while releasing the UFET dataset. With this combination, we observe a slight improvement of the F1 score for ultra-fine types for explicit mentions.

## 2 Related Work

### 2.1 Ultra-Fine Entity Typing

There are several lines of research on UFET [6] and FET [10,19]. FET contains smaller set of labels, e.g., 112 types in Ling and Weld [19], which are in an ontology, e.g., *location/city*. UFET is more diverse and finer grained, containing more than 10K labels as free-form noun phrases. Some studies investigate hierarchies/dependencies or correlations in the types in different ways [15,21,22,24,37,40], inter alia. While most attention is on English typing, several work on other languages [17].

For the challenge of the scarcity of annotated data, Li et al. [18] re-formulate the task as natural language inference (NLI) and leverage indirect supervision from NLI. Some works attempt to create more labeled data automatically and use it as distant supervision. A typical way is to obtain types from the knowledge base after linking the mention to the entity [6]. Choi et al. [6] propose to utilize head words of mentions as a distant supervision. Dai et al. [7] insert [MASK] token with a few tokens (e.g., "such as") to create an artificial Hearst pattern [11] close to a mention to retrieve the predictions from BERT masked language model [8]. Qian et al. [32] attempt to tackle FET without a knowledge base, in which they generate automatically labeled data from a large-scale unlabeled corpus and then propose a training method using this data. Since the automatically labeled data might contain noise, several studies provide solutions to denoise it [25,26]. Some others deal with also zero-shot scenarios [9,18,23,38,39] inter alia. Among them, works Zhou et al. [39] and Ding et al. [9] rely on unlabeled data. Ding et al. [9] apply prompt-learning to generate type labels using PLMs, and for zero-shot set-up, they further propose a self-supervised method relying on contrastive learning. Zhou et al. [39] utilize knowledge base information. Huang et al. [12,13] provide unsupervised solution, again using knowledge base information.

Among all, our study is more relevant to Qian et al. [32], in terms of applying a Hearst pattern to large data and applying the clustering without accessing the knowledge base. In a similar vein, Liu et al. [20] propose an NLP system that supports unsupervised FET by applying a Hearst pattern and clustering. However, both evaluate on FET task, while our focus is on UFET, which contains more fine-grained types. In Dai et al. [7], the labels are generated automatically from PLMs, and Ding et al. [9] provide also a zero-shot solution. In comparison, our study investigates particularly the usage of the JoBimText API on this task, which is a simpler scenario than their models. In comparison with Zhou et al. [39], Huang et al. [12,13], we do not use the knowledge base information.

### 2.2 JoBimText Applications

Several works utilize information provided by JoBimText in different tasks [1, 14,30], inter alia. Among them, the most similar studies might be unsupervised knowledge-free word sense disambiguation by Panchenko et al. [27,29], in which a word in a context is disambiguated using the induced senses of JoBimText. Inspired by them, we conduct a similar approach to the UFET task.

## 3    Method

### 3.1    JoBimText Framework

In our work, we generate labels relying on the JoBimText framework [4] as an end-user of the API[1] [35] provided by this framework. The underlying technology of this framework involves a holing operation as the first step, which performs the split of a term (Jo) and a contextual feature (Bim) based on structural observations of text (e.g., dependency parsing). Next, by pruning these terms and features (based, e.g., on some significance scores, like LMI [16]) and

---

[1] http://ltmaggie.informatik.uni-hamburg.de/jobimviz/#.

by aggregating terms based on their overlapping contextual features, a distributional thesaurus (DT), a graph of terms, is constructed. Furthermore, they cluster an ego/neighboring graph (i.e., a sub-graph containing similar terms to a particular term) of a DT entry (i.e., term) using the Chinese Whispers algorithm [2] to get a sense information of an entry in terms of its similar entries. Each induced sense is labeled based on the information of IS-A relationship between terms with their frequencies, collected by applying IS-A (hypernym) patterns [11] on a text collection. The API [35] allows to access the information of the JoBimText framework (for more information, see [3,4,34,35]).

**Fig. 1.** A sample prediction process: search for a mention on JoBimText (mimicked from API) to get sense clusters containing terms and labels (IS-As), vectorize clusters by averaging SBERT vectors of each term (and label), compute cosine similarities between context (and mention) vector and clusters, and obtain IS-As of the most similar cluster as a final prediction.

## 3.2   Method

In our setup, we query a mention into the JoBimText API and obtain sense clusters of this mention. The most appropriate sense cluster is selected based on the vectorial similarity between the context that the mention appeared in, and a sense cluster. To compute this similarity, each sense cluster is vectorized by using the sense terms (and labels) with Sentence BERT (SBERT) [33] (note that some clusters may not have hypernyms/terms, so we skip them). And context (and mention) is also vectorized using SBERT. The hypernym labels of the most similar sense are the final type predictions, as exemplified in the Fig. 1.

We query a mention with NN (noun) or MWE (multi-word expression) tag, depending on whether a mention contains a single token or multiple tokens. Other tags, e.g., ADJ, are not considered since the mentions in the UFET dataset [6] are pronouns, nominal expressions, and named entity mentions. For named entity mentions, some other tags, like e.g. Org, Loc, would be helpful, however, the goal of entity typing itself to produce such labels, and so, we use only NN/MWE tags. JoBimText might not provide information for all mentions, like long phrases, e.g., "the building, a violation of the Clean Air Act". Thus, we try to shorten the mentions in several ways: extraction of a head word (root word) of the mention or extraction of n-grams located close to the beginning/end of the mention due to an observation of some mentions (e.g., "shipments for the month", "the social and economic development"). Note that Choi et al. [6] also utilize a head word of the mention, however, they directly take it as a weak label, while our goal is to shorten the mention to search later on the JoBimText. Additionally, singularization is applied for the mention as an additional configuration based on the observation that the singular version of some mentions is in the JoBimText, while the plural is not, e.g., "shareholders". JoBimText might still not provide any information for the short mentions, for which we assign a *person* label as it is the most used label in the development set. The coverage for our reported results, in Table 2, is 87.74.

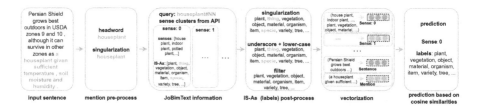

**Fig. 2.** A sample overview of the process is illustrated. In this sample, head word is used to shorten mention, and post-process is shown for labels of sense: 0, and IS-A labels are included for vector representation. The prediction is based on the cosine similarities of sentence and sense vectors, and mention and sense vectors.

In a similar vein, we apply some post-processing steps due to some mismatches between predictions (i.e., IS-As) and the type vocabulary of Choi et al. [6] (e.g., predictions may involve *people*, while the vocabulary contains *person*, not *people*). We first follow Dai et al. [7] for post-processing: the labels are singularized and filtered if they are not in the type vocabulary. In addition to them, we add underscores for the multi-token labels, e.g., *tennis_player*, and make them lower-case if not. We remove a label *thing* among our predicted labels since we consider it as a noisy label due to its high frequency. A sample overview containing pre-/post-process steps is shown in Fig. 2.

## 4    Experiments

### 4.1    Dataset

The experiments are performed on an English UFET dataset provided by Choi et al. [6]. The type vocabulary contains 10331 labels. The dataset consists of a training set, development set, and test set, each with 1998 samples.

## 4.2  Baselines

- **first cluster or random cluster** There is an order of sense clusters in JoBimText, based on the score of related terms. We either always choose the first cluster with terms and labels, or choose any sense randomly. The same pre- and post-process steps are applied as the configuration, which will be explained in Sect. 4.3.
- **Choi et al. (2018)** [6] generate representations through the pre-trained word embeddings, bi-directional LSTM, CNN, and train the model with a multitask objective.
- **Dai et al. (2021)** [7] generate labels through the BERT masked language model and leverage the generated labels in the training entity typing model.
- **Li et al. (2022)** [18] treat each sentence as a premise and generate a hypothesis through the candidate type to formulate the task as NLI. Here, the learning objective is learning-to-rank.

## 4.3  Implementation Details

JoBimText provides many DTs including different language supports from various corpora. In this study, we use the DT constructed from the DepCC corpus [28]. It is built from the web-scale data from the Common Crawl, which provides access to large amounts of data. The JBT DepCC model uses a 2016 snapshot of the Common Crawl.[2] As a Sentence BERT model, we utilize "all-mpnet-base-v2", since it is the best performing one (on average performance) among current models.[3]

We consider several parameters or features and select amongst them based on a simple manual search, in our implementation, as shown in Table 1. In Table 1, the first row represents the methodology, where we shorten the mention before searching in the JoBimText API, where n-grams are extracted using NLTK [5] and head words are extracted with the stanza library/toolkit[4] [31]. Some mentions start with "a, an, the" (or upper case of them), for which we take the tokens after the first one, if we experiment with the beginning of mention. The punctuation symbols (like, e.g. ".", ",", """", "-", as well as the tokens "-LRB-" and "-RRB-" that we think they would be used for brackets) are removed.

For some mentions (three mentions in the development set and eight mentions in the test set), there can be more than one head word, for which we use the first head word by default. The second item in the table is applying singularization to mentions before the search in API, for which we use inflect library[5], however, it might result in some mistakes for some cases as discussed in Sect. 5. To avoid the case that the term ends with "s" (e.g. "access"), we double-check its morphological property using the stanza toolkit whether it is singular or plural.

---

[2] https://commoncrawl.org.
[3] https://www.sbert.net/docs/pretrained_models.html.
[4] https://github.com/stanfordnlp/stanza.
[5] https://pypi.org/project/inflect.

**Table 1.** Parameters or features, their possible values, and the selection based on our simple manual search on the development set.

| | Parameter/Feature | Possible values | Selection |
|---|---|---|---|
| 1 | mentions shorten options | - head word<br>- n-gram beginning tokens (for n = 1 - 2 - 3)<br>- n-gram end tokens (for n = 1 - 2 - 3) | head word |
| 2 | apply singularization to mention | true - false | true |
| 3 | cluster types | 200, 200 - 200, 50 - 50, 50 | 50, 50 |
| 4 | number of terms for representation | 10 - 20 - 30 | 10 |
| 5 | number of labels for representation | false - 10 - 20 - 30 | 10 |
| 6 | weighting average | false - rank - cos. sim. | false |
| 7 | include mention similarity | true - false | true |
| 8 | include mention | true - false | false |
| 9 | number of predictions | 5 - 10 - 15 - 20 | 10 |

Here, we cross-check only the last token of the mention and we dismiss the cases of the singularized word that is located in different place. For example, "princess of Brunswick-Wolfenbuttel" is singularized as "princes of brunswick-wolfenbuttell". The third row in Table 1 is for cluster types available in the API to determine the number of some entries[6] for the Chinese Whispers algorithm. Cluster representations are created using the sense terms, and the fourth item in the table is to determine how many terms to include. Similarly, the cluster labels are optionally included while creating representations, as shown in the 5th row with the number options.

While averaging, weighting is possible with weights either from the similarity between a mention and a considered term/label or from the order present in the JoBimText with the formula, 1/its order (meaning if it is ranked first, weight becomes 1, second: 0.5, third: 0.33, fourth: 0.25..), as in the sixth entry in the table. While computing similarities for the final decision, similarities are between a context and a sense, or also including the similarities between the mention and a sense. Item eight is to determine that the context contains either only left- and right-context words or also mention words, inserted in between. The last entry is for the number of predicted labels.

We try to find the best parameters and features in the development set with a simple manual search. Among the experiments, the configuration with the best F1 score, which we can reach so far, consists of the parameters and features, as shown in the last column of Table 1.

### 4.4    Evaluation

In Table 2, we report P (precision), R (recall), and F1 by following recent works [7,18,24]. The scores are computed with the evaluation script provided by

---

[6] http://ltmaggie.informatik.uni-hamburg.de/jobimtext/documentation/sense-cluste ring.

**Table 2.** Unsupervised ultra-fine entity typing performance on UFET test set. Without pronouns: the results are for the mentions that are not pronouns (1210 samples, in the test set), 5 preds.: the results contain the first 5 predictions from Choi et al. [6] and our first 5 predictions, Ours-PRP: pronoun mentions are searched with PRP tag.

| Model | Total | | | Coarse | | | Fine | | | Ultra-Fine | | |
|---|---|---|---|---|---|---|---|---|---|---|---|---|
| | P | R | F1 | P | R | F1 | P | R | F1 | P | R | F1 |
| first cluster | 17.4 | 18.1 | 17.7 | 34.1 | 42.9 | 38.0 | 13.0 | 18.0 | 15.1 | 6.5 | 9.6 | 7.7 |
| random cluster (avg. of 5 runs) | 16.8 | 16.2 | 16.5 | 43.4 | 43.3 | 43.4 | 12.5 | 14.5 | 13.4 | 5.5 | 7.5 | 6.3 |
| Li et al. (2022) [18] | 53.3 | **56.4** | **50.6** | - | - | - | - | - | - | - | - | - |
| Dai et al. (2021) [7] | **53.6** | 45.3 | 49.1 | - | - | - | - | - | - | - | - | - |
| Choi et al. (2018) [6] | 47.1 | 24.2 | 32.0 | 60.3 | 63.4 | 61.8 | 41.2 | 38.7 | 39.9 | 42.2 | 9.4 | 15.4 |
| Ours | 20.1 | 19.3 | 19.7 | 42.3 | 41.9 | 42.1 | 16.1 | 20.0 | 17.8 | 8.9 | 11.0 | 9.8 |
| Ours-PRP | 25.6 | 22.0 | 23.7 | 58.7 | 53.3 | 55.9 | 23.2 | 20.0 | 21.5 | 9.5 | 12.0 | 10.6 |
| Choi et al. (2018) [6] + Ours | 23.2 | 33.0 | 27.3 | 49.0 | 74.4 | 59.0 | 24.4 | 46.2 | 31.9 | 13.9 | 17.7 | 15.5 |
| Choi et al. (2018) [6] + Ours (5 preds.) | 27.3 | 30.3 | 28.7 | 51.9 | 72.5 | 60.5 | 30.6 | 43.8 | 36.0 | 16.7 | 14.4 | 15.5 |
| Choi et al. (2018) [6] + Ours-PRP | 25.2 | 33.4 | 28.7 | 54.4 | 74.5 | 62.9 | 29.9 | 46.2 | 36.3 | 14.5 | 18.4 | 16.2 |
| Choi et al. (2018) [6] + Ours-PRP (5 preds.) | 29.4 | 30.4 | 29.9 | 56.7 | 72.8 | 63.8 | 34.3 | 44.0 | 38.5 | 17.4 | 14.7 | 15.9 |
| without pronouns | | | | | | | | | | | | |
| Choi et al. (2018) [6] | 46.7 | 19.6 | 27.7 | 50.3 | 50.8 | 50.5 | 44.1 | 36.0 | 39.6 | **50.2** | 7.8 | 13.4 |
| Ours | 18.8 | 25.4 | 21.6 | 46.0 | 47.4 | 46.7 | 23.1 | 34.7 | 27.8 | 12.2 | 17.7 | 14.5 |
| Choi et al. (2018) [6] + Ours | 21.1 | 33.4 | 25.9 | 43.0 | 68.6 | 52.8 | 25.2 | 48.7 | 33.2 | 13.8 | **21.1** | **16.7** |
| Choi et al. (2018) [6] + Ours (5 preds.) | 26.4 | 29.6 | 27.9 | 46.4 | 65.6 | 54.3 | 31.8 | 44.9 | 37.2 | 17.7 | 16.8 | **17.3** |

Choi et al. [6][7]. We report the results of our method on the test set by Choi et al. [6] using the features/parameters as explained in the previous section. We also report the results of the first/random cluster baseline (with the same possible parameters applied without the ones to choose the best cluster). The improvement over first and random cluster baselines suggests that our method is able to disambiguate the induced sense at some level. Additionally, the first cluster baseline scores are pretty good, so we can say that the first sense among the induced senses is prominent in the dataset. We perform an alternative experiment by searching pronoun mentions with PRP tag rather than NN/MWE and we can see some improvement there. Note that the coverage is changed to 80.73 in this experiment.

Most of the recent works are supervised (e.g., [7,18], in Table 2), and thus we cannot directly compare them with our results. For this reason, we combine our predictions with the predictions from the Choi et al. [6] model and check if additional predictions from our approach improve the scores. We directly concatenate the predictions, and then we keep unique labels (technically we make the concatenation set). They release their best model and the prediction file from this model[8], and for this experiment, we use only this prediction file.

Our solution cannot produce good hypernyms for pronouns. Therefore, we also compare the predictions for explicit mentions only, excluding pronouns. We consider pronouns: "i, me, myself, we, us, ourselves, he, him, himself, she,

---

[7] https://github.com/uwnlp/open_type/blob/master/scorer.py.
[8] http://nlp.cs.washington.edu/entity_type/model/best_model.tar.gz.

her, herself, it, itself, they, them, themselves, you, yourself" (and upper case of them), with references.[9]. Additionally, we collect our first 5 predictions, for each mention, and the first 5 predictions from the model by Choi et al. [6].

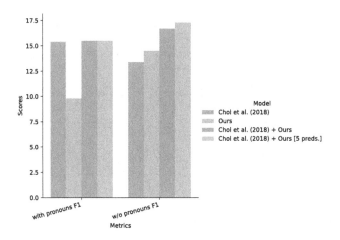

**Fig. 3.** Results on ultra-fine granularity are shown.

Based on the combination results shown in Table 2, the ultra-fine F1 scores are improved when the predictions of explicit mentions are combined in both cases all predictions and 5 predictions, which suggests our labels are complementary to the predictions by Choi et al. [6], in this set-up, as also shown in Fig. 3. Overall, we experiment with all granularities but only ultra-fine worked well for the tested dataset, which is a potential limitation of the approach.

We also take the first 1, 3, 5, 7 prediction(s) as the final prediction(s) and see the decrease of precision and the increase of recall, when the number of predictions are increased from 1 to 7, as shown in Fig. 4.

## 5  Error Analysis and Limitations

### 5.1  Error Analysis

We conduct an error analysis on 100 random samples in the test set shown in Table 3, from our predictions explained in Sect. 4.4. Based on this analysis, we classify errors into five categories:

1. Context-dependent or pronoun mentions, where the induced senses of JoBim-Text are not that useful for pronoun mentions, and mentions, where the labels are expected to be generated context dependently, e.g., for a given name as shown in Table 3.

---

[9] https://github.com/HKUST-KnowComp/MLMET/blob/main/prep.py#L9, https://en.wikipedia.org/wiki/English_pronouns#Full_list.

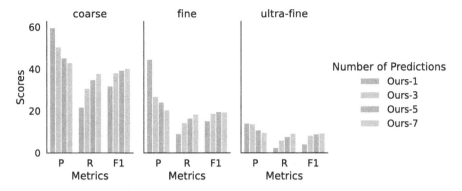

**Fig. 4.** Our results with different number of predictions on test set are shown. Ours-{1, 3, 5, 7}: the results contain the first {1, 3, 5, 7} prediction(s).

2. JoBimText does not contain the referred sense information, although it can provide many different and fine-grained induced senses. For the sample mention "Binaural", most senses are related to Binaural beats or headsets. Note that there is one sense (sense 14), which contains person names from the group Pearl Jam as terms, however there is no IS-As, and thus we cannot take into account.
3. The labels are not matching perfectly.
4. Some pre-processing issues, which are discussed in detail, in limitations paragraph below. For the sample data in Table 3, if "place" is searched, more relevant information can be collected.
5. The labels of JoBimText are not that relevant, though the correct sense (or one of the correct senses) is selected.
6. Wrong sense selection, for example, for "Orleans", induced sense 0 is the right sense in JoBimText, however, the method selects another cluster. Note that some samples included in one group of a category can also be included in another category, e.g., some pronouns (category-1) do not have the referred sense information from JoBimText (category-2). Note also that some labels can be mixed for the sense.

### 5.2   Limitations

One of the goals of ultra-fine entity typing is to generate context-dependent labels for a mention, e.g., the label for "Leonardo DiCaprio" could be *passenger* depending on its context [7]. JoBimText may not produce context-dependent labels, as discussed. In UFET, mention types can be nominals, named entities, and pronouns, and for pronouns, JoBimText is unable to produce good labels and clusters. There are some mistakes or limitations specifically due to the pre-processing steps. Since head word extraction returns only one token, named entities cannot be taken properly, e.g., for the mention "Los Angeles", the head

**Table 3. A sample per category:** Mentions are marked as red, the predictions, and gold labels are examplified for each category. 1: Context-dependent or pronoun mentions, 2: JoBimText does not contain the referred sense information, 3: The labels are not the matching perfectly, 4: Some preprocessing issues, 5: The JoBimText labels are not that relevant, 6: Wrong sense selection.

| Cat. | Context | Search Mention | Predictions | True Labels |
|---|---|---|---|---|
| 1 | "They need to allow the international humanitarian organizations full and unobstructed access because they are obstructing access right now", -Sollom- said | Sollom | no information (assigned person) | person |
| 2 | Following a full-scale tour in support of its previous album, -Binaural- -LRB- 2000 -RRB-, Pearl Jam took a year-long break | Binaural | company, format, feature, technology, brand, variety, mode, track, stuff, film | object, album |
| 3 | " People are getting deported for -even minor offenses- like not having an ID or a driver's license", said Cesar Espinosa of America for All, a group that helps immigrants in Houston | offense | offense, **crime**, activity, charge, incident, matter, case, law, act, **violation** | **violation**, difficulty, wrongdoing, error, consequence, problem, trouble, event, **crime** |
| 4 | The devastating 21 September earthquake of 1999 left Taiwan's landscape covered in scars, but researchers discovered that -the places least damaged by the quake- were areas of natural forest | damaged | no information (assigned person) | city, area, location, town, region, space |
| 5 | Jack Byrne, chairman of Fireman's Fund, said this disaster will test the catastrophe reinsurance market, causing -these rates- to soar | rate | disease, illness, condition, information, cancer, side_effect, event, number, effect, rate | share, value, capital, price, stock |
| 6 | It was formerly located at Six Flags -New Orleans before it was relocated to Six Flags Fiesta Texas- and rethemed to Goliath | Orleans | venue, stadium, attraction, **place** | town, city, placement, space, state, location, **place**, park |

word is "Los". Sometimes, the head word loses the main information, e.g. for the mention "Perhaps the biggest of those factors", the head word is "biggest", although "factors" might be a better token for this mention. There are also some limitations due to the singularization step. Sometimes it can singularize the name entities, for example for the mention "the Cleveland Browns" with "Browns" head word, after singularization the word becomes "Brown". If the plural word is not in the last token, the singularization might fail for compounds. As explained earlier, we double-check whether the token is plural using features of stanza, however, sometimes this check causes a mistake. For instance, "works" is labeled as a verb by stanza, and so it is not singularized. The induced senses might be so fine-grained for some terms, e.g., "plan" has 14 senses (with cluster 200, 200).

In some cases, the labels might be mixed and produce noise information. Therefore, the predictions are far from being usable directly in real-world and the misuse of the predictions might result in wrong information, as also discussed in Ding et al. [9].

# 6    Conclusion

In this study, we generated ultra-fine entity type labels using the JoBimText framework in an unsupervised way. We observed a slight improvement when we combine our predictions with the predictions from Choi et al. [6] for the mentions that are not pronouns, and this suggests that the labels produced through JoBimText contain helpful information. The improvement is due to the drop of the precision in favor of recall. That means, JoBimText has good lexical coverage with numerous labels, but they are also noisy.

## 7    Future Work

There are several promising further directions, such as, we consider an unsupervised solution in this work, yet the produced labels can be used as a weak label of some supervised models as in some previous models. Our produced unsupervised labels might help when supervised labels are not sufficient. We try to find good features and parameters based on a manual search, however, their combinatorial behavior is open for research, and further improvement over the search space by better tuning parameters is possible[10].

**Acknowledgements.** The work was partially supported by a Deutscher Akademischer Austauschdienst (DAAD) doctoral stipend and the DFG funded JOIN-T project BI 1544/4.

## References

1. Anwar, S., Shelmanov, A., Panchenko, A., Biemann, C.: Generating lexical representations of frames using lexical substitution. In: Proceedings of the Probability and Meaning Conference (PaM 2020), Gothenburg, pp. 95–103 (2020). https://aclanthology.org/2020.pam-1.13
2. Biemann, C.: Chinese whispers - an efficient graph clustering algorithm and its application to natural language processing problems. In: Proceedings of TextGraphs: the First Workshop on Graph Based Methods for Natural Language Processing, New York City, pp. 73–80 (2006). https://aclanthology.org/W06-3812
3. Biemann, C., Coppola, B., Glass, M.R., Gliozzo, A., Hatem, M., Riedl, M.: JoBimText visualizer: a graph-based approach to contextualizing distributional similarity. In: Proceedings of TextGraphs-8 Graph-based Methods for Natural Language Processing, Seattle, WA, USA, pp. 6–10 (2013). https://aclanthology.org/W13-5002
4. Biemann, C., Riedl, M.: Text: now in 2D! A framework for lexical expansion with contextual similarity. J. Lang. Modell. **1**(1), 55–95 (2013). https://doi.org/10.15398/jlm.v1i1.60
5. Bird, S., Klein, E., Loper, E.: Natural Language Processing with Python. O'Reilly Media Inc. (2009). https://www.nltk.org/book
6. Choi, E., Levy, O., Choi, Y., Zettlemoyer, L.: Ultra-fine entity typing. In: Proceedings of the 56th Annual Meeting of the Association for Computational Linguistics (Volume 1: Long Papers), Melbourne, Australia, pp. 87–96 (2018). https://www.aclweb.org/anthology/P18-1009

---

[10] Our code can be found at: https://github.com/uhh-lt/unsupervised-ultra-fine-entity-typing.

7. Dai, H., Song, Y., Wang, H.: Ultra-fine entity typing with weak supervision from a masked language model. In: Proceedings of the 59th Annual Meeting of the Association for Computational Linguistics and the 11th International Joint Conference on Natural Language Processing (Volume 1: Long Papers), pp. 1790–1799 (2021). https://doi.org/10.18653/v1/2021.acl-long.141

8. Devlin, J., Chang, M.W., Lee, K., Toutanova, K.: BERT: pre-training of deep bidirectional transformers for language understanding. In: Proceedings of the 2019 Conference of the North American Chapter of the Association for Computational Linguistics: Human Language Technologies, Volume 1 (Long and Short Papers), Minneapolis, MN, USA, pp. 4171–4186 (2019). https://doi.org/10.18653/v1/N19-1423

9. Ding, N., et al.: Prompt-learning for fine-grained entity typing. In: Findings of the Association for Computational Linguistics: EMNLP 2022, Abu Dhabi, United Arab Emirates, pp. 6888–6901 (2022). https://aclanthology.org/2022.findings-emnlp.512

10. Gillick, D., Lazic, N., Ganchev, K., Kirchner, J., Huynh, D.: Context-dependent fine-grained entity type tagging (2016)

11. Hearst, M.A.: Automatic acquisition of hyponyms from large text corpora. In: COLING 1992 Volume 2: The 14th International Conference on Computational Linguistics, pp. 539–545 (1992). https://aclanthology.org/C92-2082

12. Huang, L., May, J., Pan, X., Ji, H.: Building a fine-grained entity typing system overnight for a new x (x = language, domain, genre) (2016)

13. Huang, L., et al.: Liberal entity extraction: rapid construction of fine-grained entity typing systems. Big Data 5(1), 19–31 (2017). https://doi.org/10.1089/big.2017.0012

14. Jana, A., Goyal, P.: Can network embedding of distributional thesaurus be combined with word vectors for better representation? In: Proceedings of the 2018 Conference of the North American Chapter of the Association for Computational Linguistics: Human Language Technologies, Volume 1 (Long Papers), New Orleans, LO, USA, pp. 463–473 (2018). https://doi.org/10.18653/v1/N18-1043

15. Jiang, C., Jiang, Y., Wu, W., Xie, P., Tu, K.: Modeling label correlations for ultra-fine entity typing with neural pairwise conditional random field. In: Proceedings of the 2022 Conference on Empirical Methods in Natural Language Processing, Abu Dhabi, United Arab Emirates, pp. 6836–6847 (2022). https://aclanthology.org/2022.emnlp-main.459

16. Kilgarriff, A., Rychly, P., Smrz, P., Tugwell, D.: ITRI-04-08 the sketch engine. Inf. Technol. 105(116), 105–116 (2004)

17. Lee, C., Dai, H., Song, Y., Li, X.: A Chinese corpus for fine-grained entity typing. In: Proceedings of the Twelfth Language Resources and Evaluation Conference, Marseille, France, pp. 4451–4457 (2020). https://aclanthology.org/2020.lrec-1.548

18. Li, B., Yin, W., Chen, M.: Ultra-fine entity typing with indirect supervision from natural language inference. Trans. Associat. Computat. Linguist. 10, 607–622 (2022). https://doi.org/10.1162/tacl_a_00479

19. Ling, X., Weld, D.: Fine-grained entity recognition. In: Proceedings of the AAAI Conference on Artificial Intelligence, vol. 26, no. 1, pp. 94–100 (2021). https://doi.org/10.1609/aaai.v26i1.8122

20. Liu, L., et al.: TexSmart: a system for enhanced natural language understanding. In: Proceedings of the 59th Annual Meeting of the Association for Computational Linguistics and the 11th International Joint Conference on Natural Language Processing: System Demonstrations, pp. 1–10 (2021). https://doi.org/10.18653/v1/2021.acl-demo.1

21. Liu, Q., Lin, H., Xiao, X., Han, X., Sun, L., Wu, H.: Fine-grained entity typing via label reasoning. In: Proceedings of the 2021 Conference on Empirical Methods in Natural Language Processing, Online and Punta Cana, Dominican Republic, pp. 4611–4622 (2021). https://doi.org/10.18653/v1/2021.emnlp-main.378

22. López, F., Heinzerling, B., Strube, M.: Fine-grained entity typing in hyperbolic space. In: Proceedings of the 4th Workshop on Representation Learning for NLP (RepL4NLP-2019), Florence, Italy, pp. 169–180 (2019). https://doi.org/10.18653/v1/W19-4319

23. Obeidat, R., Fern, X., Shahbazi, H., Tadepalli, P.: Description-based zero-shot fine-grained entity typing. In: Proceedings of the 2019 Conference of the North American Chapter of the Association for Computational Linguistics: Human Language Technologies, Volume 1 (Long and Short Papers), Minneapolis, MN, USA, pp. 807–814 (2019). https://doi.org/10.18653/v1/N19-1087

24. Onoe, Y., Boratko, M., McCallum, A., Durrett, G.: Modeling fine-grained entity types with box embeddings. In: Proceedings of the 59th Annual Meeting of the Association for Computational Linguistics and the 11th International Joint Conference on Natural Language Processing (Volume 1: Long Papers), pp. 2051–2064 (2021). https://doi.org/10.18653/v1/2021.acl-long.160

25. Onoe, Y., Durrett, G.: Learning to denoise distantly-labeled data for entity typing. In: Proceedings of the 2019 Conference of the North American Chapter of the Association for Computational Linguistics: Human Language Technologies, Volume 1 (Long and Short Papers), Minneapolis, MN, USA, pp. 2407–2417 (2019). https://doi.org/10.18653/v1/N19-1250

26. Pan, W., Wei, W., Zhu, F.: Automatic noisy label correction for fine-grained entity typing. In: Raedt, L.D. (ed.) Proceedings of the Thirty-First International Joint Conference on Artificial Intelligence, IJCAI-22, pp. 4317–4323 (2022). https://doi.org/10.24963/ijcai.2022/599

27. Panchenko, A., et al.: Unsupervised, knowledge-free, and interpretable word sense disambiguation. In: Proceedings of the 2017 Conference on Empirical Methods in Natural Language Processing: System Demonstrations, Copenhagen, Denmark, pp. 91–96 (2017). https://doi.org/10.18653/v1/D17-2016

28. Panchenko, A., Ruppert, E., Faralli, S., Ponzetto, S.P., Biemann, C.: Building a web-scale dependency-parsed corpus from Common Crawl. In: Proceedings of the Eleventh International Conference on Language Resources and Evaluation (LREC 2018), Miyazaki, Japan, pp. 1816–1823 (2018). https://aclanthology.org/L18-1286

29. Panchenko, A., Ruppert, E., Faralli, S., Ponzetto, S.P., Biemann, C.: Unsupervised does not mean uninterpretable: the case for word sense induction and disambiguation. In: Proceedings of the 15th Conference of the European Chapter of the Association for Computational Linguistics: Volume 1, Long Papers, Valencia, Spain, pp. 86–98 (2017). https://aclanthology.org/E17-1009

30. Pelevina, M., Arefiev, N., Biemann, C., Panchenko, A.: Making sense of word embeddings. In: Proceedings of the 1st Workshop on Representation Learning for NLP, Berlin, Germany, pp. 174–183 (2016). https://doi.org/10.18653/v1/W16-1620

31. Qi, P., Zhang, Y., Zhang, Y., Bolton, J., Manning, C.D.: Stanza: a python natural language processing toolkit for many human languages. In: Proceedings of the 58th Annual Meeting of the Association for Computational Linguistics: System Demonstrations, pp. 101–108 (2020). https://doi.org/10.18653/v1/2020.acl-demos.14

32. Qian, J., et al.: Fine-grained entity typing without knowledge base. In: Proceedings of the 2021 Conference on Empirical Methods in Natural Language Processing, Online and Punta Cana, Dominican Republic, pp. 5309–5319 (2021). https://doi. org/10.18653/v1/2021.emnlp-main.431
33. Reimers, N., Gurevych, I.: Sentence-BERT: sentence embeddings using Siamese BERT-networks. In: Proceedings of the 2019 Conference on Empirical Methods in Natural Language Processing and the 9th International Joint Conference on Natural Language Processing (EMNLP-IJCNLP), Hong Kong, Hong Kong, pp. 3982–3992 (2019). https://doi.org/10.18653/v1/D19-1410
34. Riedl, M., Biemann, C.: Scaling to large[3] data: an efficient and effective method to compute distributional thesauri. In: Proceedings of the 2013 Conference on Empirical Methods in Natural Language Processing, Seattle, WA, USA, pp. 884–890 (2013). https://aclanthology.org/D13-1089
35. Ruppert, E., Kaufmann, M., Riedl, M., Biemann, C.: JoBimViz: a web-based visualization for graph-based distributional semantic models. In: Proceedings of ACL-IJCNLP 2015 System Demonstrations, Beijing, China, pp. 103–108 (2015). https:// doi.org/10.3115/v1/P15-4018
36. Sui, X., et al.: Improving zero-shot entity linking candidate generation with ultra-fine entity type information. In: Proceedings of the 29th International Conference on Computational Linguistics, pp. 2429–2437. Gyeongju, Republic of Korea (2022). https://aclanthology.org/2022.coling-1.214
37. Xiong, W., et al.: Imposing label-relational inductive bias for extremely fine-grained entity typing. In: Proceedings of the 2019 Conference of the North American Chapter of the Association for Computational Linguistics: Human Language Technologies, Volume 1 (Long and Short Papers), Minneapolis, MN, USA, pp. 773–784 (2019). https://doi.org/10.18653/v1/N19-1084
38. Zhang, T., Xia, C., Lu, C.T., Yu, P.: MZET: memory augmented zero-shot fine-grained named entity typing. In: Proceedings of the 28th International Conference on Computational Linguistics, Barcelona, Spain, pp. 77–87 (2020). https://doi. org/10.18653/v1/2020.coling-main.7
39. Zhou, B., Khashabi, D., Tsai, C.T., Roth, D.: Zero-shot open entity typing as type-compatible grounding. In: Proceedings of the 2018 Conference on Empirical Methods in Natural Language Processing, Brussels, Belgium, pp. 2065–2076 (2018). https://doi.org/10.18653/v1/D18-1231
40. Zuo, X., Liang, H., Jing, N., Zeng, S., Fang, Z., Luo, Y.: Type-enriched hierarchical contrastive strategy for fine-grained entity typing. In: Proceedings of the 29th International Conference on Computational Linguistics, Gyeongju, Republic of Korea, pp. 2405–2417 (2022). https://aclanthology.org/2022.coling-1.212

# Static, Dynamic, or Contextualized: What is the Best Approach for Discovering Semantic Shifts in Russian Media?

Veronika Nikonova[1] and Maria Tikhonova[1,2(✉)]

[1] HSE University, Moscow, Russia
m_tikhonova94@mail.ru
[2] SberDevices, Moscow, Russia

**Abstract.** This paper is focused on discovering diachronic semantic shifts in Russian news and social media using different embedding methods. Namely, in our work, we explore the effectiveness of static, dynamic, and contextualized approaches. Using these methods, we reveal social, political, and cultural changes through semantic shifts in the News and Social media corpora; the latter was collected and released as a part of this work. In addition, we compare the performance of these three approaches and highlight their strengths and weaknesses for this task.

**Keywords:** Language Modeling · Word Embeddings · BERT model · Semantic Shifts · Diachronic Shifts

## 1 Introduction

Changes in language can happen at different levels: grammatical, phonetic, syntactic, and semantic. In this work, we focus on semantic shifts, in other words, changes in words' meaning that happen throughout time.

Studying diachronic semantic shifts is important for both theoretical and practical reasons. From a scientific point of view, it helps researchers in historical linguistics investigate word meaning evolution across different time periods, provides linguists with data-driven evidence for updating and improving dictionaries and language resources, and contributes to interdisciplinary research in the digital humanities and social sciences. Such studies enable researchers to compare the embedding models with one another and evaluate their quality, assessing how well they can capture word semantics. In practical terms, such models can be applied in *Natural Language Processing (NLP)* tasks, such as information retrieval, sentiment analysis, and machine translation, helping to improve the accuracy and relevance of these systems, especially when dealing with historical texts or multilingual data.

A few works are devoted to this subject in English, while Russian, the focus of the current research, needs to be more thoroughly studied, and there are substantial volumes of data for research and analysis.

D. I. Ignatov et al. (Eds.): AIST 2023, LNCS 14486, pp. 141–153, 2024.
https://doi.org/10.1007/978-3-031-54534-4_10

This work focuses on discovering semantic shifts in the Russian language. Namely, we explore changes in words' meanings in Russian news and social media. We use three different approaches that have proved to be efficient for text semantic analysis:

- **Static approach:** Word2vec;
- **Dynamic approach:** Dynamic word embeddings model based on PPMI;
- **Contextualized approach:** pre-trained BERT model.

Using these models, we get a representation of each particular word in a vector space and then trace its movement throughout time. In this work, we deal with two tasks: 1) discovering task (what semantic changes the models can reveal from data) and 2) the diachronic semantic shift detection, which is regarded as a classification task (how well the models can detect whether a shift happened or not). In the first task, we use collected data to evaluate models' performance through the semantic shifts they highlight. In contrast, in the second task, we test the models' ability to detect semantic changes using an annotated dataset.

We use news and social media datasets for our analysis; the latter corpus was collected and released as a part of this research. These data sources are chosen because they are exposed to changes in shorter periods than fiction literature, for example, and can capture subtle social and cultural changes.

The main goal of this research is to reveal semantic shifts in the selected datasets and compare the performance of the described approaches in solving this task. Thus, the contribution of our paper is three-fold: (I) we present a new Social media corpus, (II) we compare different approaches (static, dynamic, and contextualized) to semantic shift discovering problem, and (III) conduct a comprehensive analysis of the discovered semantic shifts.

The rest of the paper is structured as follows: in Sect. 2, we discuss previous works on this topic, Sect. 3 describes the datasets, Sect. 4 introduces the models we used, Sect. 5 presents the experimental setup, Sect. 6 describes the results of our work, in Sect. 7 we discuss the limitations of the implemented approaches, and finally Sect. 8 concludes the paper.

## 2   Related Work

One of the most important works on this subject is [4], in which the authors compare three embedding algorithms, namely PPMI, SVD on top of PPMI and Word2vec (SGNS, first presented by [12]). According to the results of [4], Word2vec has shown the best performance in discovering real semantic shifts. And in our work, we use Word2vec as an example of the static approach. Other works that employ static word embeddings include [6,7], and [17].

In [20], the authors move from static to dynamic word representations by simultaneously training the model on all the periods. They propose a joint optimization problem that comprises both embedding learning and alignment problems. We use this approach and compare its performance with other approaches.

Contextualized models are gaining more and more popularity for solving NLP tasks showing state-of-the-art performance. A recent publication [16] applies contextualized approach to semantic shifts problem in Russian, namely ELMo [15] and BERT [9] models. The authors compare aggregation methods: averaging and clustering of word embeddings, and demonstrate that the first method performs better in most cases. In our work, we use this method for the contextualized approach due to its effectiveness and lower complexity. Other works that use ELMo and BERT models based on the Transformer architecture [18] for semantic shifts analysis include [2,3,5] and [11].

As a part of the research [1], one of the pioneers for diachronic semantic shifts detection in Russian, the authors released a dataset for semantic shift detection and provided baseline results: trained static and incremental Word2vec CBOW models, applied and compared different alignment techniques for micro and macro datasets.

One of the most recent works [8] dedicated to the diachronic semantic changes in Russian summarizes the results of RuShiftEval[1], in which semantic shifts analysis was conducted on three time periods (pre-Soviet, Soviet, post-Soviet). In this work, BERT [9] and ELMo [15] showed better performance than static models. Another recent work [13] that is dedicated to semantic detection problem focuses on contextualized embedding models. This survey provides a classification of the solutions, compares the previous results, and highlights existing scalability, interpretability, and robustness problems.

The previous works often compare the performance of several approaches. In our research, following their methodology, we compare classic examples of the static, dynamic, and contextualized approaches for semantic shift discovering task. The comparison of methods in such combination is performed for the first time in Russian.

## 3  Data

For this work, we use three datasets: News[2] and Social media[3] corpora for discovering semantic shifts and diachronic semantic shift detection dataset. For the News corpus, we use already collected data from lenta.ru from 2000 to 2019[4]. Social media corpus was collected as a part of this work from users' walls in VKontakte[5] social network from the year 2007 up to 2019[6]. These datasets intersect in time; however, there is no such task to compare the results between datasets, and the analyzed periods differ. Moreover, for different datasets, the

---

[1]  https://www.dialog-21.ru/evaluation/2021/rushifteval/.

[2]  https://github.com/yutkin/lenta.ru-news-dataset.

[3]  https://github.com/VeronikaNikonova/hse_thesis.

[4]  The period is chosen based on the available data and is bounded by 2019 before the Covid period.

[5]  https://m.vk.com.

[6]  The period is chosen based on the available data and is bounded by 2019 before the Covid period.

embeddings are learned in different vector spaces and, therefore, cannot be directly compared.

We use these datasets to identify social, cultural, and political semantic shifts[7] rather than linguistic ones: the time period is too small for considerable linguistic changes, and the nature of news and social media implies reflecting social, cultural, and political events and processes. Another reason for choosing these datasets is fewer studies on semantic shift discovering in Russian news and social media compared to Russian classical and fiction literature.

### 3.1 News Corpus

News articles allow us to glance at the cultural, social, and political context of a particular period. The language used in the articles typically reflects changing attitudes, beliefs, values, etc. It is highly contextualized, and news articles are consistent in style and grammar, which is beneficial for identifying patterns in them. This dataset contains 797 884 articles, on average about 36 000 articles per year. These articles correspond to over 20 various topics, with topics like Russia, World, Economy, Sports, and Culture being the most popular.

### 3.2 Social Media Corpus

As a part of our research we collected social media corpus, which includes texts posted on users' walls in the Russian social network VKontakte[8]. The users for the analysis have been chosen in two ways: 1) randomly from the first 10 000 registered users (around 2000 users) and 2) the most popular[9] users (both male and female) from the largest Russian cities: Moscow, St. Petersburg, Novosibirsk, Ekaterinburg, and Kazan (about 5200 users). Overall, there are 6 944 032 posts with an average of almost 500 000 posts per year. In such a way we obtained a completely new Social media corpus which can be used not only for semantic shift discovering but for dynamic topic modelling, training language models and other NLP tasks. The dataset is available in our repository[10].

### 3.3 Diachronic Semantic Shifts Classification Dataset

To compare model performance, we conduct additional experiments on diachronic semantic shift detection, regarding it as a classification task. For this task, we take the dataset[11] [1], which comprises 280 manually annotated Russian

---

[7] In other words, changes in associations with a word driven by social, cultural an political events and processes.
[8] https://m.vk.com.
[9] Popularity depends on the user that is making the request, that is why the dataset may be biased.
[10] https://github.com/VeronikaNikonova/hse_thesis.
[11] https://github.com/wadimiusz/diachrony_for_russian/blob/master/datasets/micro.csv.

adjectives. The annotation includes three different classes: *no meaning change* (220 words), *moderate change in meaning* (42 words), and *significant change in meaning* (18 words).

It should be noted that the classification dataset includes 21 words, which are not encountered in the News corpus we use. That is why our results cannot be directly compared to the baseline results of the original research, but we can compare the order of the resulted quality measures.

## 4    Method

In this work, we use three different models: Word2vec, Dynamic word embeddings, and BERT. These models represent three different approaches to the problem: static, dynamic, and contextualized, respectively.

– **Word2vec** Skip-gram with negative sampling [12], (further referred to simply as *Word2vec*), which is the classic example of the static approach. We train a separate Word2vec model for each year, align word embeddings using orthogonal Procrustes and then use cosine similarity measure to find the words that changed their meaning during the analyzed period the most.
– **Dynamic word embeddings** model [20] takes into account the temporal dimension of language. It is based on PPMI and captures the temporal evolution of word co-occurrence statistics. The main advantage of this model is that the embeddings are aligned during training, and the alignment is performed for all the embeddings simultaneously.
– **BERT** [9] is based on a Transformer [18] encoder, which uses a multi-head attention mechanism with scaled dot-product attention. In our work, we fine-tune the ruBert-base model[12] from HuggingFace library. Since BERT uses WordPiece tokenization [19], a subword tokenization algorithm, to obtain a word embedding from token embeddings, we take the sum of the embeddings of the tokens in which the word is tokenized.

## 5    Experiments

### 5.1    Evaluation Setup

We conduct two series of experiments: 1) main experiments on discovering semantic shifts aimed at revealing semantic changes in language; 2) an additional series of experiments on diachronic semantic shift detection to analyze how well models can detect semantic shifts and compare their performance. We analyze semantic shifts for the first series of experiments for the whole period. And in the second series of experiments, we deal with year-to-year shifts.

---

[12] https://huggingface.co/ai-forever/ruBert-base.

For discovering semantic shifts task, we assume that the shift has effectively taken place if 1) one of the existing meanings of a word comes to the foreground or 2) the historical context in which the word is used changes while the meaning of the word itself remains constant[13].

For discovering semantic shifts, we use precision scores as a quality metric. For the diachronic semantic shift detection, we use macro-averaged F1-score as the main metric and additionally compute precision, recall, and balanced accuracy scores.

## 5.2   Experimental Setup

**Semantic Shift Discovering Task**
For this task we use the pipeline:

1. Train (of fine-tune) embedding model on the training data (for BERT transform token embeddings into word embeddings).
2. Align embeddings and reduce them to the 2019 vector space (only for Word2vec).
3. Calculate cosine similarity for each of the eligible words (occurrence at least 50 times for the News corpus and 10 times for the Social media corpus and present in the corresponding periods) for word embeddings of 2000 (2007[14]) and 2019 years (for BERT, we use word prototype embeddings: averaged embeddings of all the occurrences of the eligible words in the appropriate year).
4. Obtain the top 20 words[15] with the lowest cosine similarity between these periods (which indicates the biggest semantic shift).
5. Analyze the revealed semantic shifts. Namely, for each word obtained in the previous step, we find the closest neighbors to understand the context in which they are used in each period and decide on the actual presence/absence of the semantic shift[16].

---

[13] The change of the president in some country is a vivid example of such change.
[14] For the Social media dataset.
[15] We chose 20 for the analysis as the bigger number includes more "noise" words.
[16] The analysis was performed by experts with a degree in social sciences. The final decision was made on a binary scale based on the closest semantic neighbors, however, broader textual contexts were also available.

Table 1 shows hyperparameters used for model training.

**Table 1.** Hyperparameters used for model training for semantic shift discovering task.

| Word2vec | Dynamic word embeddings | BERT |
|---|---|---|
| Vector size: 300<br>Window size: 5<br>Minimal<br>frequency: 50 (for<br>the News corpus)<br>and 10 (for the<br>Social media corpus)<br>Negative samples: 5<br>Exponent: 0.75 | Vocabulary size: 19 121<br>(for the News corpus) and 6 453 (for the Social<br>media corpus)<br>Rank (embedding dimension): 50<br>Penalty regularizer (lambda, ensures low-rank<br>word fidelity): 10<br>Time regularizer (tau, responsible for align-<br>ment<br>among time periods): 50<br>Symmetry regularizer (gamma): 50<br>Emphasize parameter (emphasizes non-zero):<br>1 | Batch size: 16<br>Learning rate: $2e-5$<br>Weight decay (for<br>weights<br>regularization): 0.01<br>Masking<br>probability: 0.15 |

**Diachronic Semantic Shifts Detection**

For this task, we use the following pipeline:

1. Take embedding models from the previous task and retrain them if neces-sary[17].
2. Calculate the cosine similarity measure for each word in the classification list (for BERT, we use word prototypes: averaged embeddings of all the occur-rences of the words in the classification list in the appropriate years).
3. Train Random Forest Classifier[18,19] using cosine similarity as an input.
4. Evaluate[20] the models with quality metrics (F1 score as the main metric, and balanced accuracy, precision, and recall as additional metrics).

To include as many words from the diachronic semantic shifts classification dataset as possible, we lowered the frequency threshold to 5. And to improve the performance of the Word2vec model, we increased the size of the window (maximum distance between the context and the target words) to 10.

---

[17] To include as many words from the classification dataset as possible, we had to retrain the Word2vec and Dynamic word embeddings models since the latter is initialized with static word2vec embeddings. For retraining, we lowered the frequency threshold from 50 to 5 and increased the window size to 10.

[18] We train Random Forest Classifier with the following parameters, which were selected on cross-validation: 1) for Word2vec: number of trees - 120, maximum depth of the tree - 4, minimum number of samples for a split - 4; 2) for Dynamic word embeddings: number of trees - 100, minimum number of samples for a split - 2; 3) for BERT: number of trees - 105, maximum depth of the tree - 5, minimum number of samples for a split - 5; 4) for combined features: number of trees - 120, maximum depth of the tree - 4, minimum number of samples for a split - 12, class weight - balanced.

[19] We use a Random Forest Classifier since it has proved efficient for classification tasks and is not as sensitive to the hyperparameters' choice as gradient boosting algorithms.

[20] We use 5-fold stratified cross-validation for evaluation.

# 6  Results

## 6.1  Discovering Task

**Table 2.** Top 20 words with the most significant semantic change for the News corpus. Words that actually experienced semantic changes are highlighted in bold.

| Word2vec | альберт, **видео**, родригес, рогозин, кстати, **гора**, титов, подъём, рубин, зато, **наряд**, кристалл, смена, **григорий**, **марат**, быков, красота, кольцо, миротво-рец, **чайка** (albert, **video**, rodriguez, rogozin, kstati, **gora**, titov, pod'em, rubin, zato, **naryad**, cristall, smena, **grigoriy**, **marat**, bykov, krasota, kol'tso, mirotvorets, **chaika**) |
|---|---|
| Dynamic word embeddings | лауреат, **свадьба**, франциско, уважение, турист, **аргумент**, законодатель, ясно, **повстанец**, хиллари, грабитель, **тайный**, **торговец**, долго, **егор**, **частота**, успех, обломок, достигнутый, **полотно** (laureat, **svad'ba**, francisco, uvazhenie, turist, **argument**, zakonodatel', yasno, **povstanets**, hillary, grabitel', **tainyi**, **torgovets**, dolgo, **egor**, **chastota**, uspekh, oblomok, dostignytyi, **polotno**) |
| BERT | **зелёный**, **цик**, **наряд**, напомнить, **лента**, **боец**, **ру**, четвёрка, **дон**, **кадр**, **правый**, подъём, рада, правда, **барак**, **миля**, редакция, **яблоко**, **корт**, **единство** (**zelenyi**, **tsik**, **naryad**, napomnit', **lenta**, **boets**, **ru**, chetverka, **don**, **kadr**, **pravyi**, pod'em, rada, **pravda**, **barak**, **milya**, redaktsiya, **yabloko**, **kort**, **yedinstvo**) |

**Table 3.** Top 20 words with the most significant semantic change for the Social media corpus. Words that actually experienced semantic changes are highlighted in bold.

| Word2vec | популярный, десятый, **экономика**, ца, **мышь**, шестой, металлический, ом, напомнить, норма, включая, **пробежать**, ооо, **заложить**, господин, отправляться, **база**, **космос**, баланс, украина (populyarnyi, desyatyi, **economica**, tsa, **mysh'**, shestoy, metallicheskiy, om, **napomnit'**, norma, vklyuchaya, **probezhat'**, ooo, **zalozhit'**, gospodin, otpravlyat'sya, **baza**, **kosmos**, balans, ukraina) |
|---|---|
| Dynamic word embeddings | положение, игорь, научный, **французский**, голод, думать, путы, **делить**, таракан, **старенький**, гордо, дак, поверить, увы, **дрянь**, **править**, раздел, огород, целое, аудио (polozheniye, igor', nauchnyi, **frantsuzskii**, golod, dumat', puty, **delit'**, tarakan, **staren'kii**, gordo, dak, poverit', uvy, **dryan'**, **pravit'**, razdel, ogorod, tseloye, audio) |
| BERT | **зенит** (zenit),**приведа** (priveda), **убогий** (ubogii), **защититься** (zashchit'sya), **малина** (malina), **серж** (serzh), **замешать** (zameshat'), **охарактеризовать** (okharacterizovat'), **чъ** (ch'), **кончать** (konchat'), **взаимопонимание** (vzaimoponimanie), **делить** (delit'), **полярный** (polyarnyi), **вырубить** (vyrubit'), **зевать** (zevat'), **тесный** (tesnyi), **пъ** (p'), **взрывать** (vzryvat'), **спасти** (spasti), **противный** (protivnyi) (**zenit,priveda**, **ubogii**, zashchit'sya, **malina**, serzh, zameshat', okharacterizovat', ch', konchat', vzaimoponimanie, **delit'**, **polyarnyi**, **vyrubit'**, **zevat'**, **tesnyi**, p', **vzryvat'**, **spasti**, protivnyi) |

Tables 2 and 3 contain top-20 words with the lowest cosine similarity measure for each of the models for the News and the Social media corpora, respectively. Further, we highlight the most curious semantic shifts for both corpora. The full tables containing cosine similarity measures, closest neighbors, and words' meanings that changed from one-time slice to another can be found on Github[21].

**News Corpus Semantic Shifts Analysis**
Table 4 shows the closest neighbors and words' meanings in 2000 and 2019 time slices.

*Word2vec.* The word "наряд" ("naryad") is an example of social change. At the beginning of the 21st century "наряд" ("naryad") was associated with the police, and in 2019 this word was used more as an outfit. The word "видео" ("video") (shift from "TV commercial" or a "movie" to a "video clip") is an example of a technological change that was prompted by the emergence of mobile phones cameras. In total, 13 out of 20 words effectively experienced semantic changes, which is 65% precision on the examined set of words.

---
[21] https://github.com/VeronikaNikonova/hse_thesis

**Table 4.** Examples of semantic changes with words' closest neighbors and meanings for the News corpus.

| Word | Top 3 closest neighbors 2000 | Meaning 2000 | Top 3 closest neighbors 2019 | Meaning 2019 |
|---|---|---|---|---|
| *Word2vec* | | | | |
| наряд (naryad) | патруль, патрульный, оцепить (patrol, patrol, cordon off) | Police | костюм, одетый, сумка (costume, dressed, purse) | Outfit |
| видео (video) | телевизионный, документальный, классический (television, documentary, classic) | TV commercial, movie | кадр, запись, снятой (shot, record, taken) | Clip |
| *Dynamic word embeddings* | | | | |
| свадьба (svad'ba) | мадонна, принцесса, леннон (madonna, princess, lennon) | The wedding of Madonna | принц, телеведущий, гарри (prince, TV presenter, harry) | Prince Harry's wedding |
| полотно (polotno) | цистерна, рельс, сойти (tank, rail, get off) | Railway road | галерея, замок, уличный (gallery, castle, street) | Painting |
| *BERT* | | | | |
| правда (pravda) | кстати, хотя, сам (by the way, though, himself) | Linking word (to tell the truth) | газета, комсомолец, новость (newspaper, komsomolets, news) | Newspaper |
| кадр (kadr) | кадровый, персонал, материал (personnel, staff, material) | Staff, personnel | снимок, фотография, запись (shot, photo, record) | Photo |

*Dynamic Word Embeddings.* Dynamic word embeddings model captured different social events, for example, weddings. The word "свадьба" ("svad'ba", wedding) in 2000 refers to the wedding of Madonna and in 2019 to the wedding of Prince Harry and Megan Markle so that we can trace "the wedding of the year" with our model. Overall, 11 out of 20 words effectively experienced semantic changes, which is 55% precision on the examined set of words.

*BERT.* The meaning change of the word "правда" ("pravda") is an example of a cultural shift. In 2000 it was used as a linking word "to tell the truth", and in 2019, it was used as a reference to a newspaper called "Pravda". The model also captured the technological change of the emergence of portable devices with cameras but with a different word: "кадр" ("kadr") (shift from "staff" and "personnel" in 2000 to "picture" and a "photo" in 2019). In total, 16 out of 20 detected words effectively experienced semantic changes, which is 80% precision on the examined set of words.

### Social Media Corpus Semantic Shifts Analysis
Table 5 shows closest neighbors and words' meanings in 2007 and 2019 time slices.

*Word2vec.* The word "норма" ("norma") in 2007 was frequently used in the meaning of "ok", "good" and in 2019, it was used more often as a "legal norm", or "law". The word "ооо" ("ooo") in 2007 was used as an interjection or exclamation, and in 2019 it referred to a limited responsibility company. We can suggest that

**Table 5.** Examples of semantic changes with words' closest neighbors and meanings for the Social media corpus.

| Word | Top 3 closest neighbors 2007 | Meaning 2007 | Top 3 closest neighbors 2019 | Meaning 2019 |
|---|---|---|---|---|
| *Word2vec* | | | | |
| норма (norma) | потихоньку, тихонька, пучок (slowly, quietly, ok) | Good, ok | мораль, стандарт, нарушать (moral, standard, break) | Norm, law |
| ооо (ooo) | оооо, оо, ооооо (оооо, оо, ооооо) | Interjection, exclamation | интеграл, компания, москоу (integral, company, moscow) | Limited responsibility company |
| *Dynamic word embeddings* | | | | |
| положение (polozheniye) | граф, зависимость, разный (count, dependency, different) | Position, status | ориентация, внешний, интерес (orientation, external, interest) | Orientation, position in space |
| французский (frantsuzskii) | контрольный, исключительно, какой (check, solely, which) | Subject | русский, итальянский, английский (russian, italian, english) | Language |
| *BERT* | | | | |
| взрывать (vzryvat') | жечь, привеза, баста (burn, hi, basta) | To party | взорвать, зажигать, взрыв (explode, burn, explosion) | To explode |
| полярный (polyarnyi) | сомнительный, защититься, альтернативный (dubious, defend, alternative) | Alternative | снежный, северный, земной (snowy, polar, terrestrial) | Polar |

these shifts can be connected to the growth of VKontakte users writing posts that caused some change in their language.

Overall, 10 out of 20 words effectively experienced semantic changes, which is 50% precision on the examined set of words.

*Dynamic Word Embeddings.* The word "французский" ("frantsuzskii") in 2007 meant subject at school/University, while in 2019, it is associated with other languages, such as Russian, Italian, and English, which is in line with our "aging" pattern. In general, 8 out of 20 words effectively experienced semantic changes, which is 40% precision on the examined set of words.

*BERT.* The word "взрывать" ("vzryvat'") in 2007 was used in the meaning of "to party" while in 2019, it was used in its more formal sense "to explode", which again can be regarded as a confirmation of our "aging" hypothesis. In total, 12 of 20 words effectively experienced semantic changes, which is 60% precision on the examined set of words.

Table 6 contains the resulting precision for the discovering task. It can be seen that the BERT model showed the best performance; the Word2vec model is in second place, and then goes the Dynamic word embeddings model. In general, we can say that all the models revealed different political, social, cultural, and technological events and changes, and sometimes different models captured the same trends.

**Table 6.** Model precision on semantic shift discovering task.

| | News corpus | | | Social media corpus | | |
|---|---|---|---|---|---|---|
| | Word2vec | Dynamic word embeddings | BERT | Word2vec | Dynamic word embeddings | BERT |
| Precision | 65% | 55% | 80% | 50% | 40% | 60% |

## 6.2  Diachronic Semantic Shift Detection

Table 7 shows results obtained for the diachronic semantic shift detection task. We use results reported in [1] and random choice results as a baseline. Our models show approximately equal performance, with Word2vec yielding the best result. Since our training corpus is smaller than the one used in the original research for the baselines, we cannot compare our results directly, but we can note that our F1 score indicators are close to the baseline scores. It should be noted that all tested models show better performance than the random choice strategy.

**Table 7.** Results on diachronic semantic shifts detection task.

| | F1 score | Presicion | Recall | Balanced accuracy |
|---|---|---|---|---|
| Random choice baseline | 0.333 | 0.333 | 0.333 | 0.333 |
| Procrustes baseline | 0.468 | – | – | – |
| Combined baseline | 0.503 | – | – | – |
| Word2vec | 0.451 | 0.442 | 0.482 | 0.481 |
| Dynamic word embeddings | 0.413 | 0.454 | 0.404 | 0.404 |
| BERT | 0.413 | 0.419 | 0.418 | 0.418 |
| Combined | 0.446 | 0.453 | 0.474 | 0.474 |

Even though BERT is the most powerful model, it showed poorer performance on the diachronic semantic shift detection task. We discuss the possible reasons for this it the next section.

## 7  Discussion

The Word2vec model is a comparatively simple but rather effective model. Its major drawback is that static word embeddings it outputs need to be aligned.

The dynamic approach solves this problem by simultaneously optimizing and aligning the embeddings during training. However, this approach is sensitive to the hyperparameters choice and is rather memory and time-consuming for large vocabularies and datasets.

The BERT model gives contextualized word representations that automatically solve the alignment problem. However, our research shows that the BERT model showed poorer performance on diachronic semantic shift detection task than the Word2vec model. There may be several reasons for that:

- The authors of [14] notice that contextualized models, like BERT, show state-of-the-art performance in detecting semantic shifts for high-polysemy words. In contrast, for low-polysemy words, the performance of such models drops considerably, even underperforming Word2vec models.
- In our research, we used averaged word embeddings to obtain word prototypes because it is computationally efficient, and this technique showed good performance in [16]. However, in [14] averaging performs worse than clustering of word embeddings and even underperforms the Word2vec model.
- Even fine-tuning the base version of the BERT model requires significant resources. We used its base variation (ruBert-base) and fine-tuned it on our data only for one epoch; if we had used the large version or fine-tuned the model for two epochs, the model could have shown better performance.
- Moreover, though BERT is considered the most powerful model among the analyzed, in some cases [10], Word2vec still outperforms BERT for some languages, which aligns with our findings.

## 8    Conclusion

This paper is devoted to the semantic shift discovering problem. In this work, we presented a new Social media corpus, compared different approaches (static, dynamic, and contextualized) to semantic shift discovering task, and conducted discovered semantic shifts analysis. The code, the trained models, and the dataset are available in our repository[22].

In the experiments, tested models revealed political, cultural, social, and technological changes in the Russian language, with the BERT model showing better quality of 80% for the News corpus and 60% for the Social media corpus on the discovering task and Word2vec giving the best F1-score of 0.45 on the diachronic semantic shift detection task.

As a future research we plan to use other BERT-like models, and explore bigger and more various text corpora, which cover longer time periods.

## References

1. Fomin, V., Bakshandaeva, D., Ju, R., Kutuzov, A.: Tracing cultural diachronic semantic shifts in Russian using word embeddings: test sets and baselines. In: Komp'juternaja Lingvistika i Intellektual'nye Tehnologii, pp. 213–227 (2019)
2. Giulianelli, M.: Lexical semantic change analysis with contextualised word representations. Unpublished master's thesis, University of Amsterdam, Amsterdam (2019)
3. Giulianelli, M., Del Tredici, M., Fernández, R.: Analysing lexical semantic change with contextualised word representations. In: Proceedings of the 58th Annual Meeting of the Association for Computational Linguistics, pp. 3960–3973 (2020)

---

[22] https://github.com/VeronikaNikonova/hse_thesis.

4. Hamilton, W.L., Leskovec, J., Jurafsky, D.: Diachronic word embeddings reveal statistical laws of semantic change. In: Proceedings of the 54th Annual Meeting of the Association for Computational Linguistics (Volume 1: Long Papers), pp. 1489–1501 (2016)
5. Hu, R., Li, S., Liang, S.: Diachronic sense modeling with deep contextualized word embeddings: an ecological view. In: Proceedings of the 57th Annual Meeting of the Association for Computational Linguistics, pp. 3899–3908 (2019)
6. Kim, Y., Chiu, Y.I., Hanaki, K., Hegde, D., Petrov, S.: Temporal analysis of language through neural language models. In: Proceedings of the ACL 2014 Workshop on Language Technologies and Computational Social Science, pp. 61–65 (2014)
7. Kulkarni, V., Al-Rfou, R., Perozzi, B., Skiena, S.: Statistically significant detection of linguistic change. In: Proceedings of the 24th International Conference on World Wide Web, pp. 625–635 (2015)
8. Kutuzov, A., Pivovarova, L.: RuShiftEval: a shared task on semantic shift detection for Russian. Comput. Linguist. Intellect. Technol. (2021)
9. Kenton, J.D., Chang, M.W., Toutanova, L.K.: BERT: pre-training of deep bidirectional transformers for language understanding. In: Proceedings of NAACL-HLT, pp. 4171–4186 (2019)
10. Martinc, M., Montariol, S., Zosa, E., Pivovarova, L.: Discovery team at SemEval-2020 task 1: context-sensitive embeddings not always better than static for semantic change detection. In: Proceedings of the Fourteenth Workshop on Semantic Evaluation. International Committee for Computational Linguistics (2020)
11. Martinc, M., Novak, P.K., Pollak, S.: Leveraging contextual embeddings for detecting diachronic semantic shift. In: Proceedings of the Twelfth Language Resources and Evaluation Conference, pp. 4811–4819 (2020)
12. Mikolov, T., Sutskever, I., Chen, K., Corrado, G.S., Dean, J.: Distributed representations of words and phrases and their compositionality. In: Advances in Neural Information Processing Systems, vol. 26 (2013)
13. Montanelli, S., Periti, F.: A survey on contextualised semantic shift detection. arXiv preprint arXiv:2304.01666 (2023)
14. Montariol, S.: Models of diachronic semantic change using word embeddings. Doctoral dissertation, Université Paris-Saclay (2021)
15. Peters, M.E., et al.: Deep contextualized word representations. NAACL-HLT. arXiv (2018)
16. Rodina, J., Trofimova, Y., Kutuzov, A., Artemova, E.: ELMo and BERT in semantic change detection for Russian. In: van der Aalst, W.M.P., et al. (eds.) AIST 2020. LNCS, vol. 12602, pp. 175–186. Springer, Cham (2021). https://doi.org/10.1007/978-3-030-72610-2_13
17. Rosenfeld, A., Erk, K.: Deep neural models of semantic shift. In: Proceedings of the 2018 Conference of the North American Chapter of the Association for Computational Linguistics: Human Language Technologies, Volume 1 (Long Papers), pp. 474–484 (2018)
18. Vaswani, A., et al.: Attention is all you need. In: Advances in Neural Information Processing Systems, vol. 30 (2017)
19. Wu, Y., et al.: Google's neural machine translation system: bridging the gap between human and machine translation. arXiv preprint arXiv:1609.08144 (2016)
20. Yao, Z., Sun, Y., Ding, W., Rao, N., Xiong, H.: Dynamic word embeddings for evolving semantic discovery. In: Proceedings of the Eleventh ACM International Conference on Web Search and Data Mining, pp. 673–681 (2018)

# Controllable Story Generation Based on Perplexity Minimization

Sergey Vychegzhanin$^{(\boxtimes)}$ ⓘ, Anastasia Kotelnikova ⓘ, Alexander Sergeev ⓘ, and Evgeny Kotelnikov ⓘ

Vyatka State University, Kirov, Russia
vychegzhaninsv@gmail.com

**Abstract.** Large-scale pre-trained language models have demonstrated impressive results in producing human-like texts. However, controlling the text generation process remains a challenge for researchers. Controllable text generation consists of generating sentences that satisfy desired constraints (e.g., sentiment, topic, or keywords). Recent studies that control the decoding stage of a language model have proved the high efficiency of this approach for control of generated texts. This approach, in contrast to the fine-tuning of pre-trained language models, requires much less computing resources. In this work, we propose and investigate a method that controls the process of language generation using perplexity minimization. The method is designed to create stories from a sequence of guide phrases that form a storyline and is based on the search for sequences of tokens that reduce text perplexity when generation is directed towards the guide phrase. First, we generate several arbitrary small sequences of tokens from the language model vocabulary. Then we choose the most probable subsequence - the one, the probability of following the guide phrase after which is the biggest. The proposed method induces the model to shift the content of the generated text to the guide phrase. Experiments on the Russian-language corpus of fairy tales with storylines have shown the high efficiency of the proposed method for creating stories corresponding to the user-specified storyline.

**Keywords:** Large Language Models · Text generation · Decoding strategy · GPT · LLaMA

## 1 Introduction

Natural language generation (NLG) has been one of the most important subfields in natural language processing. It combines computational linguistics and artificial intelligence to generate plausible and readable text in human language. In recent years, large-scale pre-trained language models (PLMs) have shown promising results to enhance the understanding of language and improve the generation quality, achieving state-of-the-art performance. With the emergence of Transformer neural architecture [21] and language models based on it, such as GPT-3 [2] and LLaMA [20], automatically generated texts became more realistic and difficult to distinguish from human ones.

D. I. Ignatov et al. (Eds.): AIST 2023, LNCS 14486, pp. 154–169, 2024.
https://doi.org/10.1007/978-3-031-54534-4_11

NLG technologies have been extensively applied to a wide range of applications. For example, they have been used to generate responses to user questions in dialogue systems (chatbots) and question-answer systems [11], generate stories [25], poems [7], theatre scripts and screenplays [14] and others.

One of the most difficult properties of the NLG is the controllability. Controllable text generation (CTG) is the task of generating texts that meet certain control constraints set by a human [16]. Many applications need a good control over the output text. For example, to generate stories for kids, it is important to guide the content of the story so that it is safe and understandable by children.

There are two types of CTG: soft and hard. The aim of soft CTG is, e.g., to provide the desired sentiment or topic of the generated text. Hard CTG requires ensuring that the text contains explicit constraints, e.g., certain keywords. Figure 1 shows an example of hard CTG, where the story is generated according to the key phrases provided by the storyline.

| Storyline | dragon holds the princess in a cave → princess loves the prince → prince marries the princess in the temple |
|---|---|
| Generated text | The **dragon holds the princess in a cave**. The prince slays the dragon. The **princess loves the prince**. The prince asks the king's permission. The **prince marries the princess in the temple**. |

**Fig. 1.** Example of controllable story generation with hard control.

Fine-tuning of PLMs is widely used to create various methods of CTG, such as in [5,8,26]. However, the disadvantage of this approach is the need to create a training corpora and perform a labor intensive and time consuming training procedure. This paper overcomes this problem by developing a plug-and-play method applicable to any large-scale PLM. Proposed method is tested on Russian text corpora and models capable of generating Russian texts. This choice is due to the insufficient number of studies on CTG for Russian.

The idea of the method is to generate a set of fixed length token sequences, and then estimate the probability of following the guide phrase (also called plot phrase) after each generated subsequence and choose the most probable one. This method is plug-and-play, i.e. it can be used with any autoregressive model. The experiments carried out on generating stories from a set of events that make up the plot of a story prove the effectiveness of the proposed method for creating texts from a set of plot phrases.

The contribution of the paper is as follows:

- we propose a method that generates stories in accordance with the plot;
- we apply the method to the Russian language;
- we form a text corpus containing stories with extracted storylines;
- we experiment with story generation using ruGPT-3 Large, ruAlpaca and Saiga models to confirm the effectiveness of the proposed method.

## 2     Previous Work

This section discusses the existing methods of CTG that can be applied to the problem of story generation, which is of primary research interest. Automatic story generation takes as input a short sentence (a prompt) and aims at generating a narrative from it that describes real or imaginary actors and events [4]. The complexity of the task of generating stories is due to the need to meet a number of conditions. The story must be thematically consistent throughout the document, requiring modeling very long range dependencies, the story must be creative, and the story must contain a plot [24].

CTG methods can be classified into fine-tuning, prompt engineering, retraining or refactoring, and post-processing [29]. Fine-tuning PLMs on a specialized data set is the main way to interact with models. Methods of this type fine-tune some or all of the model parameters to create texts that satisfy certain constraints. For example, works [6,26] used fine-tuning and a two-stage hierarchical approach to text generation. First, a premise was generated that determined the structure of the story. Then the premise was converted into a text passage.

Later, large language models (LLMs) based on the Transformer architecture began to be used for CTG. The prompt-based approach became widespread. The prompt engineering technique is to carefully construct precise and context-specific prompts to get the desired outputs. Yuan et al. [27] developed Wordcraft, a web application consisting of text editor and set of integrated LLM-powered controls for the purpose of writing a story. Mirowski et al. [14] developed Dramatron, a system for generating screenplays and theatre scripts that uses several hard-coded prompts to guide the LLM and prompt chaining technique.

The prompt tuning occupies an intermediate position between fine-tuning and prompt engineering. This approach forms the prompt as a floating-point-valued vector found by gradient descent to maximize the log-probability on outputs. Li and Liang [12] proposed a method called "prefix tuning" that freezes the parameters of the PLM and performs error backpropagation to optimize a small continuous task-specific vector called "prefix". A similar P-tuning method [10] differs from prefix tuning in that it does not place a prompt with the "prefix" in the input, but constructs a suitable template composed of the continuous virtual token, which is obtained through gradient descent.

Retraining or refactoring involves changing the architecture of the language model or retraining a model from scratch. This approach is limited by the insufficient amount of labeled data and the high consumption of computing resources. One of the first models in this direction was CTRL [8]. The model was trained on a set of control codes. Zhang et al. [30] proposed POINTER, an insertion-based method for hard-constrained text generation, which involves preserving of specific words.

Methods based only on using a decoder are called post-processing. Such methods require less computational resources. This group of methods includes the PPLM [5], which first trains an attribute discriminant model and then uses it to guide language model to generate the text with corresponding topic or sentiment. Another method is the Keyword2Text [15] shifts the output distribution of the language generation model to the semantic space of a given guide word in the word2vec or GloVe vector space. A similar idea is used [22], but the difference is that the score function of the autoregressive language model is modified with the score function of the autoencoding language model rather than with the cosine similarity to the target keyword.

In this paper, we propose a post-processing method that implements a decoding strategy based on heuristics. The difference from previous works [15,22], lies in the fact that at each generation step for small sequences of tokens, the probability of following the guide phrase is estimated. The method is based on the idea that choosing a sequence of tokens, after which the probability of following the guide phrase is maximum, will induce the model to generate text, shifting its content to the guide phrase.

## 3   Controllable Text Generation Method

We consider conditional probabilistic models for which the probability of the output text $X = \{x_1, ..., x_n\}$ can be factorized by tokens:

$$P(X) = \prod_{i=1}^{n} P(x_i|x_{<i}), \tag{1}$$

where $x_i$ denotes the $i$-th output token, and $x_{<i}$ denotes previous tokens $x_1, ..., x_{i-1}$.

In accordance with formula (1), the goal of conditional text generation can be formulated as follows:

$$P(X|C) = \prod_{i=1}^{n} P(x_i|x_{<i}, C), \tag{2}$$

where $C$ denotes the control conditions and $X$ is the generated text, which complies with the control conditions.

An important role in the generation process belongs to the decoding strategy using which natural language units (symbols, words, or sentences) are decoded from the probability distribution $P$. It selects tokens from the probability distribution over a model vocabulary at each time step. An example of a decoding strategy is beam search [13].

Generative language models such as GPT learn to predict the next token in a given sequence of tokens. Text generation is a natural application for such models. However, when predicting the next token of a sequence, they are not able to take into account the context following it, which is supposed to be the content of the generated text.

In this study, we propose the method, which at each generation step determines the most probable sequence of tokens for logically linking the prompt and the guide phrase that should be used in the text. The idea of the method is based on using intrinsic knowledge of a PLM to evaluate many small token sequences and select the appropriate sequence for a coherent transition to the guide phrase.

Let us consider the sequence $X = \{x_1, ..., x_{i-1}, x_i, x_{i+1}, ..., x_{i+k}, t_1, ..., t_m\}$. For a given prompt $X_{1:i-1} = \{x_1, ..., x_{i-1}\}$ and a guide phrase $T = \{t_1, ..., t_m\}$ theoretically it is possible to find the sequence $X_{i:i+k} = \{x_i, x_{i+1}, ..., x_{i+k}\}$ using exhaustive search of tokens from the model vocabulary. However, such search has an exponential dependence on the length of the connecting sequence and is not applicable in practice. Therefore, in order to reduce the number of variants we propose a heuristic technique for generating and evaluating connecting sequences (Fig. 2).

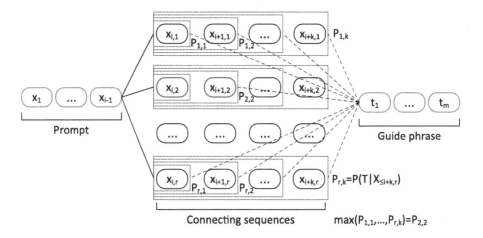

**Fig. 2.** Scheme of our controllable text generation method.

First, as continuations of the prompt $X_{1:i-1}$, $r$ different sequences of tokens of length $k+1$ are generated using some decoding strategy. Next, for each subsequence $X_{i:i+q}(q = 1, ..., k)$ of the $r$ sequences, the probability of following the guide phrase $T$ after it is determined by the formula:

$$P(X_{i:i+q}|X_{1:i-1}, T) = P(T|X_{\leq i+q}) = \prod_{j=1}^{m} P(t_j|t_{<j}, X_{\leq i+q}). \tag{3}$$

Further, at the current generation step, a subsequence is selected for which the probability (3) is maximum, and the sequences of length $k+1$ are repeatedly generated.

At each generation step, it makes sense to check the presence of the guide phrase in the generated text. It is possible that by the current step, the model has already said something about the guide phrase. To do this, we calculate the proportion of words in the guide phrase that are contained in the generated text. If the share exceeds a certain threshold value, then we consider that the target phrase is present in the text and direct the generation to the next guide phrase. However, if after the completion of the generation of a given number of tokens, the share of words of the guide phrase is less than the threshold value, then we insert this phrase in the position where it had the maximum probability for the entire time of generation, and continue the generation process.

Formula (3) makes it possible to estimate the probability of following the guiding phrase for each connecting sequence of tokens, but does not evaluate their semantic similarity. There may be cases where semantic similarity is more important than the likelihood of following the guide phrase. To assess the similarity of the connecting sequence and the guide phrase, it is proposed to use the Jaccard coefficient:

$$K_J = \frac{C}{A + B - C}, \qquad (4)$$

where $A$ is the set of words in normal form from the prompt, $B$ is the set of words in normal form from the guide phrase, $C$ is the set of common words for the prompt and the guide phrase.

Taking into account formulas (3) and (4) for connecting sequences, the average score, which establishes a balance between the two measures, can be determined by the formula:

$$Score_{X_{i:i+q}} = w_{prob}P_{norm} + w_J K_J, \qquad (5)$$

where $w_{prob}$, $w_J$ are weight coefficients, $P_{norm}$ is the normalized probability of following the guide phrase.

Thus, at each time step, the proposed method allows selecting the most logical subsequence of tokens for linking the prompt and the guide phrase, based on the knowledge of the generative model itself.

As an example of how the method works, let us consider a text at some $i$-th generation step and a guide phrase separated by a sequence of unknown tokens, for example, of length 3 (Fig. 3). In the figure, the prompt for the autoregressive model is highlighted in blue, and the guide phrase is highlighted in orange. The connecting sequence is marked with labels $\langle x_1 \rangle \langle x_2 \rangle \langle x_3 \rangle$.

Однажды в лесу, около речки, сидел мальчик с бабушкой. Вдруг в это время из-за $\langle x_1 \rangle \langle x_2 \rangle \langle x_3 \rangle$ волк напал на ребенка

Once in the forest, near the river, a boy was sitting with his grandmother. Suddenly, at this time, $\langle x_1 \rangle \langle x_2 \rangle \langle x_3 \rangle$ the wolf attacked the child

| Score | P | $K_J$ | $\langle x_1 \rangle \langle x_2 \rangle \langle x_3 \rangle$, Russian | $\langle x_1 \rangle \langle x_2 \rangle \langle x_3 \rangle$, English |
|---|---|---|---|---|
| 0.944 | 3.20E-11 | 0.200 | кустов вышли волки | wolves came out from behind the bushes |
| 0.226 | 6.50E-12 | 0.200 | поворота вышел волк, | a wolf came out from around the corner, |
| 0.105 | 1.90E-13 | 0.100 | дерева на поляну | from behind a tree to a clearing |
| 0.100 | 4.60E-18 | 0.100 | дерева выскочило | from behind a tree jumped out |
| 0.100 | 9.30E-19 | 0.100 | деревьев вышел лев, | a lion came out from behind the trees, |

**Fig. 3.** Example of prompt and connecting sequences at the $i$-th generation step.

The prompt is an input of the autoregressive model. With some decoding strategy, such as top-$k$ sampling, $r$ different sequences of 3 tokens $\langle x_1 \rangle \langle x_2 \rangle \langle x_3 \rangle$

are generated. For their subsequences, the probabilities of following the guide phrase $P$ and the Jaccard coefficients $K_J$ are calculated. The calculated values are averaged by formula (5). The subsequences of tokens are sorted in descending order of *Score*, and the subsequence with the highest value of the average score is selected. The selected subsequence is attached to the prompt, and the generation process continues until the specified number of tokens is generated.

## 4   Text Corpus

To conduct experiments, a text corpus[1] was formed from fairy tales in Russian with extracted storylines. The corpus is made up of fairy tales placed on nukadeti.ru[2] with a length of no more than 5000 characters. In total, the training corpus contains 562 fairy tales.

In each fairy tale, plot phrases were singled out, i.e. phrases that determine the main events in the story, the storyline. To do this, first, in each fairy tale keywords and phrases were selected, using the methods *yake* [3], *rakun* [19], *frake* [28], *textrank*[3], *rutermextract*[4], *keybert*[5] methods. Each method selected 15 keywords and phrases. The *yake* and *rutermextract* methods were manually selected. These methods showed the highest quality, so their results were used in the next stage to compose plot phrases.

Further, plot phrases were extracted from fairy tales according to the following algorithm:

1. Events were found. Events are syntactically related triples ⟨object, action, object⟩ (for example, "старуха, испекла, колобок" - "old woman, baked, bun"). The objects were selected from a set of keywords, and the actions was determined from the parse tree as nodes, syntactically associated with the objects. The stanza library was used to make the syntax parsing of the sentences.
2. The most important events were selected from the found events. Each selected event was assigned a weight obtained by summing the weights of the keywords extracted by the *yake* and *rutermextract* methods separately.
3. From the selected important events, a plot phrase was formed, determined by a 4-element set $(o_1, v, o_2, m)$, where $v$ is a verb, $o$ are objects related to the verb, $m$ is a modifier, prepositional object, or indirect object. Prepositions are possible before $o$ and $m$. An example of an event: "grooves in the forest spilled into whole streams", where "spilled" is $v$, "grooves" and "streams" are $o$, "forest" is $m$ ("канавки в лесу разлились в целые ручьи", $v$ – "разлились", $o$ – "канавки", "целые ручьи", $m$ – "лесу").

---

[1] https://github.com/icecreamz/MaxProb.

[2] https://nukadeti.ru.

[3] https://github.com/JRC1995/TextRank-Keyword-Extraction.

[4] https://github.com/igor-shevchenko/rutermextract.

[5] https://github.com/MaartenGr/KeyBERT.

For each of the two methods for extracting keywords, their own plot phrases were formed, the number of which, depending on the fairy tale, varied from 0 to 26. The number of sentences in fairy tales varied from 4 to 139. The distribution of the number of plot phrases extracted using the *yake* and *rutermextract* methods and the distribution of the number of sentences are shown in Appendix A.

Since the number of plot phrases should correlate with the length of the tale, the plot was assembled from the selected phrases according to the following algorithm:

1. The minimum number of phrases in the plot is 1, the maximum is the rounded-up value of the logarithm to base 2 of the number of sentences $n$ in the text: $\lceil log_2 n \rceil$.
2. If the *yake* method returned the number of plot phrases in the above range, these phrases were taken in order as a plot.
3. If the *yake* method produced fewer plot phrases, and the *rutermextract* method yielded enough, then the *rutermextract* phrases were taken in order as a plot.
4. If both methods returned the number of phrases less than the minimum value, their results were combined without repetitions in the order of the sentences in the text.
5. If the *yake* method produced more plot phrases than the maximum allowable in accordance with point 1, then a part of the fragments with maximum weights was taken for the required amount.

A test corpus of 25 plots was also formed. The distributions of numbers of plot phrases in the training and test corpora and statistics on the number of tokens received using the ruGPT-3 Large and LLaMA tokenizer in fairy tales of training corpus, depending on the number of plot phrases, are shown in Appendix B.

## 5 Experimental Setup

### 5.1 Models and Libraries

Keywords used in plot events were extracted from texts using the *yake* and *rutermextract* libraries. The initial word forms for calculating the Jaccard coefficient were determined using the pymorphy2 library [9].

Text generation experiments were carried out using three models:

- ruGPT-3 Large[6] language model (760 million parameters), which is the Russian-language version of the GPT-2 model [17];
- ruAlpaca 13B[7], Russian LLaMA-based model, trained on the Russian instructions dataset ru_turbo_alpaca[8];
- Saiga 13B[9], Russian LLaMA-based chatbot, trained on the Russian instructions dataset ru_turbo_saiga[10].

---

[6] https://huggingface.co/sberbank-ai/rugpt3large_based_on_gpt2.
[7] https://huggingface.co/IlyaGusev/llama_13b_ru_turbo_alpaca_lora.
[8] https://huggingface.co/datasets/IlyaGusev/ru_turbo_alpaca.
[9] https://huggingface.co/IlyaGusev/saiga_13b_lora.
[10] https://huggingface.co/datasets/IlyaGusev/ru_turbo_saiga.

In the experiments, fairy tales were generated according to a given sequence of events that determines the plot of the fairy tale. The top-$p$ sampling decoding strategy with parameter $p = 0.85$ was used as a decoding strategy in our method to obtain connecting sequences of tokens. The threshold value of the proportion of words from the guide phrase present in the text for its insertion into the text was equal to 0.5.

The values of the weight coefficients in formula (5) were determined empirically based on the analysis of the generated connecting sequences. The coefficients took the values $w_{prob} = 0.65$ and $w_J = 0.35$. The probability of following the guide phrase turned out to be more significant, and due to the $w_J$ coefficient, the connecting sequence that was closest in content to the guide phrase was ranked first.

The length of connecting sequences was 10 tokens. Experiments were also carried out for windows ranging in size from 5 to 30 tokens. According to the results of the experiments, a small window of connecting sequences had a better effect on shifting the content of the generated text towards the plot phrase than a large window.

The maximum length of the generated tale (in tokens) depended on the number of plot phrases and was equal to the average number $+10\%$ of the tokens (see Appendix B).

## 5.2   Competing Methods

The proposed method was compared with three methods of CTG:

1. Constrained beam search (CBS). It was used as the baseline of CTG. Its implementation was taken from the transformers library from HuggingFace[11]. Plot phrases were tokenized and used as a list of restrictions. The following prompts were used as an input for the models:
   - ruGPT-3: "Однажды" ("Once");
   - ruAlpaca: "Задание: Сочини длинный рассказ.\nВыход:" ("Task: Compose a long story.\nOutput:");
   - Saiga: "<s>user\nСочини длинный рассказ.</s>\n<s>bot\n" ("<s>user\nCompose a long story.</s>\n<s>bot\n").

   The number of beams varied from 6 to 8 to generate different stories. A prohibition on the repetition of 5-grams was also established. The length of the generated fairy tale was chosen similarly to our method.
2. Few-shot learning (FS). The following prompts were used for the models:
   - ruGPT-3: "Compose text with key phrases:\nPlot: {plot phrase 1}, ..., {plot phrase n}.\nText: {the text of fairy tale} ### Plot: {plot phrase 1}, ..., {plot phrase n}.\nText: {the text of fairy tale}";
   - ruAlpaca: "Task: Write a long story, making sure to include the following key phrases.\nInput: {plot phrase 1}, ..., {plot phrase n}\nOutput: {the text of fairy tale}";

---

[11] https://huggingface.co/blog/constrained-beam-search.

    – Saiga: "<s>user\nWrite a long story, making sure to include the following key phrases: {plot phrase 1}, ..., {plot phrase n}</s>\n<s>bot\n{the text of fairy tale}".

The number of fairy tales input to the model depended on the estimated maximum length of the generated text so that the total input sequence fit into 2048 tokens. The range of the number of input training examples is from 1 to 5, most often 3. To generate fairy tales, sampling was used with parameters $p = 0.85$ and $k = 10$. The length of the generated fairy tale was chosen similarly to our method.

3. Prompt engineering (PE). The following prompts were used for the models:
    – ruGPT-3: "Write a long story, making sure to include the following key phrases: {plot phrase 1}, ..., {plot phrase n}.\nOnce";
    – ruAlpaca: "Task: Write a long story, making sure to include the following key phrases.\nInput: {plot phrase 1}, ..., {plot phrase n}\nOutput:";
    – Saiga: "<s>user\nWrite a long story, making sure to include the following key phrases: {plot phrase 1}, ..., {plot phrase n}</s>\n<s>bot\n".

When generating texts, the same parameters as for FS were used. The length of the generated fairy tale was chosen similarly to our method.

### 5.3   Evaluation Metrics

The quality of the generated texts was evaluated using four automatic [23, 31] and three human-centric evaluation measures:

- perplexity (PPL) – is a metric to measure how well the language probability model predicts a sample. It is usually calculated as the exponential mean of the negative log-probability per token in the language model. We calculated perplexity using the ruGPT-3 Medium[12] language model (350 million parameters);
- repetition (Rep) evaluates the proportion of repeated 4-grams in the text, where the tokens belong to the vocabulary of the tested model;
- word inclusion coverage (Cov) shows the percentage of plot words included in the generated text. Plot and generated words are lemmatized;
- self-BLEU-5 (S-BLEU) evaluates the syntactic diversity of a given set of texts. It is defined as the average overlap between all generated texts;
- coherence (Coh) – whether the story is consistent in terms of causal relationships;
- relevance (Rel) – the events in the story unfold in accordance with the storyline;
- interestingness (Intr) – how the user likes the story, whether it is interesting.

---

[12] https://huggingface.co/sberbank-ai/rugpt3medium_based_on_gpt2.

# 6  Results and Discussion

All measures were calculated for fairy tales generated from 25 storylines of test corpus. For each storyline, two fairy tales were generated. A total of 50 tales were generated by each method.

Average values of automatic quality scores are shown in Table 1. The values of the Cov measure show that proposed method ensures that more than 93% of the words from the storyline events appear in the text. The texts generated by our method met the requirement of matching the storyline to the best extent. The CBS method also has rather high values of the Cov measure, however, all plot phrases are usually found in the first sentence of the generated text, and the fairy tale does not correspond to the given plot.

**Table 1.** Automatic and human-centric quality scores for generation methods.

| Model | Method | Automatic scores | | | | Human-centric scores | | |
|---|---|---|---|---|---|---|---|---|
| | | ↓ PPL±Std | ↓ Rep (%) | ↑ Cov (%) | ↓ S-BLEU | ↑ Coh | ↑ Rel | ↑ Intr |
| ruGPT-3 | CBS | $6.8 \pm 2.5$ | **4.53** | 81.86 | 0.093 | 1.78 | 2.94 | 1.60 |
| | FS | $9.9 \pm 6.1$ | 16.40 | 43.49 | **0.014** | 2.32 | 1.88 | 2.42 |
| | PE | $\mathbf{2.5 \pm 1.8}$ | 74.26 | 58.60 | 0.070 | 2.10 | 2.53 | 1.82 |
| | Our | $7.2 \pm 1.4$ | 18.33 | **93.90** | 0.067 | **2.89** | **4.44** | **2.85** |
| ruAlpaca | CBS | $15.7 \pm 9.3$ | **3.27** | 95.59 | 0.143 | 2.74 | 3.19 | 2.90 |
| | FS | $\mathbf{9.5 \pm 5.7}$ | 31.67 | 83.28 | 0.145 | **3.53** | 3.85 | **3.33** |
| | PE | $20.5 \pm 8.7$ | 12.04 | 79.82 | **0.087** | 2.83 | 3.46 | 2.90 |
| | Our | $28.8 \pm 19.3$ | 5.75 | **96.61** | 0.118 | 3.11 | **4.47** | 2.99 |
| Saiga | CBS | $11.8 \pm 5.6$ | 5.09 | 95.92 | 0.196 | 2.94 | 3.15 | 3.01 |
| | FS | $\mathbf{7.4 \pm 3.4}$ | 27.27 | 86.40 | 0.136 | 4.13 | 4.00 | 4.14 |
| | PE | $8.5 \pm 1.5$ | **3.64** | 74.42 | 0.231 | **4.29** | 3.79 | **4.22** |
| | Our | $23.6 \pm 7.2$ | 6.28 | **96.06** | **0.084** | 3.28 | **4.56** | 3.24 |

The lowest PPL value, equal to 2.5, has the PE method with the ruGPT-3 model. With the ruAlpaca and Saiga models, the FS method has the lowest PPL values of 9.5 and 7.4, respectively. A lower perplexity value corresponds to a better model, the generated texts look more natural. The method of controlled generation proposed by us shows almost the largest value of PPL. The increase in perplexity indicates that the control process is "unnatural" for the model. This causes the model to be more "surprised" by the tokens observed in the text.

The S-BLEU measure has the lowest value for FS with ruGPT-3 model. The texts generated by this method turned out to be the most syntactically diverse. With the Saiga model, the method we developed allowed us to obtain the most diverse texts among all methods used with this model. For the PE method, the chatbot responses were the most similar to each other.

To calculate human-centric measures, the generated texts were evaluated by three annotators. The assessment was carried out on a 5-point Likert scale

(1 – the worst, 5 – the best). Inter-annotator agreement was measured using the Spearman coefficient [1]. The value of this coefficient for the "coherence" criterion was 0.75, "relevance" – 0.74, "interestingness" – 0.71. The values, which are greater than 0.7 indicate high annotator agreement [18]. Average values of human-centric scores are shown in Table 1.

According to the annotators, the proposed method allowed us to generate texts that were most relevant to the storyline (Rel measure). Our method performed best when using the relatively small ruGPT-3 model, receiving the highest score on all three human evaluation measures. Although ruGPT-3 model generated less coherent and interesting texts than ruAlpaca and Saiga.

The statistical characteristics of the generated texts calculated using the GEM-metrics library[13] are shown in Appendix C: avg length - the average length of texts (in words); vocab size - the number of different words; distinct-n - the ratio of distinct n-grams over the total number of n-grams. Analysis of the results showed that the FS method with ruGPT-3 model, on average, generated tales 3 times shorter than the other methods. When generating longer tales, the first tale was often interrupted and a new tale began. Similarly, PE method with Saiga model, on average, generated tales 2 times shorter. Being a chatbot, in response to a request to compose a tale, this model generated small but complete tales that well corresponded to the given plot. The PE method with the ruGPT-3 model had low distinct-n score. Tales often had many repeating fragments.

## 7 Conclusion

The proposed method allows generating stories in accordance with a sequence of guide phrases that describe some of the key events in the story and consist of several words. The method uses a generative language model to estimate the probability of following a guide phrase after all subsequences of connected sequences of tokens generated by the model between prompt and guide phrase. The method selects the subsequence with the highest probability, prompting the model to shift the content of the text towards the guide phrase. Experiments with LLMs carried out using the Russian-language corpus of fairy tales with extracted storylines showed that the proposed method provides a high proportion of story words (from 93.90% to 96.61% in Cov) in the text and high correspondence of the text to the storyline (from 4.44 to 4.56 in Rel). The method has low S-BLEU values, which means that it generates various texts. At the same time, the method increases the perplexity of the text. Plot phrases that appear in the text cause more surprise for the model. The method performed best with the ruGPT-3 model, receiving the highest human scores among other methods. For the larger models, it can be used as a complement to other methods to increase the relevance of text to a given storyline.

**Acknowledgments.** This work was supported by Russian Science Foundation, project № 23-21-00330, https://rscf.ru/en/project/23-21-00330/.

---

[13] https://github.com/GEM-benchmark/GEM-metrics.

## Appendix A

(See Figs. 4 and 5).

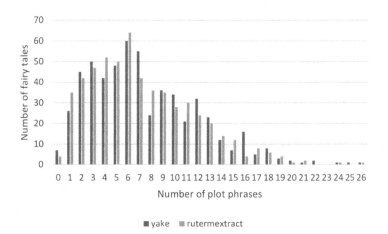

**Fig. 4.** Distribution of the number of plot phrases.

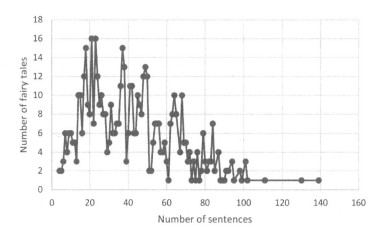

**Fig. 5.** Number of sentences in fairy tales.

## Appendix B

The first column of Table 2 contains the number of phrases in the plot, the second - the number of fairy tales with such a number of phrases, the third - the share of the total number of fairy tales in the training corpus. The fifth and sixth columns contain statistics on the number of tokens received using the ruGPT-3 Large and LLaMA tokenizer in tales of training corpus, depending on the number of plot phrases.

**Table 2.** Distribution of the number of phrases in the plot.

| # Plot phrases | # Tales in the training corpus | Share of the total number of tales (%) | # Tales in the test corpus | Avg number of tokens ruGPT-3 | LLaMA |
|---|---|---|---|---|---|
| 1 | 31 | 5.46 | 1 | 230.9 | 448.7 |
| 2 | 48 | 8.45 | 2 | 238.9 | 527.8 |
| 3 | 53 | 9.33 | 2 | 344.3 | 614.3 |
| 4 | 56 | 9.86 | 3 | 308.9 | 665.6 |
| 5 | 107 | 18.84 | 5 | 476.4 | 789.7 |
| 6 | 185 | 32.57 | 8 | 796.0 | 1,298.6 |
| 7 | 80 | 14.08 | 4 | 1,150.1 | 1,758.8 |
| 8 | 2 | 0.35 | 0 | 1,726.0 | 2,399.5 |

# Appendix C

Table 3 contains statistical characteristics of generated texts: avg length - the average length of texts (in words); vocab size - the number of different words; distinct-n - the ratio of distinct n-grams over the total number of n-grams.

**Table 3.** Statistical characteristics of generated texts.

| Model | Method | Avg length | Vocab size | Distinct-1 | Distinct-2 | Distinct-3 |
|---|---|---|---|---|---|---|
| ruGPT-3 | CBS | 447 | 3,149 | 0.11 | 0.49 | 0.85 |
| | FS | 158 | 1,998 | 0.19 | 0.57 | 0.77 |
| | PE | 504 | 1,289 | 0.05 | 0.16 | 0.23 |
| | Our | 577 | 1,521 | 0.12 | 0.48 | 0.75 |
| ruAlpaca | CBS | 571 | 4,683 | 0.16 | 0.57 | 0.84 |
| | FS | 461 | 2,878 | 0.12 | 0.39 | 0.56 |
| | PE | 575 | 5,480 | 0.19 | 0.57 | 0.80 |
| | Our | 581 | 5,714 | 0.20 | 0.59 | 0.84 |
| Saiga | CBS | 561 | 4,344 | 0.15 | 0.55 | 0.83 |
| | FS | 502 | 3,133 | 0.12 | 0.41 | 0.60 |
| | PE | 282 | 2,685 | 0.19 | 0.58 | 0.82 |
| | Our | 578 | 3,710 | 0.20 | 0.63 | 0.88 |

# References

1. Amidei, J., Piwek, P., Willis, A.: Agreement is overrated: a plea for correlation to assess human evaluation reliability. In: Proceedings of the 12th International Conference on Natural Language Generation, pp. 344–354 (2019)
2. Brown, T.B., Mann, B., Ryder, N., Subbiah, M., Kaplan, J., et al.: Language models are few-shot learners. Adv. Neural. Inf. Process. Syst. **33**, 1877–1901 (2020)
3. Campos, R., et al.: YAKE! Keyword extraction from single documents using multiple local features. Inf. Sci. J. **509**, 257–289 (2020)
4. Cavazza, M., Pizzi, D.: Narratology for interactive storytelling: a critical introduction. In: Göbel, S., Malkewitz, R., Iurgel, I. (eds.) TIDSE 2006. LNCS, vol. 4326, pp. 72–83. Springer, Heidelberg (2006). https://doi.org/10.1007/11944577_7
5. Dathathri, S., et al.: Plug and play language models: a simple approach to controlled text generation. arXiv preprint arXiv:1912.02164 (2020)
6. Fan, A., Lewis, M., Dauphin, Y.: Hierarchical neural story generation. arXiv preprint arXiv:1805.04833 (2018)
7. Hejazi, H.D., Khamees, A.A., Alshurideh, M., Salloum, S.A.: Arabic text generation: deep learning for poetry synthesis. In: Hassanien, A.-E., Chang, K.-C., Mincong, T. (eds.) AMLTA 2021. AISC, vol. 1339, pp. 104–116. Springer, Cham (2021). https://doi.org/10.1007/978-3-030-69717-4_11
8. Keskar, N.S., McCann, B., Varshney, L., Xiong, C., Socher, R.: CTRL - a conditional transformer language model for controllable generation. arXiv preprint arXiv:1909.05858 (2019)
9. Korobov, M.: Morphological analyzer and generator for Russian and Ukrainian languages. In: Khachay, M.Y., Konstantinova, N., Panchenko, A., Ignatov, D.I., Labunets, V.G. (eds.) AIST 2015. CCIS, vol. 542, pp. 320–332. Springer, Cham (2015). https://doi.org/10.1007/978-3-319-26123-2_31
10. Lester, B., Al-Rfou, R., Constant, N.: The power of scale for parameter-efficient prompt tuning. In: Proceedings of the 2021 Conference on Empirical Methods in Natural Language Processing, pp. 3045–3059 (2021)
11. Li, Y., Li, K., Ning, H., Xia, X., Guo, Y., Wei, C., Cui, J., Wang, B.: Towards an online empathetic chatbot with emotion causes. In: Proceedings 44th International ACM SIGIR Conference on Research and Development in Information Retrieval, pp. 2041–2045 (2021)
12. Li, X.L., Liang, P.: Prefix-tuning: optimizing continuous prompts for generation. In: Proceedings of the 59th Annual Meeting of the Association for Computational Linguistics and the 11th International Joint Conference on Natural Language Processing, pp. 4582–4597 (2021)
13. Meister, C., Vieira, T., Cotterell, R.: If beam search is the answer, what was the question? In: Proceedings of the 2020 Conference on Empirical Methods in Natural Language Processing, pp. 2173–2185 (2020)
14. Mirowski, P., Mathewson, K.W., Pittman, J., Evans, R.: Co-writing screenplays and theatre scripts with language models: an evaluation by industry professionals. arXiv preprint arXiv:2209.14958 (2022)
15. Pascual, D., Egressy, B., Meister, C., Cotterell, R., Wattenhofer, R.: A plug-and-play method for controlled text generation. In: Findings of the Association for Computational Linguistics, pp. 3973–3997 (2021)
16. Prabhumoye, S., Black, A.W., Salakhutdinov, R.: Exploring controllable text generation techniques. In: Proceedings of the 28th International Conference on Computational Linguistics, pp. 1–14 (2020)

17. Radford, A., Wu, J., Child, R., Luan, D., Amodei, D., Sutskever, I.: Language models are unsupervised multitask learners. In: OpenAI Blog, vol. 1, no. 8 (2019)
18. Rosenthal, J.A.: Qualitative descriptors of strength of association and effect size. J. Soc. Serv. Res. **21**(4), 37–59 (1996)
19. Škrlj, B., Repar, A., Pollak, S.: *RaKUn*: rank-based keyword extraction via unsupervised learning and meta vertex aggregation. In: Martín-Vide, C., Purver, M., Pollak, S. (eds.) SLSP 2019. LNCS (LNAI), vol. 11816, pp. 311–323. Springer, Cham (2019). https://doi.org/10.1007/978-3-030-31372-2_26
20. Touvron, H., et al.: LLaMA: open and efficient foundation language models. arXiv preprint arXiv:2302.13971 (2023)
21. Vaswani, A., et al.: Attention is all you need. In: Proceedings of the 31st Conference on Neural Information Processing Systems (NeurIPS), vol. 30, pp. 6000–6010 (2017)
22. Vychegzhanin, S., Kotelnikov, E.: Collocation2Text: controllable text generation from guide phrases in Russian. In: Computational Linguistics and Intellectual Technologies: Proceedings of the International Conference "Dialogue-2022", vol. 21, pp. 564–576 (2022)
23. Welleck, S., Kulikov, I., Roller, S., Dinan, E., Cho, K., Weston, J.: Neural text generation with unlikelihood training. In: Proceedings of the 8th International Conference on Learning Representations, pp. 1–18 (2020)
24. Wiseman, S., Shieber, S., Rush, A.: Challenges in data-to-document generation. In: Proceedings of the 2017 Conference on Empirical Methods in Natural Language Processing, pp. 2253–2263 (2017)
25. Yang, K., Tian, Y., Peng, N., Klein, D.: Re3: generating longer stories with recursive reprompting and revision. In: Proceedings of the 2022 Conference on Empirical Methods in Natural Language Processing (EMNLP 2022), pp. 4393–4479 (2022)
26. Yao, L., Peng, N., Weischedel, R., Knight, K., Zhao, D., Yan, R.: Plan-and-write: towards better automatic storytelling. In: Proceedings of the AAAI Conference on Artificial Intelligence, vol. 33, no. 01, pp. 7378–7385 (2019)
27. Yuan, A., Coenen, A., Reif, E., Ippolito, D.: Wordcraft: story writing with large language models. In: 27th International Conference on Intelligent User Interfaces, pp. 841–852 (2022)
28. Zehtab-Salmasi, A., Feizi-Derakhshi, M.R., Balafar, M.A.: FRAKE: fusional real-time automatic keyword extraction. arXiv preprint arXiv:2104.04830 (2021)
29. Zhang, H., Song, H., Li, S., Zhou, M., Song, D.: A survey of controllable text generation using transformer-based pre-trained language models. arXiv preprint arXiv:2201.05337 (2022)
30. Zhang, Y., Wang, G., Li, C., Gan, Z., Brockett, C., Dolan, B.: POINTER: constrained progressive text generation via insertion-based generative pre-training. In: Proceedings of the 2020 Conference on Empirical Methods in Natural Language Processing (EMNLP), pp. 8649–8670 (2020)
31. Zhu, Y., et al.: Texygen: a benchmarking platform for text generation models. In: The 41st International ACM SIGIR Conference on Research & Development in Information Retrieval, pp. 1097–1100 (2018)

# Automatic Detection of Dialectal Features of Pskov Dialects in the Speech of Native Speakers

Ekaterina Zalivina[✉]

The National Research University Higher School of Economics, Moscow, Russia
zalivina01@gmail.com

**Abstract.** In this work, we made an attempt to solve the problem of detecting dialectal features in the speech of informants who have dialectal features characteristic of the Pskov dialects found on the territory of the Opochetsky district of the Pskov region and the Zapadnodvinsky district of the Tver region. The task is divided into two parts: speech recognition and features detection. First of all, we developed a model, the functionality of which include transcription of the interview collected during expeditions to the Pskov dialects. In order to find the most suitable architecture for the task, we compare several recently proposed systems. The obtained transcriptions of the interview can be used to update and expand the dialect corpus. The next step was to develop a system that can help the researcher pay attention to possible dialectal features in the annotation. The study considered several approaches to detecting dialect features, and identified the advantages and disadvantages of each. Ultimately, we developed a unified algorithm that incorporates the best of the approaches considered, it takes a .wav file as input and returns .TextGrid and .eaf files with annotations and detected dialect features.

**Keywords:** Automatic speech recognition · Russian dialects · Detecting dialect features

## 1 Introduction

With the development of new speech recognition architectures researchers also faced the problem of a shortage of transcribed audio data. Over time datasets have grown in volume, now they have a wide variety of data: multilingual datasets including low-resource languages have been collected. Despite the fact that in recent years oral corpora of Russian dialect speech has been collected, they are still not used for speech recognition tasks. The purpose of this work is to develop a model for transcribing dialect speech, namely Pskov dialects, as well as to provide researchers with a tool to identify dialectal features characteristic

---

I express my gratitude to my supervisor R. V. Ronko. I would also like to thank A.S. Vyrenkova for comments and corrections.

D. I. Ignatov et al. (Eds.): AIST 2023, LNCS 14486, pp. 170–184, 2024.
https://doi.org/10.1007/978-3-031-54534-4_12

of these dialects in the speech of informants. The paper discusses approaches to fine-tuning models in conditions of a limited amount of dialect data, identifies the best of them, and, based on the results, proposes a new approach for future research. Although we consider only Pskov dialects in this paper, this approach can be scaled to other dialects for which oral corpus data are collected.

## 2    Related Work

Dialect speech recognition has already been proposed, for example, for English and Arabic dialects [1,2]. For dialects of the English language two approaches were used: dialect-dependent and dialect-independent models. In the case of the first approach the original English model (American English) was chosen, then the model was trained on dialect data (for example, Australian English). A language model trained on dialect data was additionally used for decoding. The second approach was to train the model on a mixture of all dialects (Australian, Canadian, British). For Australian data, the first approach showed the best result, for other dialects, the model trained on the entire collected corpus turned out to be better [1]. For Arabic dialects, many models have been presented and trained for various dialects: Algerian, Moroccan, Tunisian, Egyptian, etc. Common to such works is the preference for MFCC as a feature extraction method, acoustic models are presented in a greater variety, for example, Hidden Markov Model, LSTM, Mixed Gaussian Model [2].

A characteristic feature of research on the automatic detection of dialect features is the focus on textual data. For example, in [3], the model was trained to extract 22 dialectal features (e.g., omitting articles or prepositions, lack of agreement) based on Indian English corpus data. The paper used minimal pair approaches and also assumed regular expressions, but the second approach was used for only five of the twenty-two features. However, within the framework of this approach, the data format does not allow detecting some dialect features. For example, it will be more productive to use data in phonetic transcription to highlight dialectal phonetic realizations. In addition, detecting syntactic features can be improved by using data from standard orthography.

## 3    Data

### 3.1    Dialect Classification

According to the classification [4], the Opochetsky dialect and the Zapadnodvin-sky dialect belong to the Western Central Russian dialects with akanye, but the second one is distributed on the border with the Smolensk dialects, which belong to the Southern dialect. According to the classification of N. Pshenichnova [5], the considered dialects belong to the dialects of the West Russian dialect type (P), namely, to the North-Western dialect type (P2), but they are in different groups. The dialect of the villages of the Opochetsky district belongs to the dialects of the South-Pskov dialect type (P211), the dialect of the villages of the Zapadnodvinsky district belongs to the heterogeneous dialects of the fourth rank among the dialects P2P (P2PP).

Thus, the considered dialects are classified quite closely. In [4] they are included in one Pskov group of dialects. N. Pshenichnova [5] suggests a more fractional classification: at the second level of classification dialects are already divided into different subgroups, as they are found to have some dialectal features that have different realizations. The closeness of these dialects makes it possible to expect better results from the same methods.

## 3.2   Dialect Features

To identify speech and highlight features, materials from several years of expeditions to Pskov dialects were analyzed: expeditions to the Opochetsky district of the Pskov region and the Zapadnodvinsky district of the Tver region. A grammatical sketch of the Opochetsky dialect is presented in [6,7]. The articles briefly describe the phonetic, morphological and syntactic systems, and also provide some transcripts of interview recordings. A grammatical sketch of the Zapadnodvinsky dialect is contained in the work [8]. We present a brief comparison of the realizations of dialect features that are described in the articles and materials from expeditions to Pskov dialects. Further in the work, we will consider approaches to the detection of these dialect features. This list of features is not exhaustive, this will be discussed in Sect. 4.2.

- **Realizations of the phoneme /v/**
  In the dialect of the Zapadnodvinsky district, realizations of [u] and [w] were found, less often [v]/[f] in prepositions, [w], [v]/[f] inside the word form. Allophones [v]/[f] predominate in prepositions in Opochechetsky dialect, [w] occurs, [u] almost does not occur. Inside the word form, it occurs mainly [v]/[f], but the allophone [w] is also rare. Statistically significant differences are observed in the intervocalic position and in the position between consonant and vowel [9].
- **The unstressed vocalism**
  In both dialects, there is variability in the first prestressed position, including the implementation of [a] after palatalized consonants. Vowels [ɛ], [ě], [a] positioned after palatalized consonants (and before the stressed syllable) are pronounced as [a] regardless of which vowel is stressed, e.g., [dʲaˈrʲɛvnʲa] 'village-NOM.SG', [zʲaˈmlʲa] 'ground-NOM.SG'. In standard Russian, vowels [ɛ], [ě], [a] positioned after palatalized consonants (and immediately before the stressed syllable) are pronounced as [i]: [dʲiˈrʲɛvnʲ a] 'village-NOM.SG', [zʲiˈmlʲa] 'ground-NOM.SG'. As noted in the work [6], perhaps it depends on the pace of speech, phrasal conditions of utterance and individual characteristics.
- **Instrumental plural forms**
  In both dialects, it is found that the forms of the dative case and instrumental coincide while in standard Russian, the forms have different affixes: *мололи (лён) рукам* 'ground (flax) by hand' instead of *мололи (лён) руками* - in standard Russian.

– **Forms of deictic pronouns and adverbs**
  In the Opochetsky dialect, as well as in the Zapadnodvinsky dialect, there are forms of deictic pronouns that have been influenced by the adjectival declension: *тая* instead of *та* in standard Russian, *этая* instead of *эта* in standard Russian and others.
– **Third person verb forms in present tense**
  The systems of third person verb affixes in present tense are different. The affixes of plural form in these dialects are the same. The main differences are singular forms: in the Opochetsky dialect, the verbs of the I conjugation, regardless of the stress, show variability between the affixes without the velarized final [t] and the literary variant (with velarized final [t]). Verbs of the second conjugation have three types of affixes: with a final palatalized [t], without a final [t] and a literary variant. In the Zapadnodvinsky dialect in these forms there are only variants of palatalized final [t] and standard Russian variant. Dialectal variants of affixes in both dialects prevail among the older generation; among younger informants, they are replaced by a standard variant.
– **Reflexive verbs**
  The competition of syllabic (*-ся*) and non-syllabic (*-сь*) variants after a vowel in reflexive verbs is found in both studied dialects, while there are only non-syllabic variant in standard Russian: *боюсь* (standard) and *боюся* (dialect).
– **Participle forms**
  In the Zapadnodvinsky dialect, the active past participles are formed according to the following model: *-дчи* for verbs of motion, *-ши* for all other verbs. Such forms are more often formed from intransitive verbs, but they are also allowed from transitive ones, but much less often. In Opochetsky dialects, the suffix *-лши* occurs, which is attached to the stem with a vowel. Such forms of participles are not found in the Russian standard language.

– **Lack of agreement of passive participles in the presence of a auxiliary verb and in the absence**
  In the Opochetsky dialect, the auxiliary verb agrees in gender and number with the subject; reduction of the final vowel is rare, but this is possible only in the presence of a non-canonical subject or in impersonal sentences. In the Zapadnodvinsky dialect, an implication is derived: only if the participle is agreed, then the auxiliary verb will also be agreed in gender and number. In the materials of the Opochetsky and Zapadnodvinsky dialect, there is also a lack of agreement between the subject and the passive participle in the absence of a auxiliary verb.

– **Object purpose construction with verbs of motion and a preposition *в* 'in'**
  In the materials of both dialects, there are object purpose constructions "X in Y", where X is a verb of movement, and Y are nouns that name objects of forest gathering: *ходить в ягоды* 'go to the forest to pick berries', *пойти в клюкву* 'go to the forest to pick cranberries'.

- **Marking the Addressee of Speech**
  Such constructions are found in both dialects. In general, they are found on
  the territory of the western part of the Central Russian dialects and the South
  Russian dialect. Constructions have can express an impulse to action or the
  meaning of scolding, disapproval [10].

### 3.3   Preprocessing of Corpus Data

The complete collection of interviews is presented in the corpus of Opochetsky
dialects [11], where the annotation is presented in standard orthography [12].
Thus, the corpus does not reflect the phonetic features of the dialect, dialectical affixes are replaced with standard variant. «The dialectical features of the
composition of roots have been saved (particularities of the phonemic composition of roots, phonemes, and suffixes), along with the syntax and, in part,
dialectical features of control. Among the preserved grammatical features are
the participle forms -*ш*, -*вш*, and -*лш* which in this dialect have acquired resul-
tative and perfective semantics». Recordings of interviews and their transcripts
from Zapadnodvinsky district are also in the corpus [10], the methodology for
annotation is similar to that presented in the corpus of Opochetsky dialects.

The corpus data were filtered: the dataset did not include interviewers' repli-
cas. Such data are characterized by noisiness, the presence of breaks in words and
phrases, the presence of several speakers at the same time. The total duration of
the dataset for the villages of the Zapadnodvinsky district is 6.67 h (7922 utterances, the average audio duration is 3.03 s), for the villages of the Opochetsky
district - 2.64 h (2256 utterances, the average audio duration is 4.22).

As part of the work with models that were trained to detect dialect features,
several manual annotations were made. For methods that are based on the classification of the entire sentence each fragment was marked for the presence of a
dialectal feature (0 marked the absence, 1 - the presence). For token classification methods, a markup, similar to IOB2, was used. This kind of format is used
to train models, highlighting entities both named [13] and unnamed [14]. Each
annotation was divided into tokens and one of the tags was manually assigned
to each token. Tags are of two types: 1) X-Y, where X is either B (Beginning)
or I (Inside) and Y is the language level to which the dialect feature belongs.
For example, a token containing an implementation of the phoneme /v/ as [w]
was marked as B-PHON; 2) The O tag marked the absence of dialectal features
within the token.

## 4   Methods

### 4.1   Speech Recognition

In this work, several models were used, all of them were already pretrained on
the data of the standard Russian language. It is assumed that the combination of
the pretrained model on the data of standard Russian and fine-tuned with Pskov

dialect materials can show a high result as in [1] (dialect-dependent method). The use of this method has been also demonstrated in [15]. In this study, data from standard Japanese and six dialects was used for speech recognition and dialect identification tasks. The use of such a dataset led to an increase in the quality of recognition of both standard Japanese and its dialects.

**Wav2vec2.** For speech recognition task the wav2vec2 model pretrained on data from the Golos dataset [16] was chosen. A word error rate of 12.12 was achieved by this model on Common Voice data [22]. The basic version of wav2vec [17] and the later version of wav2vec2 [18] were evaluated on the TIMIT dataset [19] in which transcriptions are presented as phonemes. Thus, it is assumed that the quality of dialect speech recognition may be higher, since the characteristic features of the considered dialect include, among other things, different realizations of the phoneme /v/ [9], the coincidence of non-close vowel phonemes in the first prestressed syllable after a palatalized consonant, which is not observed in standard Russian.

After fine-tuning the model on half of the data a brief analysis of the results was done. Common mistakes include: combining two tokens into one (*онавючилась* 'she had learned' instead of *она вючилась*), splitting one token into two parts (*это го* 'that' instead of *этого*), inserting extra characters (*четверьо* 'four' (collective) instead of *четверо*). The transcripts also revealed realizations of dialectal features that negatively affected quality metrics: reflection of the system I and II conjugation verbs with unstressed affixes (*помнют* '(they) remember' instead of *помнят*), realizations of the phoneme /v/ (*часоуже* 'hours' instead of *часов же*, *девоки* 'girls' instead of *девки*), The unstressed vocalism after palatalized consonant (*дявали* '(they) gave' instead of *надевали*), palatalized affixes in the third person present tense (*вырастить* 'will grow up' instead of *вырастет*). As a result of the analysis, it was suggested that some of the errors can be corrected by a spellchecker. Since in the end it is supposed to combine the best approaches into one tool, it was decided to use a spellchecker instead of a language model, since the inclusion of another model will affect the performance of the final tool that is intended to be used in field research. Moreover, language models can replace dialectal syntactic features, such as the lack of agreement between the passive participle and the target, which need to be preserved in transcription. To choose a spellchecker, the Precision, Recall and F1-score metrics were taken into account, in the work [20] a comparison was made on these metrics on the SpellRuEval dataset. After fine-tuning the wav2vec2 model on dialect data, Yandex.Speller was applied to the transcripts, which gave a strong increase in quality as expected.

With the help of a spellchecker it was possible to correct spelling errors that occurred during recognition. In some cases corrections of the spellchecker did not agree with the annotation rules adopted in the corpus. For example, changing the participial forms with affixes -ши / -чи with literary forms: replacing *оставши* 'have stayed' with *оставшиеся*, *замерзши* 'have freezed over' with *замерзшие*, *пришодчи* 'have come' with *пришедшие*, *одичавши* 'have gone

**Table 1.** Percentage of spellchecker corrections on recognized annotations, total number of corrections was 571.

| Correct | | | Incorrect | Other |
|---|---|---|---|---|
| insertion or deletion | within the token | insertion/deletion and within the token | | |
| 8.06% | 37.30% | 0.70% | 22.07% | 31.87% |

wild' with *одичавшие*. Spellchecker corrections were classified: correction within the token, fixing gaps (insertion/deletion), correction that replaced the correct token with an incorrect one or an incorrect one was replaced with an incorrect one, and a block Other, which included editions for tokens that were partially recognized model or recognized without the possibility of restoring the correct word form. As a result, the following distribution was obtained (Table 1).

**QuartzNet.** The second approach to the speech recognition task was the QuartzNet $15 \times 5$ architecture. This architecture was trained on the Golos dataset [21] and a word error rate of 8.06 was achieved on Common Voice data [22] with the use of the language model. One of the features of this model is the use of the transfer learning [21].

**Conformer-Transducer.** The model (STT Ru Conformer-Transducer Large[1]) for the Russian language was trained on the Common Voice (28 h) and Golos (1181 h), as well as LibriSpeech [23] (92 h) and SOVA (75 h) subcorpuses. As a result, high WER results were obtained (the best for some Russian-language datasets): Common Voice - 4.0, Golos - 2.7 (for the Crowd subcorpus), LibriSpeech - 12.0. As part of the fine-tuning of the model on dialect data, parameters similar to those of QuartzNet were changed. During the fine-tune process, it was found that the model overfits faster on such a small amount of data comparing with previous approaches.

### 4.2 Dialect Features Detection

**Rule-Based Method.** First of all, to extract dialect features from text fragments, a method that operates with rules was developed. Initially, a list of dialect features was taken from a grammatical essay [8], as well as some features described during the expeditions. In the process of forming the rules some dialectal features were omitted such as third person forms of the lexeme *мочь* (can). In the dialect this lexeme in the meaning of 'to be able' has a palatalized affix like other verbs of both conjugations. If the lexeme acts as an indicator of epistemic modality then the verb forms do not have consonant at the end.

---

[1] https://catalog.ngc.nvidia.com/orgs/nvidia/teams/nemo/models/stt_ru_ conformer_transducer_large.

The rule-based method is not applicable to this type of dialect features, the selection of all lexemes could lead to deterioration in the quality of the tool. The following order of rules was defined for annotation processing: phonetic, morphological, syntactic. If one token, for example, has both phonetic and morphological features then it gets only tag MORPH.

For each rule the function was written; here we present some examples of rules. To determine the first prestressed position and the realization of phonemes in it, as well as the palatalization of the previous consonant, two libraries were used. To obtain transcriptions in standard Russian, the wiktionaryparser library was used, this library provides transcriptions in IPA notation. We also use the ipapy library to parse transcriptions. We separate the prestressed part and with the help of the ipapy library determine the backness and height of the prestressed vowel and the presence of a palatalized consonant before this vowel. Morphological dialectal features were extracted based on morphological parsing, which was obtained using the pymorphy2. For example, to detect verb forms in the third person present tense, the following conditions were used: POS (Part of speech) = VERB, Person = 3, Tense = Pres. Syntactic feature extraction was implemented using the natasha library, the output of which is in ConLL format. This format contains forms, lemmas, indices for each token and indices of the vertix. Such parsing, for example, has been used to detect mismatch between the controller (the passive participle in this case) and the target. Initially, a participle was detected in a text fragment, then for this participle, all vertices that were children of the vertex. Among the children, those that had the relation nsubj:pass (passive nominal subject) and had a gender category in their set of grammatical categories were selected. The result was a list of indices at the positions of which the required syntactic construction was found.

**Binary Classification Based on MFCC.** To apply this method, binary markup of audio fragments was used. Mel Frequency Cepstral Coefficients (MFCC) were calculated for each audio fragment. MFCC is often used as input to a classifier when working with audio data [24–26]. As part of the experiment, three models LGBMClassifier, XGBClassifier and GaussianNB were used.

**Entity Extraction Method: XLM-RoBERTa.** XLM-RoBERTa refers to models of the Transformer type [27] and is used for entity extraction, in particular for extracting personalities, locations, organizations, as well as for narrower tasks, for example, extracting terms and their definitions from texts [28]. A feature of the model is that the training dataset also includes low-resource languages: Uighur, Khmer, Swahili, etc. As a result, the accuracy of such a model is higher than mBERT [29] for these low-resource languages. The choice of this model is due to the fact that such an architecture successfully copes with token classification for the task of extracting entities. Entities in this case are meant as dialectological features, that is, they are referents to a specific level of implementation of the feature. Transfer learning assumes that token labeling will allow detection of more than one dialect feature in a single annotation. In this work, dialect features were taken as entities, depending on which language level they are implemented at. To train the model, interviews with the oldest consultant were annotated. In the speech of the consultant, the studied features of the dialect are preserved and consistently implemented. As a result, 1444 annotations were marked up, 1010 of which were included in the training set, 434 in the test set.

## 5   Results

Several approaches to fine-tune models were used: training and testing on data of only one dialect; training on the data of the first dialect, then additional training on the data of the second dialect and testing on the data of the first; training on the data of one dialect, then additional training on the data of the second dialect and testing on the data of the second one.

The results shown in Table 2. Corpus$_1$ is the dataset of the villages of the Zapadnodvinsky district, corpus$_2$ is the dataset of the villages of the Opochetsky district. The highest values for the test data are bolded in the table. The numbering in the second column shows the sequence of fine-tuning steps.

In the case of the wav2vec2 and QuartzNet models, approaches using a spellchecker proved to be the most effective: transcriptions were improved on average by 8% in the WER metric for each of the mentioned models, while the CER metric for both models remained almost unchanged. The use of a spellchecker for the Conformer-Transducer architecture, namely for the STT Ru Conformer-Transducer Large model, fine-tuned on dialect data, did not show such an increase in metrics.

**Table 2.** Comparison of the metrics for the best speech recognition models within each architecture.

| Approach | Data for fine-tuning | Test data | Using the spellchecker | WER | CER |
|---|---|---|---|---|---|
| wav2vec2 | 1. 70% corpus$_1$ | 1. 30% corpus$_1$ | + | **0.34** | **0.18** |
| QuartzNet Golos | | | + | 0.46 | 0.25 |
| Conformer-Transducer | | | − | 0.35 | 0.21 |
| wav2vec2 | 1. 70% corpus$_2$ | 1. 30% corpus$_2$ | + | 0.53 | 0.32 |
| QuartzNet Golos | | | + | 0.60 | 0.36 |
| Conformer-Transducer | | | − | 0.39 | 0.27 |
| wav2vec2 | 1. 70% corpus$_1$ | 1. 30% corpus$_1$ | + | 0.54 | 0.31 |
| QuartzNet Golos | 2. 70% corpus$_2$ | | + | 0.56 | 0.31 |
| Conformer-Transducer | | | − | 0.48 | 0.40 |
| wav2vec2 | 1. 70% corpus$_1$ | 1. 30% corpus$_2$ | + | 0.50 | 0.30 |
| QuartzNet Golos | 2. 70% corpus$_2$ | | + | 0.60 | 0.35 |
| Conformer-Transducer | | | − | **0.37** | **0.25** |

During the experiments, it was found that the models that were first fine-tuned on the Zapadnodvinsky data and then on the Opochetsky data performed worse on the Zapadnodvinsky data. The following hypothesis is proposed for further research in this area: it is necessary to combine data from close dialects into one dataset in order to obtain a model that shows high results on each of them. The possible forgetting of previous datasets by the model during fine-tuning was mentioned in [29] for audio recordings of African Americans. Therefore, for a large sample of dialects and dialect material, respectively, it may be more efficient to conduct fine-tuning on the entire sample at once. Thus, if there is a need to fine-tune the model for one dialect, then it is possible to use those already pretrained on a close dialect and fine-tune on a small amount of data of the target dialect. It is assumed that combining all possible datasets will provide the highest quality if the goal of creating a tool is to make it universal across all dialects.

As a result of the dialect features recognition experiments, it was determined that the considered approaches differ in their tasks. The rule-based approach is rather more suitable for highlighting all possible positions where this or that feature can be found. The algorithm for extracting dialect features using rules was applied to all manually marked data. Table 3 presents the results of this approach.

**Table 3.** Results of applying the rule-based approach for PHON, MORPH and SYNT tags.

| Data | Precision | Recall | F1-score |
|------|-----------|--------|----------|
| Corpus$_1$ | 0.26 | 0.56 | 0.33 |
| Corpus$_2$ | 0.39 | 0.63 | 0.45 |

Binary classification solves the problem of splitting the dataset into two groups: the presence of dialect features and their absence. This approach is rather an auxiliary step either before the formation of a dataset for training or for applying subsequent methods for extracting dialect features in order to reduce the sample and thereby reduce the tool's run time. Table 4 shows the results of model training only on the Zapadnodvinsky data and testing on the data of the corresponding dialect, Table 5 shows training on the entire sample of Pskov dialects presented in this work and testing on the Zapadnodvinsky data.

**Table 4.** Results for the LGBMClassifier, XGBClassifier, and GaussianNB models on the Zapadnodvinsky data.

| Classifier | Precision | Recall | F1-score |
|------------|-----------|--------|----------|
| LGBMClassifier | 0.43 | 0.55 | 0.48 |
| XGBClassifier | **0.55** | 0.44 | 0.49 |
| GaussianNB | 0.47 | **0.66** | **0.55** |

**Table 5.** Results of binary audio classification for the LGBMClassifier, XGBClassifier, and GaussianNB models on the entire sample of Pskov dialects.

| Data | Precision | Recall | F1-score |
|---|---|---|---|
| LGBMClassifier | 0.63 | 0.78 | 0.70 |
| XGBClassifier | 0.64 | **0.85** | **0.73** |
| GaussianNB | **0.65** | 0.84 | **0.73** |

Based on the results, it is assumed that, in general, to train the classifier from scratch, a larger amount of manually marked data is needed, while it is possible to include data from different dialects and corpora, in order to obtain audio of various characteristics. Such a task is rather more voluminous, and it can be a full-fledged study involving a large amount of dialect material, which is not limited to Pskov dialects.

The entity extraction method is more accurate than the rule-based method, as it takes into account the presence of variability and the predominance of dialect or literary variants, which is reflected in the results (Table 6).

**Table 6.** Results of applying the XLM-RoBERTa model for PHON, MORPH, LEX, and SYNT tags.

| Data for fine-tuning | Test data | Precision | Recall | F1-score |
|---|---|---|---|---|
| 70% corpus$_1$ | 30% corpus$_1$ | 0.53 | 0.36 | 0.43 |
| 70% corpus$_1$ | 30% corpus$_2$ | 0.42 | 0.05 | 0.08 |
| 70% corpus$_1$ and 70% corpus$_2$ | 30% corpus$_2$ | 0.59 | 0.57 | 0.58 |

The model without fine-tuning on Opochetsky data successfully classifies tags, but at the same time it allocates a very small proportion of the total data volume. Despite the similarity of implementations of syntactic features, the model does not cope with their detection. Results for lexical features are expected, since, in general, archaic lexis is less common in interviews (3% of LEX tags in the labeled data of the dialect of the villages of the Zapadnodvinsky district, 8% in the Opochetsky labeled data) than other features.

## 6   Creating Tool

One of the stages of work was the integration of the developed methods into a single tool. The tool accepts an audio recording in wav format as input. All audio recordings are converted to a single-channel format, since the model can work with data of only one dimension. There are both single-channel and two-channel recordings in the cases, therefore, in order to unify, a choice was made in favor of a smaller number of channels.

On being converted, the audio is divided into fragments. Fragmentation and conversion are done using the pydub[2] library. Fragmentation occurs on the basis of the selection of sections of silence - fragments of audio, below the specified decibel threshold - the check occurs in 5 millisecond steps, sections of silence must last at least 700 milliseconds. The transcription is done using the Conformer-Transducer model, since the model much less often produces pseudowords as transcriptions, which make it difficult for the annotator to understand the context as a whole during the annotation review process. Moreover, this model does not require any additional tools to be used, since the quality is generally comparable to other models that showed low CER and WER metrics where the spellchecker was used. After receiving the transcripts, a TextGrid formatted file was generated. Each transcript was fed to the input of the pre-trained XLM-RoBERTa model, the transcriptions were divided into tokens, and a tag was received for each token. The ELAN file was generated using the pympi-ling[3] library. The layer structure was as follows: 1) the original layer with annotation; 2) layer with tokens, the first layer is the parent layer; 3) a layer with dialect features, the parent layer is the second. In total, the result of the tool was two files: in the TextGrid format for working with the Praat program [30] and in the eaf format for the ELAN program [31].

# 7 Conclusion

First of all, it should be noted that the field data differs from the existing datasets on which modern architectures are trained. Such data is noisier, more often there is a quantitative reduction of sounds, hesitation, the presence of several speakers in one time interval, etc.

Three models were used for recognition, which were already pretrained on standard Russian. Each of them was fine-tuned on the dialect material. For two of the three models a spellchecker was also used. Based on the analysis of the results, it was found that:

1. If the goal is to get the model to recognize one selected dialect, fine-tuning a model that have already seen a close dialect will give a better result than fine-tuning a model pretrained only in standard Russian. Thus, phonetic, morphological, syntactic and lexical differences between close dialects do not impair the quality of recognition.
2. The last data to fine-tune should be the target dialect (for which the model is being trained), and not some close one, otherwise the quality will be lower.

A new hypothesis has been put forward for further research: to create a universal dialect speech recognition model, it is necessary to fine-tune the model

---

[2] https://pypi.org/project/pydub/.
[3] https://pypi.org/project/pympi-ling/.

on the entire sample of dialects at the same time. It can be related with dialect-independent approach [1] which may be more appropriate for Russian dialects altogether.

The objectives of this work also included the automatic extraction of dialect features in recognized annotations. The binary classifier took MFCC as input and classified as 1 the data in which the presence of dialect features was assumed, as 0 - the absence. This approach has its drawbacks, since the sample of audio recordings is not representative in terms of the number of informants, their gender and age, as well as the quality of the recordings due to such a distribution within the corpora under consideration. In spite of the method's high quality, it should be viewed more as an intermediate step because of the above-mentioned features. The other two methods, rule-based and entity extraction, classify each token in the annotation. The rule-based method can be expanded by adding functions that can be based on the Dialectological Atlas of the Russian Language [32]. The entity extraction method was applied for such a task for the first time, but it has already shown a result higher than the rule-based approach, since it takes into account variability, unlike the previous method.

As a result of the experiments, the best models for each of the stages of the declared tool were determined. A single pipeline was created that takes an audio file in wav format as input and as result provides a file in TextGrid format for further work with annotations in the Praat program, as well as an eaf file with layers of annotations, tokens and tags of dialect features for work in the ELAN program.

# References

1. Das, A., Kumar, K., Wu, J.: Multi-dialect speech recognition in English using attention on ensemble of experts. In: ICASSP 2021–2021 IEEE International Conference on Acoustics, Speech and Signal Processing (ICASSP), pp. 6244–6248 (2021)
2. Alsayadi, H.A., Abdelhamid, A.A., Hegazy, I., Alotaibi, B., Fayed, Z.T.: Deep investigation of the recent advances in dialectal Arabic speech recognition. IEEE Access **10**, 57063–57079 (2022)
3. Demszky, D., Sharma, D., Clark, J., Prabhakaran, V., Eisenstein, J.: Learning to recognize dialect features. In: North American Chapter of the Association for Computational Linguistics, pp. 2315–2338 (2021)
4. Zakharova, K.F., Orlova, V.G.: Dialect Division of the Russian Language. Editorial URSS, Moscow (2004)
5. Pshenichnova, N.N.: Linguistic Geography. Azbukovnik, Moscow (2008)
6. Dyachenko, S.V., Zhidkova, E.G., Malysheva, A.V., Ronko, R.V., Ter-Avanesova, A.V.: Dialectological expedition to Opochka district, Pskov region. Russ. Lang. Linguist. Theory **2**(36), 257–312 (2018)
7. Ronko, R.V.: Expedition to Opochka district of the Pskov region in 2019: texts and a brief commentary. Russ. Lang. Linguist. Theory **1**(39), 234–258 (2020)
8. Ronko, R.V., et al.: The dialect of the villages Shetnevo, Makeevo, Sh'elkino, Terekhovo and Stepan'kovo of the Zapadnodvinskij district, Tver Oblast'. A grammar sketch and field notes. Russ. Lang. Linguist. Theory **1**(45), 216–263 (2023)

9. Zambrzhitskaya, M.S.: Description of the implementation of the phoneme /v/ in Pskov dialects. In: Gerasimov, D.V. (ed.) Nineteenth Conference on Typology and Grammar for Young Researchers. St. Petersburg, 24–26 November 2022. ILS Russian Academy of Sciences, St. Petersburg (2022)

10. Ronko, R.V.: Na with the accusative: marking the addressee of speech in some western and southern Russian dialects. Slověne **10**(2), 277–296 (2022)

11. Ronko, R., et al.: Corpus of Opochetsky dialects 2019. Linguistic Convergence Laboratory, HSE University; V.V. Vinogradov Russian Language Institute Russian Academy of Science, Moscow (2019)

12. Von Waldenfels, R., Daniel, M., Dobrushina, N.: Why standard orthography? Building the Ustya River Basin corpus, an online corpus of a Russian dialect. In: Selegey, V.P. (ed.) Kompjuternaja lingvistika i intelektual'nyje tehnologii, pp. 720–728 (2014)

13. Cho, H.C., Okazaki, N., Miwa, M., Tsujii, J.I.: Named entity recognition with multiple segment representations. Inf. Process. I& Manage. **49**(4), 954–965 (2013)

14. Nakashole, N.: Unnamed entity recognition of sense mentions. arXiv preprint arXiv:1811.07092 (2018)

15. Imaizumi, R., Masumura, R., Shiota, S., Kiya, H.: End-to-end Japanese multi-dialect speech recognition and dialect identification with multi-task learning. APSIPA Trans. Signal Inf. Process. **11**(1), (2022). https://doi.org/10.1561/116.00000045

16. Bondarenko, I.: XLSR Wav2Vec2 Russian with 2-gram language model. Hugging Face Hub. (2022). https://huggingface.co/bond005/wav2vec2-large-ru-golos-with-lm

17. Schneider, S., Baevski, A., Collobert, R., Auli, M.: wav2vec: unsupervised pre-training for speech recognition. arXiv preprint arXiv:1904.05862 (2019)

18. Baevski, A., Zhou, Y., Mohamed, A., Auli, M.: Wav2vec 2.0: a framework for self-supervised learning of speech representations. In: Advances in neural information processing systems, vol. 33, pp. 12449–12460 (2022)

19. Garofolo, J.S., et al.: TIMIT acoustic-phonetic continuous speech corpus LDC93S1. Web download. Linguistic Data Consortium, Philadelphia (1993)

20. Burtsev, M., et al.: DeepPavlov: open-source library for dialogue systems. In: Proceedings of ACL 2018, System Demonstrations, pp. 122–127 (2018)

21. Karpov N., Denisenko A.A., Minkin F. Golos: Russian dataset for speech research. arXiv preprint arXiv:2106.10161 (2021)

22. Ardila R., et al.: Common voice: a massively-multilingual speech corpus. In: International Conference on Language Resources and Evaluation. arXiv preprint arXiv:1912.06670 (2019)

23. Panayotov, V., Chen, G., Povey, D., Khudanpur, S.: LibriSpeech: an ASR corpus based on public domain audio books. In: 2015 IEEE International Conference on Acoustics, Speech and Signal Processing (ICASSP), pp. 5206–5210. IEEE (2015)

24. Suksri, S., Yingthawornsuk, T.: Speech recognition using MFCC. In: International Conference on Computer Graphics, Simulation and Modeling, vol. 9, pp. 135–138 (2012)

25. Milton, A., Roy, S.S., Selvi, S.T.: SVM scheme for speech emotion recognition using MFCC feature. Int. J. Comput. Appl. **69**, 34–39 (2013)

26. Aggarwal, G., Singh, L.: Characterization between child and adult voice using machine learning algorithm. In: International Conference on Computing, Communication I& Automation, pp. 246–250 (2015)

27. Conneau, A., et al.: Unsupervised cross-lingual representation learning at scale. arXiv preprint arXiv:1911.02116 (2020)

28. Jin, Y., Kan, M., Ng, J., He, X.: Mining scientific terms and their definitions: a study of the ACL anthology. In: Conference on Empirical Methods in Natural Language Processing, pp. 780–790 (2013)
29. Huang, J., et al.: Cross-Language transfer learning, continuous learning, and domain adaptation for end-to-end automatic speech recognition. arXiv Audio and Speech Processing (2019)
30. Boersma, P., Weenink, D.: PRAAT: doing phonetics by computer (1992–2022). [Computer program]
31. Brugman, H., Russel, A.: Annotating multimedia/multi-modal resources with ELAN. In: Proceedings of LREC 2004, Fourth International Conference on Language Resources and Evaluation, pp. 2065–2068 (2004)
32. Marchenko, I.A., et al.: Database of the dialectological atlas of the Russian language. https://da.ruslang.ru/

# Needle in a Haystack: Finding Suitable Idioms Based on Text Descriptions

Dmitrii Zhernokleev[1] and Pavel Braslavski[1,2(✉)] [iD]

[1] HSE University, Moscow, Russia
ddzhernokleev@edu.hse.ru, pbras@yandex.ru
[2] Nazarbayev University, Astana, Kazakhstan

**Abstract.** Idioms are an important part of natural languages and are often used to enhance expressiveness and fluency of speech. However, it can be difficult to find a contextually appropriate idiom when writing an essay or crafting a headline for a news article, especially for non-native speakers. This gives rise to the idea of an automated system that is able to recommend an idiom for an input sentence. The goal of this study is to develop and compare methods that would make such a system possible. We used an existing collection of idioms and employed several configurations based on models from the Sentence-BERT family. We also automatically expanded the initial dataset and fine-tuned a pre-trained Sentence-BERT model on the idiom/context matching task. This approach achieved the highest MRR score of 0.507. The data and the trained model are publicly available.

**Keywords:** idioms · idiomatic expressions · writing assistance · semantic search · Sentence-BERT

## 1 Introduction

Idiomatic expressions are multi-word expressions that are commonly used in both written and oral speech in order to convey ideas succinctly and to enhance the expressiveness and fluency of speech. Idiomatic expressions are non-compositional, i.e. their meaning cannot be derived from the constituent words [13]. For example, the idiom *water under the bridge* is used to refer to past events, usually unfortunate ones, that cannot be rectified; the meaning of this idiomatic expression as a whole cannot be derived from the literal meanings of the individual words. Moreover, so-called "potentially idiomatic expressions" (PIE) [7] can have literal or figurative meaning in different contexts. For example, in the sentence *You will fail your exams, unless you start working harder, so it's time to wake up and smell the coffee* the underlined idiom is used in a figurative sense, while in the sentence *It was so great to wake up and smell the coffee this morning* – in a literal sense. These characteristics of idioms pose a challenge in such tasks as information retrieval, sentiment analysis, text summarization, question answering, and machine translation [5,9,16,21].

D. I. Ignatov et al. (Eds.): AIST 2023, LNCS 14486, pp. 185–196, 2024.
https://doi.org/10.1007/978-3-031-54534-4_13

In recent years, writing assistants have become very popular, especially among non-native speakers. Modern writing assistants not only correct spelling and grammatical errors, but also improve style and suggest continuations of the entered text [4]. The use of idioms can be extremely appropriate in some cases – for example, when writing an essay or crafting a headline for a news article. However, finding a suitable idiomatic expression might be complicated, especially when writing in a foreign language. This naturally gives rise to the idea of an automatic writing assistant that would be able to recommend an idiom for a given sentence.

In our study, we address the task of idiom recommendation based on the provided context. We used an existing corpus of idioms along with their usage contexts and experimented with several configurations based on Sentence-BERT models. Furthermore, we automatically expanded the dataset and fine-tuned a Sentence-BERT model specifically for the idiom/context matching task. This approach achieved the highest MRR score of 0.507.

The contributions of our study are twofold: 1) we propose and evaluate a novel approach to the task of idiom recommendation based on semantic similarity search and 2) we publish a Sentence-BERT model fine-tuned specifically for this task along with the automatically compiled dataset of idioms in their contexts used for fine-tuning.[1]

## 2   Related Work

### 2.1   Translation and Detection of Idiomatic Expressions

In their study, Dankers et al. [5] analyze the processing of idiomatic expressions by pre-trained Transformer models in the context of machine translation. The authors demonstrate that interference into the model's processing of idioms can change non-compositional translations into compositional ones. According to the observations made in this research, figurative PIEs tend to be grouped as one lexical unit in the encoder compared to the literal usages of PIEs. They conclude that Transformer models "typically translate idioms in a too compositional manner, providing a word-for-word translation".

Gamage et al. [6] approach the task of idiom detection. The authors treat the task as a binary token classification problem, where each token is classified as either idiomatic or non-idiomatic using the DistilBERT model. The authors used Bag-of-Words and word2vec representations as baselines, which achieved accuracy of 0.44 and 0.55, respectively. The pre-trained DistilBERT model was fine-tuned using data from the Idioment [21] and EPIE datasets, as well as additional data collected from an online idiom dictionary. The best model in this study achieved an accuracy of 94%.

---

[1] https://github.com/archimedes1515/idiom_search.

## 2.2   Collections of Idiomatic Expressions

For the current study, we needed a dataset containing English idiomatic expressions and their usage contexts. Table 1 summarizes statistics of five publicly available datasets presented in previous studies. The size of the collection is an important factor, which affects the robustness of models trained on them. Therefore, we focus on detailed analysis of the three largest datasets – PIE-English, EPIE, and MAGPIE.

**Table 1.** Comparison of existing corpora of idiomatic expressions.

| Corpus | # unique idioms | # samples |
|---|---|---|
| Li and Sporleder [18] | 17 | 3,964 |
| IDIX [19] | 78 | 5,836 |
| PIE-English [1] | 1,197 | 20,174 |
| **EPIE** [17] | 717 | 25,206 |
| MAGPIE [7] | 1,756 | 56,622 |

Adewumi et al. [1] present PIE-English – a collection of potentially idiomatic expressions. For the idioms selected from a dictionary, the authors collected sentences from two sources: the British National Corpus (BNC) [3] and web pages. The corpus contains around 1,200 unique idioms with definitions. PIE-English dataset contains 20,174 samples, which were classified by two annotators into 10 categories: euphemism, literal, metaphor, personification, simile, oxymoron, parallelism, paradox, hyperbole, and irony.

Saxena et al. [17] introduced the English Possible Idiomatic Expressions (EPIE) dataset, which contains 25,206 sentences for 717 unique idioms. The list of unique idioms was taken from the IMIL dataset [2]. The authors divide all idioms into two groups: *static* and *formal* idioms. *Static* idioms do not undergo lexical changes (e.g. "rule of thumb", "rat race", "on cloud nine", etc.), while *formal* idioms can be modified (e.g. "keep an eye on" can occur as "keep a keen eye on"). For static idioms, the authors collected sentences that contain the idiomatic expression as a substring, while for formal idioms the StringNET [20] tool was used to account for possible variations of potentially idiomatic expressions. The final corpus contains 21,891 sentences for 359 static idioms and 3,135 sentences for 358 formal idioms.

Haagsma et al. [7] proposed the MAGPIE dataset, which exceeds the size of analogues. The presented corpus contains 56,622 sentences for 1,756 different idiomatic expressions, of which 1,035 are used exclusively in the idiomatic sense. Sentences containing PIEs were extracted from BNC and annotated via crowdsourcing with three class labels: idiom is used figuratively, literally, or in a way that does not fit the binary distinction.

## 2.3  Recommendation of Idiomatic Expressions

Liu et al. [11] focus on the task of recommending Chinese idioms. Chinese idioms originate from ancient legends and usually consist of only four characters. The authors approach the task of idiom recommendation as a context-to-idiom machine translation problem. The proposed algorithm consists of three stages: BiLSTM-based encoding and attention layers are used to encode the original text, then the encoded context attention vector is decoded into an intermediate sequence, and finally, the translated character sequences received as an output are further mapped into the standard idioms from the fixed set using edit distance and the most similar idiom is selected as the final recommendation.

## 3  Data

We chose the EPIE dataset as the main dataset in our study for several reasons: firstly, this corpus is the second-largest among all publicly available datasets of idiomatic expressions; secondly, idioms in this dataset are initially divided into *static* and *formal* ones (we focus on static idioms exclusively); thirdly, EPIE contains definitions for the idiomatic expressions, which we also use in our experiments. However, the EPIE dataset lacks the classification of sentences into literal and figurative uses of idioms, and this corpus required pre-processing for the idiom recommendation task. Pre-processing included the following steps:

1. Removing duplicate sentences and idioms.
2. Filtering of non-English idioms:
   (a) Latin idioms (e.g. "mea culpa", "carpe diem").
   (b) Idioms not present in the IBM Debater dataset [8] (e.g., "absence makes the heart grow fonder", "thank goodness", "calendar year").

After the filtering described above, 14,794 sentences were left for 173 unique idioms. Examples of idioms and their contexts are given in Table 2.

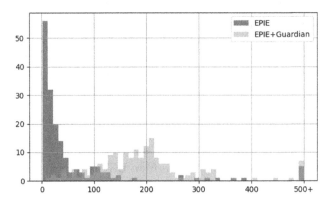

**Fig. 1.** Distribution of the number of different contexts per idiom in the initial EPIE training set and after adding examples from *The Guardian*.

**Table 2.** Examples of idioms and their contexts.

| Idiom | Sentence |
|---|---|
| a breath of fresh air | Comes to corner of the street for <u>a breath of fresh air</u> |
| cutting edge | From the same source come innovations which keep the company in the lead and form the <u>cutting edge</u> of competition |
| water under the bridge | That was a long time ago, there's been a lot of <u>water under the bridge</u>, why not talk again? |

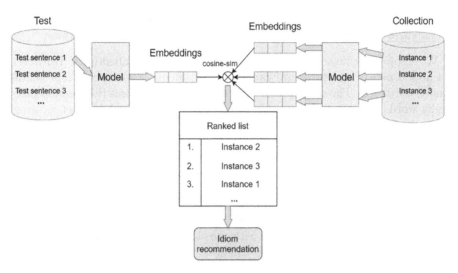

**Fig. 2.** Scheme of the proposed approach.

To evaluate different configurations of the method proposed in this study, we split the data in an 80:20 ratio with stratification. Then we removed idioms from all sentences in the test set. The test set is fixed for all the configurations that will be discussed further.

We also collected additional contexts for idioms presented in the EPIE dataset from *The Guardian*, one of the UK's largest newspapers. We queried *The Guardian* API[2] with idioms as phrase queries, then extracted idiom-bearing sentences from the returned documents. This way we obtained 24,660 additional sentences. The distribution of different contexts per idiom in the training set is illustrated in Fig. 1.

## 4   Method

Our approach is illustrated in Fig. 2. The idiom recommendation is based on the semantic similarity search: we obtain embeddings for the input sentence ('query')

---

[2] https://open-platform.theguardian.com/documentation/.

with a missing idiom and all collection items ('documents'). Then, we rank the collection items based on cosine similarity to the 'query' and consider the corresponding idioms as recommendations. The algorithm has two key parameters that can be varied: the *collection* on which the semantic search is performed and the *model* used to obtain the embeddings. Using different collections and models in our research, we obtain various configurations of the main approach.

In our experiments, we use the following collections (ordered by size):

1. <u>Idioms</u> – unique idiomatic expressions themselves;
2. <u>Idioms+Defs</u> – idiomatic expressions with corresponding definitions;
3. <u>Sentences</u> – sentences from the training set;
4. <u>Sentences++</u> – <u>Sentences</u> extended with examples from *The Guardian*.

We employ word2vec as a baseline model for sentence embeddings. Word2vec is a neural network architecture capable of producing static vector representations of words [12]. The core idea behind this group of models is that semantically similar words occur in similar contexts. To obtain sentence embeddings, we average embeddings of the words that make up the sentence. We employ word2vec model trained on the Google News dataset from the Gensim library [14].[3]

SBERT [15] is a family of models that address the problem of sentence-level semantic similarity. SBERT implements siamese or triplet-loss network architecture trained on a large collection of sentence pairs. The original study describes the implementation of classification, regression, and triplet objective functions. For the classification objective function, the sentence embeddings of a pair of sentences $u$ and $v$ are concatenated with the element-wise difference $|u - v|$. The resulting vector is multiplied by the trainable weight matrix, then the result of the multiplication is fed into softmax function and cross-entropy loss is optimised. In case of the regression objective function the authors use MSE loss and compute cosine similarity between two sentence embeddings $u$ and $v$. The triplet objective function fine-tunes the network so that the distance between an anchor sentence and a positive sentence is smaller than the distance between an anchor sentence and the negative sentence. Since the publication of the original paper, the SBERT family has been greatly expanded: it now includes models with different underlying Transformer models trained on different datasets. We use several Sentence-BERT (SBERT) models from the Sentence-Transformers framework in the main body of experiments.[4]

In our experiments, we use the following models to obtain vector representations of sentences (ordered by size):

1. word2vec-google-news-300 – <u>word2vec</u>.
2. all-MiniLM-L12-v2 – <u>MiniLM</u>.
3. all-distilroberta-v1 – <u>distilroberta</u>.
4. Fine-tuned all-distilroberta-v1 – <u>distilroberta+</u>.
5. all-mpnet-base-v2 – <u>mpnet</u>.

---

[3] https://radimrehurek.com/gensim/models/word2vec.html.
[4] https://www.sbert.net/docs/pretrained_models.html.

For convenience, we will further refer to collections and models using the corresponding underlined notations in the lists above, and we will use a paired notation to refer to configurations: for instance, the combination of sentences from the training set as the search collection and all-mpnet-base-v2 model is referred to as *mpnet@Sentences*.

# 5 Experiments

## 5.1 Metrics

Since our approach involves ranking, we find it reasonable to use Mean Reciprocal Rank (MRR) as the main metric, following [10]:

$$MRR = \frac{1}{|Q|} \sum_{i=1}^{|Q|} \frac{1}{rank_i}, \tag{1}$$

where $rank_i$ in our case is the position of the first correct idiom for $i$-th query context; $Q$ is the set of all test contexts. MRR has a straightforward interpretation: e.g. $MRR = \frac{1}{2}$ means that the correct answer is in the second position on average; in case $MRR = \frac{1}{5}$ we can say that the correct answer is in the fifth position on average, etc.

## 5.2 Jaccard Similarity

The Jaccard similarity coefficient is a measure used to estimate the word-level similarity of two strings:

$$J(A, B) = \frac{|A \cap B|}{|A \cup B|} \tag{2}$$

In this study, we use the Jaccard coefficient to ensure that there is no 'leakage' between test and train sets, i.e. context matching doesn't occur on the lexical level.

## 5.3 Baseline

Recently, transformer-based language models have shown impressive performance in a wide range of natural language processing applications, establishing themselves as the state-of-the-art approach. This is one of the reasons why SBERT was selected for our study. However, to establish a starting point, we decided to take a simpler neural network architecture as a baseline against which we could evaluate the performance of SBERT models. The pretrained word2vec model was chosen as such an architecture. So we consider the *word2vec@Sentences* configuration to be a baseline, following [6].

## 5.4   Fine-Tuning

The first stage of the experimental part of our study involved the use of pre-trained models. At the second stage, in order to improve the quality of the model, we fine-tune the distilroberta model using the EPIE training set augmented with contexts collected using *The Guardian* API. First of all, we split this new set into (new) training and validation set in a 90:10 ratio. Then, we created positive and negative pairs for training. To create a positive pair for a sentence, we matched a given sentence with another random sentence from the new training set containing the same idiomatic expression, after that we removed the idiom itself from the first sentence in the pair. The process of creating negative pairs was identical, except we paired a sentence with a random sentence with a different idiom, see examples in Table 3. The same process was applied for the validation set.

**Table 3.** Examples of positive and negative pairs.

| | Sentence 1 | Sentence 2 |
|---|---|---|
| + | You were able to drop all your own commitments ~~at the drop of a hat~~? | He's flown out from Michigan at the drop of a hat to help distribute food to flood victims |
| − | You were able to drop all your own commitments ~~at the drop of a hat~~? | You have to face the music when it comes your way; meet the challenge |

Using this data, we fine-tuned *distilroberta* model for five epochs, batch size was set to 16. Figure 3 shows the accuracy score on the validation set during the fine-tuning process. It can be seen that after the fourth epoch the accuracy reaches a plateau, so we can conclude that further training is unlikely to improve the model.

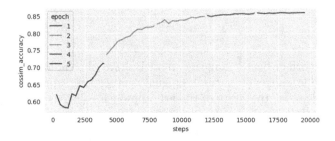

**Fig. 3.** Model performance on the validation set during fine-tuning.

**Table 4.** Experimental results.

|  | Idioms | Idioms+Defs | Sentences | Sentences++ |
| --- | --- | --- | --- | --- |
| Jaccard | 0.0505 | 0.0329 | 0.2330 | 0.2106 |
| word2vec | 0.1207 | 0.1207 | 0.2821 | 0.2563 |
| MiniLM | 0.0754 | 0.0870 | 0.3296 | 0.2975 |
| distilroberta | 0.1018 | 0.1029 | 0.3274 | 0.3021 |
| mpnet | 0.1493 | 0.1110 | 0.3479 | 0.3141 |
| distilroberta+ | **0.5071** | 0.4564 | 0.4594 | 0.4482 |

## 5.5 Experimental Results

Table 4 shows MRR scores for different configurations. The first row corresponds to the idiom recommendation based on Jaccard similarity and can be seen as a sanity check of the data. Low MRR scores suggest that sentences in train and test subsets corresponding to the same idiom don't have a high lexical similarity, thus the task is not trivial.

Based on the obtained results, we can conclude that, firstly, the use of SBERT models with all collections except Idioms and Idioms+Defs allowed us to improve the result of the baseline configuration word2vec@Sentences; secondly, the fine-tuned distilroberta model (*distilroberta+*) achieved the highest score compared to other configurations. The MRR score of the *distilroberta+@Idioms* is slightly greater than 0.5, which means that on average the correct idiomatic expression is ranked second. Interestingly, the fine-tuned model achieves the best result on bare *Idioms* in contrast to pre-trained models that score best on *Sentences*. To ensure that the fine-tuned *distilroberta+* doesn't simply memorize idiom contexts in an extended dataset, we also calculated Jaccard similarity scores between test prompts and top-ranked sentences from *Sentences++*. It can be seen from Fig. 4 that all distributions are in the lower range. We can conjecture that the model doesn't rely on lexical similarity and has learned a higher-level representation of idioms' contexts. The configuration *distilroberta+@Idioms* showed a 80% gain compared to the baseline and a 46% gain over *mpnet@Sentences*, which achieved the highest MRR score among all configurations without fine-tuning.

We can see from the results that increasing the size of the collection is only beneficial up to a certain extent: thus, the MRR scores for the *MiniLM, distilroberta,* and *mpnet* models improve from the *Idioms* to the *Sentences* (except for the decrease observed between *mpnet@Idioms* and *mpnet@Idioms+Defs*), but decrease for *Sentences++*. We also cannot state unequivocally that using a larger model to obtain sentence embeddings guarantees a higher MRR score, since in the case of the *Sentences* collections, the *distilroberta* model performs on par with the *MiniLM* model. On the other hand, the largest model (*mpnet*) consistently outperforms both the *MiniLM* and *distilroberta* models across all five collections.

Table 5 provides examples of 'simple' and 'difficult' idioms for our best configuration *distilroberta+@Idioms*. 'Simple' idioms are characterized by a high

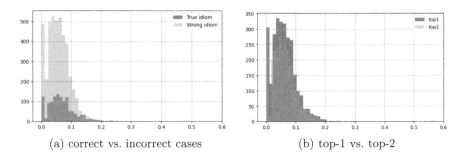

(a) correct vs. incorrect cases          (b) top-1 vs. top-2

**Fig. 4.** Distribution of Jaccard scores between the 'query' and its closest (top-1 and top-2) contexts from the *Sentences++* according to *distilroberta+* model.

reciprocal rank (RR) averaged over all corresponding test sentences. On the other hand, average RR of 'difficult' idioms is close to zero.

A possible explanation of these performance variations is that some idiomatic expressions are used in wider or more complicated contexts. Another reason could be related to the evaluation data and protocol: we assume that there is only one suitable idiom for each sentence, which is overly strict. Firstly, there are synonymous idiomatic expressions that can be interchangeable, such as "sheathe the sword" and "bury the hatchet" or "make hay while the sun shines" and "strike while the iron is hot". Secondly, some contexts might be suitable for multiple idiomatic expressions, due to their ambiguity, for example, in the sentence *Sorry guys, I have to hit the hay now!* the underlined idiomatic expression can be replaced with other idioms such as *eat like a bird, give it a shot*, etc. In this case different idiom may change the meaning of the sentence, but the idioms themselves are still suitable substitutes. Moreover, some sentences contain idiomatic expressions used in the literal sense, and some idioms (e.g. *on the table, on the same page*) appear in the EPIE dataset almost exclusively in the literal sense. This can also affect the capability of algorithm to find the correct idiom.

**Table 5.** Examples of simple and difficult idioms.

| 'simple' idioms | | 'difficult' idioms | |
|---|---|---|---|
| Idiom | Average RR | Idiom | Average RR |
| dead ringer | 1.0 | icing on the cake | $0.94 \cdot 10^{-4}$ |
| fish out of water | 1.0 | the upper crust | $0.99 \cdot 10^{-4}$ |
| inside job | 1.0 | on the same page | $1.18 \cdot 10^{-4}$ |
| greener pastures | 1.0 | dressed up to the nines | $1.29 \cdot 10^{-4}$ |
| in the wake of | 0.912 | in the blink of an eye | $1.68 \cdot 10^{-4}$ |
| all the way | 0.826 | silver bullet | $2.56 \cdot 10^{-4}$ |
| a fly on the wall | 0.750 | under the wire | $2.56 \cdot 10^{-4}$ |

# 6   Conclusion and Future Work

In this paper, we presented a new approach to the idiom recommendation task based on semantic similarity search and application of the Sentence-BERT model. We examined the suitability of different neural models, including word2vec and Sentence-BERT, for the task of idiom recommendation. The fine-tuned *distilroberta* model showed the best performance and achieved $MRR = 0.5071$ (80% gain compared to the word2vec baseline configuration). In addition, we automatically expanded an existing idiom dataset, increasing the size of the corpus by more than 2.5 times. We made the data, code, and fine-tuned model publicly available.

In the future, we plan to:

- experiment with existing tools for detection of literal/metaphoric contexts in order to filter instances containing idiomatic expressions used in literal sense;
- further expand the dataset to improve the quality of the models;
- add filters to prevent recommendations of inappropriate idioms (for instance, *kick the bucket* for the sentence *Queen Elizabeth II, Britain's longest-reigning monarch, dies aged 96*);
- analyze the impact of context length on the performance of our approach and experiment with contexts longer than one sentence.

**Acknowledgments.** The study develops ideas partially derived from Anna Vysheslavova's 2020 summer internship under Pavel Braslavski's supervision. We would like to express gratitude to Yulia Badryzlova for fruitful discussion of the paper draft.

# References

1. Adewumi, T., Vadoodi, R., Tripathy, A., Nikolaido, K., Liwicki, F., Liwicki, M.: Potential idiomatic expression (PIE)-English: corpus for classes of idioms. In: LREC, pp. 689–696 (2022)
2. Agrawal, R., Kumar, V.C., Muralidharan, V., Sharma, D.M.: No more beating about the bush: a step towards idiom handling for Indian language NLP. In: LREC (2018)
3. BNC Consortium, et al.: British national corpus. Oxford Text Archive Core Collection (2007)
4. Dale, R., Viethen, J.: The automated writing assistance landscape in 2021. Nat. Lang. Eng. **27**(4), 511–518 (2021)
5. Dankers, V., Lucas, C., Titov, I.: Can transformer be too compositional? Analysing idiom processing in neural machine translation. In: ACL, pp. 3608–3626 (2022)
6. Gamage, G., De Silva, D., Adikari, A., Alahakoon, D.: A BERT-based idiom detection model. In: HSI, pp. 1–5 (2022)
7. Haagsma, H., Bos, J., Nissim, M.: MAGPIE: a large corpus of potentially idiomatic expressions. In: LREC, pp. 279–287 (2020)
8. Jochim, C., Bonin, F., Bar-Haim, R., Slonim, N.: SLIDE - a sentiment lexicon of common idioms. In: LREC (2018)
9. Liu, P., Qian, K., Qiu, X., Huang, X.J.: Idiom-aware compositional distributed semantics. In: EMNLP, pp. 1204–1213 (2017)

10. Liu, Y., Liu, B., Shan, L., Wang, X.: Modelling context with neural networks for recommending idioms in essay writing. Neurocomputing **275**, 2287–2293 (2018)
11. Liu, Y., Pang, B., Liu, B.: Neural-based Chinese idiom recommendation for enhancing elegance in essay writing. In: ACL, pp. 5522–5526 (2019)
12. Mikolov, T., Sutskever, I., Chen, K., Corrado, G.S., Dean, J.: Distributed representations of words and phrases and their compositionality. In: NIPS, pp. 3111–3119 (2013)
13. Nunberg, G., Sag, I.A., Wasow, T.: Idioms. Language **70**(3), 491–538 (1994)
14. Řehůřek, R., Sojka, P.: Software framework for topic modelling with large corpora. In: Workshop on New Challenges for NLP Frameworks, pp. 45–50 (2010)
15. Reimers, N., Gurevych, I.: Sentence-BERT: sentence embeddings using siamese BERT-networks. In: EMNLP, pp. 3982–3992 (2019)
16. Sag, I.A., Baldwin, T., Bond, F., Copestake, A., Flickinger, D.: Multiword expressions: a pain in the neck for NLP. In: Gelbukh, A. (ed.) CICLing 2002. LNCS, vol. 2276, pp. 1–15. Springer, Berlin (2002). https://doi.org/10.1007/3-540-45715-1_1
17. Saxena, P., Paul, S.: EPIE dataset: a corpus for possible idiomatic expressions. In: Sojka, P., Kopeček, I., Pala, K., Horák, A. (eds.) TSD 2020. LNCS, vol. 12284, pp. 87–94. Springer, Cham (2020). https://doi.org/10.1007/978-3-030-58323-1_9
18. Sporleder, C., Li, L.: Unsupervised recognition of literal and non-literal use of idiomatic expressions. In: EACL, pp. 754–762 (2009)
19. Sporleder, C., Li, L., Gorinski, P., Koch, X.: Idioms in context: the IDIX corpus. In: LREC (2010)
20. Wible, D., Tsao, N.L.: StringNet as a computational resource for discovering and investigating linguistic constructions. In: Proceedings of the NAACL HLT Workshop on Extracting and Using Constructions in Computational Linguistics, pp. 25–31 (2010)
21. Williams, L., Bannister, C., Arribas-Ayllon, M., Preece, A., Spasić, I.: The role of idioms in sentiment analysis. Expert Syst. Appl. **42**(21), 7375–7385 (2015)

# Computer Vision

# DeepLOC: Deep Learning-Based Bone Pathology Localization and Classification in Wrist X-Ray Images

Razan Dibo[1], Andrey Galichin[1], Pavel Astashev[3], Dmitry V. Dylov[1], and Oleg Y. Rogov[1,2(✉)]

[1] Skolkovo Institute of Science and Technology, Moscow, Russia
`o.rogov@skoltech.ru`
[2] Artificial Intelligence Research Institute (AIRI), Moscow, Russia
[3] Pirogov National Medical and Surgical Center, Moscow, Russia

**Abstract.** In recent years, computer-aided diagnosis systems have shown great potential in assisting radiologists with accurate and efficient medical image analysis. This paper presents a novel approach for bone pathology localization and classification in wrist X-ray images using a combination of YOLO (You Only Look Once) and the Shifted Window Transformer (Swin) with a newly proposed block. The proposed methodology addresses two critical challenges in wrist X-ray analysis: accurate localization of bone pathologies and precise classification of abnormalities. The YOLO framework is employed to detect and localize bone pathologies, leveraging its real-time object detection capabilities. Additionally, the Swin, a transformer-based module, is utilized to extract contextual information from the localized regions of interest (ROIs) for accurate classification.

**Keywords:** Pathology localization · transformers · medical imaging

## 1 Introduction

The wrist bones are important as they allow for a wide range of movements in the hand and wrist, and they are necessary for everyday activities such as grasping objects, writing, typing, and using tools [21]. Additionally, the wrist bones are important for transmitting forces from the forearm to the hand and fingers, which is necessary for activities such as lifting and carrying objects [1]. Injuries or conditions affecting the wrist bones can greatly impact the function of the hand and wrist, making it important to maintain their health and integrity.

Bone fractures are a significant global public health concern that greatly affects people's well-being and quality of life. In pediatric patients, they are the leading cause of wrist trauma [9,22]. According to [17], fractures of the wrists were found to have the highest frequency of missed diagnoses, accounting for 32.4% of cases. The manual analysis of X-ray images is the current method

© The Author(s), under exclusive license to Springer Nature Switzerland AG 2024
D. I. Ignatov et al. (Eds.): AIST 2023, LNCS 14486, pp. 199–211, 2024.
https://doi.org/10.1007/978-3-031-54534-4_14

employed for fracture detection, but it is time-consuming, requiring experienced radiologists to review and report their findings to clinicians. Hence, a deficiency in the number of radiologists could significantly impact the timely provision of patient care [24]. Extensive experiments are conducted on the modified GRAZPEDWRI-DX [18] dataset of wrist X-ray images, demonstrating the effectiveness of the proposed approach. The results indicate significant improvements in both bone pathology localization and classification accuracy compared to state-of-the-art methods.

This paper encompasses the development of an advanced model capable of accurately detecting various bone diseases in X-ray images by employing bounding boxes for localization and subsequent disease classification. Unlike previous research that primarily concentrated on fractures as a classification problem, our model extends its scope to encompass a wider range of bone pathologies. To enhance the classification performance, a new approach incorporating multi-scale feature fusion and attention mechanisms is introduced, effectively capturing both local and global contextual information, enabling comprehensive representation learning for improved classification accuracy. Through the utilization of the proposed model, significant improvements in both speed and efficiency of the detection process have been achieved.

The contributions of this research exposition can be summarized as follows. *(1) Improved architecture for medical vision using GAM attention*: we enhance the YOLOv7 architecture through the integration the GAM attention mechanism. The study identifies that incorporating the attention mechanism before the detection heads yields the best results in bone pathology localization. *(2) Swin Transformer Integration:* We also explore the integration of a Swin Transformer along with the GAM mechanism, leading to further improvements in model performance. By identifying the optimal positions for incorporating these components before the detection heads, the study achieves favorable outcomes, advancing the state-of-the-art in bone pathology detection and recognition surpassing the current state-of-the-art results by 6%.

## 2    Related Work

Bone fracture detection and classification using DL techniques has been developed during the last few years on both open source and clinical bone image datasets gathered from various medical devices.

Pathare et al. used preprocessing techniques to reduce the noise of the X-ray images; the median filter is used to specifically remove the salt and pepper noise. Edge detection methods are then used to find the edge in the images. Finally, segmentation methods that used the Hough transform were applied to analyze the edges and detect the fracture. The accuracy of this system was 75%, and it was able to detect the main fractures only [20].

A deep neural network model was developed to classify healthy and fractured bones [30]. Data augmentation techniques were used to address overfitting caused by the small dataset. The model employed a convolutional neural network architecture, with three experiments testing its performance. The proposed model achieved an accuracy of 92.44% through 5-fold cross-validation. However, further validation on a larger dataset is needed to confirm its performance.

A set of 10 convolutional neural networks was developed to detect fractures in radiographs [11]. Each network had variations in architecture and output structure but used the Dilated Residual Network [31] as the basis. They were individually optimized on a training set to predict the probability of a radiograph containing a fracture and, for some networks, the probability of each pixel being part of a fracture site. The ensemble output was obtained by averaging the individual network outputs. The dataset used for training consisted of 16 anatomical regions, with some regions over-represented due to imbalanced data.

In [19], the authors proposed a deep learning-based method using YOLACT++ and the CLAHE algorithm for fracture detection in X-ray images of arm bones and trained YOLOv4 on a small dataset, incorporating data augmentation techniques to enhance the detector's performance. The method achieved an AP result of 81.91%. However, the detection of fractures in more complex wrist bones was not addressed. Similarly, the study of [7] aimed to detect wrist fracture areas in X-ray images using deep learning-based object detection models with different backbone networks. The study also featured five ensemble models to develop a unique detection system that achieved an average precision of 0.864, surpassing the individual models. However, the study focused solely on fractures, and the final models were complex.

*You Only Look Once (YOLOv7)* model is considered one of the fastest and most accurate real-time object detection model for computer vision tasks comparing to other YOLO versions and other object detection models [2,10,26]. YOLO directly predicts the bounding boxes and class probabilities from the entire image in a single feedforward pass of the neural network. Therefore, it is considered as a one-stage detection model. As opposed to two-stage models as Faster R-CNN [6], which first propose regions of interest (RoIs) in the image before classifying and refining them. In a YOLO model, image frames are featured through a backbone, which is then combined and MIXed in the neck.

*The Transformer* architecture enables information exchange among all entity pairs, such as regions or patches in computer vision. It has achieved remarkable performance in tasks like machine translation [29], object detection [8], multimodality reasoning [16], image classification [12,25,32] and image generation [23]. However, the scalability of the Transformer poses challenges due to its quadratic complexity concerning the length of input sequences [4]. To tackle this computation bottleneck, we suggested to use Swin transformer [15] which utilize non-overlapping local windows for self-attention computation while enabling cross-window consections. This way, the shifted window technique enhances the efficiency.

## 3   Methods

**Dataset.** The GRAZPEDWRI-DX dataset was used in this research [18]. It is open source with 20327 annotated pediatric trauma wrist radiograph images of

6091 patients, treated at the Department for Pediatric Surgery of the University Hospital Graz. Several pediatric radiologists annotated the images by placing bounding boxes to mark nine different classes, which are: bone anomaly, bone lesion, foreign body, fracture, metal, periosteal reaction, pronator sign, soft-tissue and text. Each X-ray image can be with more than one class. After discussion with radiologists, it was suggested to mix the classes ("foreign body" and "metal" were treated as one class name "Foreign body", "Bone anomaly" and "Bone lesion" were treated as one class called "Bone lesion") and to keep fracture and periosteal reaction classes which were considered the most important classes for detection.

**YOLOv7.** Object detector models feature a "neck" consisting of additional layers inserted between the backbone and head. The backbone is crucial for the model's accuracy and speed, composed of multiple convolutional layers that extract features at various scales. These features are then processed in the head and neck before being used to predict object locations and classes. The ELAN and ELAN-H modules, employing the CBS base convolution module, are significant parts of the Backbone component, responsible for feature extraction. In YOLOv7 architecture, MP (Max-Pooling), UP (Upsample), and RepConv (Reparameterized Convolution) are essential for the model's performance. MP downsamples feature maps to increase the receptive field, while UP upsamples to recover spatial information for better object localization. RepConv combines depthwise separable convolution and pointwise convolution to enhance model efficiency without sacrificing performance. These components enable YOLOv7 to detect objects accurately and efficiently in real-time applications. The model also incorporates the SPPCSPC module, a CSPNet with an additional SPP block that utilizes multiple pooling layers to extract region features at different scales. The inclusion of CSPNet reduces computational complexity, improves memory efficiency, and enhances inference speed and accuracy by addressing redundant gradient information.

**Transformer.** For processing the patch tokens, multiple Swin Transformer blocks are employed, consisting of a shifted window-based multi-head self-attention module (MSA) and two layers of MLP, are employed. Layer normalization (LN) is applied before each MSA and MLP, with residual connections facilitating information flow. For a hierarchical representation, the number of tokens is progressively reduced using patch merging layers as the network deepens. The first patch merging layer concatenates the features of neighboring patches grouped in $2 \times 2$ grids. A linear layer is then applied to the concatenated features, which are $4C$-dimensional features. This downsamples the number of tokens by a factor of $2 \times 2 = 4$ (equivalent to a $2\times$ reduction in resolution), and the output dimension is set to $2C$. Then, Swin Transformer blocks are applied to transform features while maintaining a resolution of $\frac{H}{8} \times \frac{W}{8}$. This process is repeated for subsequent stages and resulting in a hierarchical representation of the feature maps.

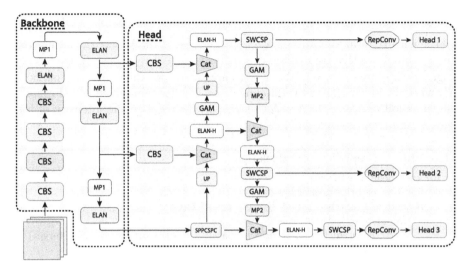

**Fig. 1.** The proposed architecture with the GAM attention block and Swin transformer.

Different attention processes that employ either channel attention or spatial attention were recently proposed. The convolutional block attention module (CBAM) considered both spatial and channel attention sequentially [28], but it ignored the channel-spatial interaction and as a result lose the cross-dimension information. Therefore, GAM was proposed [14] where 3D permutation before channel attention was applied to preserve information across 3 dimensions and a feed-forward network which consists of two linear transformations with a ReLU activation in between to amplify cross-dimension channel-spatial dependencies. Then two convolutional layers for spatial information fusion were used with batch normalization and Relu in between.

**Proposed Approach.** Based on the previous explanation for each block and method that will be used in the final architecture, the final model for bone pathology detection is denoted as DeepLOC (**Deep L**earning-based Bone Pathology **LO**calization and **C**lassification), which is obtained by inserting Swin transformer block (SWCSP) along with GAM attention block before each detection head of YOLOv7 architecture, and it is illustrated in Fig. 1 and the pseudocode for training and computing the loss is illustrated in Algorithm 1. Experimental evidence in Sect. 4 indicates that the best placement of Swin and GAM blocks is the head part before the RepConv. Anchor- based models employ a Feature-Pyramid-Network (FPN) which is presented in [13] to assist with the detection of objects of different scales. YOLOv7 also uses FPN to find the candidate bounding boxes. Center Prior consists of anchor boxes whose anchor is sufficiently close to the target box center and whose dimensions closely align with the target box size.

**Loss Function.** Following the work of [33], we utilize the complete IoU metric and the penalty term for CIoU as $\mathcal{R}_{CIoU} = \frac{\rho^2(\mathbf{b},\mathbf{b}^{gt})}{c^2} + \alpha v$. Where the parameter $\alpha$ represents a positive trade-off and is defined as: $\alpha = \frac{v}{(1-IoU)+v}$. Therefore, it assigns greater importance to the overlap area factor for the regression, particularly when dealing with non-overlapping cases. The loss function can be formally defined as follows $\mathcal{L}_{CIoU} = 1 - IoU + \mathcal{R}_{CIoU}$. To regulate the intensity of the edges $\lambda$ parameter is used [27].

---

**Algorithm 1** DeepLOC loss calculation

---

1: **for** each head of FPN **do**
2:     find Center Prior anchor boxes
3:     **if** there is a matched target "ground truth" **then**
4:         calculate objectness_loss $\mathcal{L}_O$ score as:
            **BCE** (predicted objectness probability, CIoU with the matched target)
5:     **else** $\mathcal{L}_O = 0$
6:     **end if**
7:     **if** there are any selected anchor boxes candidates **then**
8:         calculate localization_loss $\mathcal{L}_{\mathrm{Loc}}$ score as:
            **mean(1 - CIoU)**
            between all candidate anchor boxes and their matched target
9:         calculate classification_loss $\mathcal{L}_C$ score as:
            **BCE** (predicted class probabilities for each anchor box,
            one-hot encoded vector of the true class of the matched target)
10:    **else**  $\mathcal{L}_{\mathrm{Loc}} = \mathcal{L}_{\mathrm{Loc}} = 0$
11:    **end if**
12:    Multiply $\mathcal{L}_O$ by corresponding FPN head weight "predefined hyperparameter"
13: **end for**
14: Multiply each loss component:
        $\mathcal{L}_O, \mathcal{L}_{\mathrm{Loc}}, \mathcal{L}_C$ by their contribution weights (predefined $\lambda_O$, $\lambda_{\mathrm{Loc}}$, $\lambda_C$).
15: Sum the already weighted loss components.
        $\mathcal{L}_{\mathrm{sample}} = \lambda_{\mathrm{Loc}} . \mathcal{L}_{\mathrm{Loc}} + \lambda_O . \mathcal{L}_O + \lambda_C . \mathcal{L}_C$
16: Multiply the final loss value by the batch size.
17: **return** $\mathcal{L}_{total}$

---

## 4  Experiments

All the models were trained from scratch using the GRAZPEDWRI-DX dataset. The different architectures were trained for 100 epochs with a batch size of 32. The image size was set to $640 \times 640$, and the confidence threshold to 1e-3. In the feature-extraction phase, the YOLO family networks were trained using the stochastic gradient descent (SGD) optimizer with a learning rate of 1e-2, weight decay of 5e-4 and momentum of 0.937. All models were trained on an 4-Nvidia-V100s machine. Images were labeled with bounding box coordinates and class labels, which were then normalized for use in YOLOv7. The dataset was divided into 70%, 20%, and 10% splits for training, validation, and testing respectively.

YOLOv7 was trained using data preprocessing techniques like CLAHE, Unsharp masking with median and Gaussian filters, and a proposed preprocessing method called MIX. The MIX method involves resizing images to $640 \times 640$, applying mosaic and mixup with a 0.15 probability, rotation between $-15°$ and $15°$, and horizontal flipping with a 0.5 probability. Training was performed using the SGD optimizer and a batch size of 32. Performance was assessed with Average Precision (AP) computed at an Intersection over Union (IOU) of 0.5.

A total of nine anchor boxes are utilized in the training process. The width and height of each anchor box for each scale are illustrated in Table 1.

**Table 1.** The dimensions (width and height) of each anchor box for every scale/head in YOLOv7, which were determined based on the sizes of the bounding boxes in the annotations.

| Scale\Size | Anchor_1 (Width, Height) | Anchor_2 (Width, Height) | Anchor_3 (Width, Height) |
|---|---|---|---|
| Small | 12,16 | 19,36 | 40,28 |
| Medium | 36,75 | 76,55 | 72,146 |
| Large | 142,110 | 192,243 | 459,401 |

Patch size for the first stage of Swin transformer was set to $4 \times 4$, window_size was 7, number of heads in MHSA is 4, The image size was set to $672 \times 672$ to match the used window_size if Swin block is added. The query dimension of each head is $d = 32$, and the expansion layer of each MLP is $\alpha = 4$ with SilU activation [5]. CBAM and GAM attention mechanisms were tested while applied to different positions of the YOLOv7 architecture (backbone, neck, and before detection heads). The reduction ratio in MLP in the channel attention submodule which was set to 4. The same reduction ratio was used in the spatial attention submodule.

## 5    Results and Discussion

The results of the trained models are illustrated in Table 2.

Based on the obtained results, the "MIX" method, which demonstrated the highest mean average precision value, will be adopted as the preferred preprocessing and data augmentation technique for the subsequent sections.

We investigate the effects of inserting the GAM attention mechanism before each individual head and before all three heads collectively. The results of these experiments are presented and summarized in the subsequent Table 3.

Additionally, we explore the consequences of incorporating the Swin Transformer in the backbone of the architecture and before the detection heads. Finally, we examine the effects of incorporating both the Swin Transformer and

**Table 2.** The results of training YOLOv7 model with different preprocessing and augmentation techniques with SGD optimizer and batch size 32.

| Method | AP @0.5 | | | | |
|---|---|---|---|---|---|
| | F | FB | PR | BL | mAP |
| UM (M) | 0.929 | 0.572 | **0.648** | 0.049 | 0.550 |
| UM (G) | 0.932 | 0.891 | 0.605 | 0.04 | 0.617 |
| CLAHE | 0.933 | 0.870 | 0.617 | 0.044 | 0.621 |
| MIX | **0.938** | **0.912** | 0.606 | **0.057** | **0.628** |

**Note.** Here UM is the unsharp masking (UM), (M) stands for (using median filter), (G) for (Gaussian filter). WS is window size, and CL is clip limit for CLAHE. The classes are: fractures (F), foreign body (FB), Periosteal reaction (PR), Bone lesion (BL). mAP is calculated as total. CLAHE is used with parameters WS = 8 and CL = 4.

**Table 3.** The results of training YOLOv7 model with CBAM and GAM attention mechanisms in different positions of the network with SGD optimizer and batch size 32 on the test dataset.

| Model | P | R | F1 | mAP |
|---|---|---|---|---|
| Yolov7 | 0.556 | 0.582 | 0.569 | 0.628 |
| GAM (bh 1) | 0.745 | 0.574 | 0.646 | 0.634 |
| GAM (bh 2) | 0.743 | 0.543 | 0.627 | 0.636 |
| GAM (bh 3) | 0.749 | 0.614 | 0.675 | 0.638 |
| GAM (ba) | 0.751 | 0.621 | 0.680 | 0.638 |
| CBAM (ba) | 0.709 | 0.593 | 0.646 | 0.633 |

**Note.** Here (bh1), (bh2), (bh3) stand for inserting before head 1, head 2 and head 3, respectively; (ba) stands for inserting before all detection heads.

attention module in the YOLOv7 architecture, specifically before the detection heads. Our aim is to assess the impact of this combination on the overall performance of the system. The results of these experiments are presented and summarized in the following Table 6.

The proposed model DeepLOC yielded the most favorable outcomes, surpassing YOLOv7 alone by 2.6% and outperforming the insertion of Swin blocks before the detection heads by 0.3%. This model was chosen due to its superior average precision (AP) in the bone lesion class, exhibiting a notable improvement of 5.8% compared to YOLOv7 alone. Moreover, it exhibited a 2.6% enhancement in the foreign body class. Figure 2 displays precision-recall curves obtained during the testing process.

## 5.1 Ablation Study

Additional experiments were conducted to scrutinize the influence of image rotation at angles of 5°, 15°, 30°, and 60° on the model's quality. This assessment involved the calculation of mAP (mean Average Precision) in each experiment, followed by a meticulous comparison of the outcomes. As delineated in Table 4, it is evident that rotations of less than 30° yield mAP values closely aligned with those achieved by the well-trained DeepLOC model. This alignment can be attributed to our model's training process, which incorporates data augmentation techniques, including the use of rotated images as outlined in our preprocessing and augmentation methodology denoted as "MIX". In contrast, the utilization of 60° rotations for input images resulted in a notably inferior mAP score of 0.536.

The rotation of wrist images by 60° likely led to lower results due to the increased degree of distortion and deformation introduced into the image data. Such a substantial rotation can cause the model to encounter challenges in recognizing and interpreting the features, patterns, and spatial relationships within the images (Table 5).

**Table 4.** In-Depth Ablation Study: Assessing the impact of various orientation degrees on DeepLOC model performance across all classes

| Degrees | F | FB | PR | BL | mAP |
|---|---|---|---|---|---|
| 5° | 0.945 ± 0.001 | 0.917 ± 0.021 | 0.676 ± 0.026 | 0.139 ± 0.006 | 0.671 ± 0.008 |
| 15° | 0.952 ± 0.001 | 0.924 ± 0.009 | 0.688 ± 0.015 | 0.139 ± 0.022 | 0.676 ± 0.006 |
| 30° | 0.943 ± 0.002 | 0.916 ± 0.030 | 0.633 ± 0.015 | 0.117 ± 0.031 | 0.654 ± 0.008 |
| 60° | 0.823 ± 0.010 | 0.851 ± 0.012 | 0.412 ± 0.035 | 0.07 ± 0.06 | 0.536 ± 0.003 |

**Note.** The classes are: fractures (F), foreign body (FB), Periosteal reaction (PR), Bone lesion (BL). mAP is calculated as total. The results were calculated with 90% confidence level.

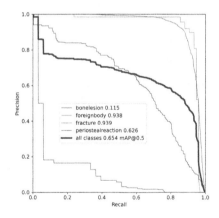

**Fig. 2.** Precision-Recall curves using DeepLOC model.

**Table 5.** Precision, Recall, mAP@.5 and mAP@.5-0.95 for all classes using DeepLOC model.

| Class | Labels | P | R | mAP@.5 | mAP@.5-0.95 |
|---|---|---|---|---|---|
| All | 2161 | 0.64 | 0.632 | **0.654** | 0.398 |
| BL | 33 | 0.165 | 0.108 | 0.115 | 0.0418 |
| FB | 76 | 0.871 | 0.934 | 0.938 | 0.76 |
| F | 1735 | 0.877 | 0.889 | 0.939 | 0.535 |
| PR | 317 | 0.648 | 0.599 | 0.626 | 0.257 |

Furthermore, experiments were performed to compare various models based on parameters, GFLOPs, layers, inference time (including pre-processing and post-processing time), and batch 32 inference time. The resulting comparisons are presented in the Table 7, illustrating the outcomes of the analysis. In addition, Table 3 illustrates the detailed results in terms of precision, recall, mAP

**Table 6.** Performance comparison for the difference models.

| Model | AP | | | | |
| --- | --- | --- | --- | --- | --- |
| | Fracture | Foreign body | Periosteal reaction | Bone lesion | mAP |
| Yolov7 | 0.938 | 0.912 | 0.606 | 0.057 | 0.628 |
| Yolov7+CBAM | 0.932 | 0.922 | 0.608 | 0.07 | 0.633 |
| Yolov7+GAM | 0.933 | 0.932 | 0.607 | 0.08 | 0.638 |
| Yolov7+SW(ba) | 0.936 | 0.928 | **0.636** | 0.106 | 0.651 |
| Yolov7+SW(B+ba) | 0.932 | 0.926 | 0.589 | 0.02 | 0.651 |
| Yolov7+SW+CBAM(ba) | 0.938 | 0.932 | 0.635 | 0.07 | 0.646 |
| DeepLOC | **0.939** | **0.938** | 0.626 | **0.115** | **0.654** |

**Note.** Here B stands for inserting the block in the Backbone, (ba) stands for inserting before all the detection heads. SW is the Swin transformer block. Reproduced with the official code under fine-tuning setting for fair comparison with SGD and batch 32.

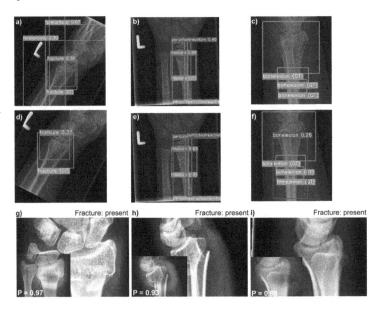

**Fig. 3.** Original X-Ray images and feature localization heatmaps. Each sub-figure presents an input image with GradCAM++ heatmaps as insets with the prediction probability inside.

for all classes using the DeepLOC model. In Fig. 3, with the ground truth represented by the blue bounding boxes, it can be observed that the YOLOv7 **(a)-(c)** architecture alone exhibited misclassifications where the model incorrectly identified the edges of the image as instances of the foreign body class. Additionally, the model accurately predicted only one out of the two instances of periosteal reaction present in the ground truth. Here YOLOv7 model fails to detect any instances of the bone lesion classes while DeepLOC correctly localizes the pathology (e.g., Fig. 3 **(d)-(f)**).We additionally visualized the GradCAM++

**Table 7.** Comparison of different model performances.

| Model | Param(M) | GFLOPs | Layers | $t_i$ | Batch average time(ms) |
|---|---|---|---|---|---|
| Yolov7+CBAM | 36.54 | 103.3 | 350 | 20.9 | 8 |
| Yolov7+GAM | 37.63 | 107.4 | 347 | 21.4 | 4.1 |
| Yolov7+SW(ba) | 42.02 | 317.8 | 395 | 18.2 | 4.1 |
| Yolov7+SW(B+ba) | 76.08 | 1614.6 | 449 | 26.7 | 15.8 |
| Yolov7+SW+CBAM(ba) | 42.04 | 317.8 | 431 | 20.3 | 5.5 |
| **DeepLOC** | 43 | 321.9 | 428 | 24.8 | 4.4 |

**Note.** Here B stands for inserting the block in the Backbone, (ba) stands for inserting before all the detection heads. SW is Swin transformer block. $t_i$ is the inference time in ms.

[3] heatmaps in Fig. 3. Images **g)** and **h)** show examples of true positive cases from the test set with easily visible fractures while in **(i)** the pathology can hardly be seen with the naked eye. Remarkably, for the True Positive cases in the dataset, on subplots **(g)-(i)**, DeepLOC identified the correct areas, where the distal radius fractures are present. In contrast, when employing our proposed model DeepLOC, notable improvements are observed. The model successfully addresses the issue of erroneous edge prediction, accurately detects both instances of periosteal reaction, and demonstrates the ability to predict one out of three instances of the Bone lesion class.

# 6    Conclusion

We emphasize the potential of incorporating new transformer architectures, specifically designed for object detection tasks, to further enhance performance. The exploration of novel attention mechanisms and the refinement of training processes to address false positives and false negatives show promising results. Optimizing confidence thresholds dynamically, expanding the dataset, and conducting ablation studies to identify influential architectural elements are suggested as important areas for further research. These efforts aim to refine and optimize object detection models for pediatric wrist trauma X-ray images, ultimately advancing the state-of-the-art in this domain.

**Acknowledgements.** The work was supported by Ministry of Science and Higher Education grant No. 075-10-2021-068.

# References

1. Berger, R.A.: The anatomy and basic biomechanics of the wrist joint. J. Hand Ther. **9**(2), 84–93 (1996)
2. Bochkovskiy, A., Wang, C.Y., Liao, H.Y.M.: Yolov4: optimal speed and accuracy of object detection. arXiv preprint arXiv:2004.10934 (2020)

3. Chattopadhay, A., Sarkar, A., Howlader, P., Balasubramanian, V.N.: Grad-CAM++: generalized gradient-based visual explanations for deep convolutional networks. In: 2018 IEEE Winter Conference on Applications of Computer Vision (WACV), pp. 839–847 (2018)
4. Chekalina, V., Novikov, G., Gusak, J., Oseledets, I., Panchenko, A.: Efficient gpt model pre-training using tensor train matrix representation. arXiv preprint arXiv:2306.02697 (2023)
5. Elfwing, S., Uchibe, E., Doya, K.: Sigmoid-weighted linear units for neural network function approximation in reinforcement learning. Neural Netw. **107**, 3–11 (2018). special issue on deep reinforcement learning
6. Girshick, R.: Fast r-cnn. In: Proceedings of the IEEE International Conference on Computer Vision, pp. 1440–1448 (2015)
7. Hardalaç, F., et al.: Fracture detection in wrist x-ray images using deep learning-based object detection models. Sensors **22**(3), 1285 (2022)
8. He, L., Todorovic, S.: DESTR: object detection with split transformer. In: Proceedings of the IEEE/CVF Conference on Computer Vision and Pattern Recognition (CVPR), pp. 9377–9386, June 2022
9. Hedström, E.M., Svensson, O., Bergström, U., Michno, P.: Epidemiology of fractures in children and adolescents: increased incidence over the past decade: a population-based study from Northern Sweden. Acta Orthop. **81**(1), 148–153 (2010)
10. Jocher, G., et al.: ultralytics/yolov5: v3.1 - Performance Improvements, October 2022
11. Jones, R.M., et al.: Assessment of a deep-learning system for fracture detection in musculoskeletal radiographs. NPJ Digit. Med. **3**(1), 1–6 (2020)
12. Lanchantin, J., Wang, T., Ordonez, V., Qi, Y.: General multi-label image classification with transformers. In: Proceedings of the IEEE/CVF Conference on Computer Vision and Pattern Recognition (CVPR), pp. 16478–16488, June 2021
13. Lin, T.Y., Dollár, P., Girshick, R., He, K., Hariharan, B., Belongie, S.: Feature pyramid networks for object detection. In: Proceedings of the IEEE Conference on Computer Vision and Pattern Recognition, pp. 2117–2125 (2017)
14. Liu, Y., Shao, Z., Hoffmann, N.: Global attention mechanism: retain information to enhance channel-spatial interactions. arXiv preprint arXiv:2112.05561 (2021)
15. Liu, Z., et al.: Swin transformer: hierarchical vision transformer using shifted windows. In: Proceedings of the IEEE/CVF International Conference on Computer Vision, pp. 10012–10022 (2021)
16. Lu, P., et al.: Learn to explain: multimodal reasoning via thought chains for science question answering. In: Koyejo, S., Mohamed, S., Agarwal, A., Belgrave, D., Cho, K., Oh, A. (eds.) Advances in Neural Information Processing Systems, vol. 35, pp. 2507–2521. Curran Associates, Inc. (2022)
17. Mounts, J., Clingenpeel, J., McGuire, E., Byers, E., Kireeva, Y.: Most frequently missed fractures in the emergency department. Clin. Pediatr. **50**(3), 183–186 (2011)
18. Nagy, E., Janisch, M., Hržić, F., Sorantin, E., Tschauner, S.: A pediatric wrist trauma x-ray dataset (grazpedwri-dx) for machine learning. Sci. Data **9**(1), 222 (2022)
19. Nguyen, H.P., Hoang, T.P., Nguyen, H.H.: A deep learning based fracture detection in arm bone x-ray images. In: 2021 International Conference on Multimedia Analysis and Pattern Recognition (MAPR), pp. 1–6. IEEE (2021)
20. Pathare, S.J., Solkar, R.P., Nagare, G.D.: Detection of fractures in long bones for trauma centre patients using hough transform. In: 2020 International Conference on Communication and Signal Processing (ICCSP), pp. 088–091. IEEE (2020)

21. Rainbow, M., Wolff, A., Crisco, J., Wolfe, S.: Functional kinematics of the wrist. J. Hand Surg. (Eur. Vol.) **41**(1), 7–21 (2016)

22. Randsborg, P.H., et al.: Fractures in children: epidemiology and activity-specific fracture rates. JBJS **95**(7), e42 (2013)

23. Razzhigaev, A., et al.: Pixel-level BPE for auto-regressive image generation. In: Proceedings of the First Workshop on Performance and Interpretability Evaluations of Multimodal, Multipurpose, Massive-Scale Models, pp. 26–30. International Conference on Computational Linguistics, Virtual, October 2022

24. Rimmer, A.: Radiologist shortage leaves patient care at risk, warns royal college. BMJ Br. Med. J. (Online) **359** (2017)

25. Selivanov, A., Rogov, O.Y., Chesakov, D., Shelmanov, A., Fedulova, I., Dylov, D.V.: Medical image captioning via generative pretrained transformers. Sci. Rep. **13**(1) (2023). https://doi.org/10.1038/s41598-023-31223-5

26. Wang, C.Y., Bochkovskiy, A., Liao, H.Y.M.: Scaled-YOLOv4: scaling cross stage partial network. In: Proceedings of the IEEE/CVF Conference on Computer Vision and Pattern Recognition, pp. 13029–13038 (2021)

27. Wang, C.Y., Bochkovskiy, A., Liao, H.Y.M.: Yolov7: Trainable bag-of-freebies sets new state-of-the-art for real-time object detectors. arXiv preprint arXiv:2207.02696 (2022)

28. Woo, S., Park, J., Lee, J.Y., Kweon, I.S.: CBAM: convolutional block attention module. In: Ferrari, V., Hebert, M., Sminchisescu, C., Weiss, Y. (eds.) Computer Vision–ECCV 2018. ECCV 2018. LNCS, vol. 11211, pp. 3–19. Springer, Cham (2018). https://doi.org/10.1007/978-3-030-01234-2_1

29. Xiao, F., et al.: Lattice-based transformer encoder for neural machine translation. In: Proceedings of the 57th Annual Meeting of the Association for Computational Linguistics, pp. 3090–3097. Association for Computational Linguistics, Florence, Italy, July 2019

30. Yadav, D., Rathor, S.: Bone fracture detection and classification using deep learning approach. In: 2020 International Conference on Power Electronics & IoT Applications in Renewable Energy and its Control (PARC), pp. 282–285. IEEE (2020)

31. Yu, F., Koltun, V., Funkhouser, T.: Dilated residual networks. In: 2017 IEEE Conference on Computer Vision and Pattern Recognition (CVPR), pp. 636–644 (2017)

32. Yu, J., Wang, Z., Vasudevan, V., Yeung, L., Seyedhosseini, M., Wu, Y.: Coca: contrastive captioners are image-text foundation models. Trans. Mach. Learn. Res. (2022). https://openreview.net/forum?id=Ee277P3AYC

33. Zheng, Z., et al.: Enhancing geometric factors in model learning and inference for object detection and instance segmentation. IEEE Trans. Cybern. **52**(8), 8574–8586 (2022)

# MiVOLO: Multi-input Transformer for Age and Gender Estimation

Maksim Kuprashevich$^{(\boxtimes)}$ and Irina Tolstykh

Layer Team, SaluteDevices, Moscow, Russia
mvkuprashevich@gmail.com

**Abstract.** Age and gender recognition in the wild is a highly challenging task: apart from the variability of conditions, pose complexities, and varying image quality, there are cases where the face is partially or completely occluded. We present MiVOLO (Multi Input VOLO), a straightforward approach for age and gender estimation using the latest vision transformer. Our method integrates both tasks into a unified dual input/output model, leveraging not only facial information but also person image data. This improves the generalization ability of our model and enables it to deliver satisfactory results even when the face is not visible in the image. To evaluate our proposed model, we conduct experiments on five popular benchmarks and achieve state-of-the-art performance, while demonstrating real-time processing capabilities. Additionally, we introduce a novel benchmark based on images from the Open Images Dataset. The ground truth annotations for this benchmark have been meticulously generated by human annotators, resulting in high accuracy answers due to the smart aggregation of votes. Furthermore, we compare our model's age recognition performance with human-level accuracy and demonstrate that it significantly outperforms humans across a majority of age ranges. Finally, we grant public access to our models, along with the code for validation and inference. In addition, we provide extra annotations for used datasets and introduce our new benchmark. The source code and data can be accessed at https://github.com/WildChlamydia/MiVOLO.git.

**Keywords:** computer vision · age recognition · gender recognition · visual transformer · multi-task model · multi-input multi-output model · human vs machine · age and gender benchmark · regression · feature fusion

## 1 Introduction

Age and gender prediction of a person in a photo is a highly important and complex task in computer vision. It is crucial for various real-world applications, including retail and clothes recognition, surveillance cameras, person identification, shopping stores and more. Additionally, this task becomes even more challenging in uncontrolled scenarios. The significant variability of all conditions such as image quality, angles and rotations of the face, partial facial occlusion,

D. I. Ignatov et al. (Eds.): AIST 2023, LNCS 14486, pp. 212–226, 2024.
https://doi.org/10.1007/978-3-031-54534-4_15

or even its absence in the image, coupled with the necessary speed and accuracy in real-world applications, makes the task quite challenging.

Our objective was to develop a simple and easy to implement approach capable of simultaneously recognizing both age and gender, even in situations where the face is not visible. We aimed for scalability and speed in our solution.

In this paper, "gender recognition" or "gender prediction" refers to a well-established computer vision problem, specifically the estimation of biological sex from a photo using binary classification. We acknowledge the complexity of gender identification and related issues, which cannot be resolved through a single photo analysis. We do not want to cause any harm to anyone or offend in any way.

Many popular benchmarks and research papers [7,14] consider age prediction as a classification problem with age ranges. However, this approach can be inaccurate because age, by its nature, is a regression problem. Moreover, it is inherently imbalanced [25]. Treating it as a classification causes several issues. For a classification model, it makes no difference whether it misclassifies into a neighboring class or deviates by several decades from the ground truth. Additionally, as stated in [25], classification models cannot approximate age ranges from unseen classes, while regression models can.

In this paper, we consider the following popular benchmarks: IMDB-Clean [18], UTKFace [28], Adience [7], FairFace [14], AgeDb [19]. These are some of the most famous datasets containing both age and gender ground truth. IMDB-Clean is the largest available dataset for this task, but it primarily consists of celebrity images and exhibits significant bias. This bias poses challenges for age and gender prediction in real-world conditions. To address these limitations, our work introduces a completely new benchmark called LAGENDA, which is well-balanced and designed to overcome these issues. For further information, please refer to Sect. 3.

While most existing works focus on estimating age and/or gender solely from face images, this work introduces the MiVOLO model, which is built upon the visual transformer model VOLO [26]. The MiVOLO allows for the simultaneous prediction of age and gender by incorporating both face and body features.

The main contributions of our work can be summarized as follows:

- We provide publicly available models that achieved SOTA results in 5 benchmarks.
- We have developed a readily implementable architecture called MiVOLO, capable of simultaneously handling faces and bodies. It enables accurate age and gender prediction, even in cases where humans may struggle. The architecture supports predictions with and without face input. MiVOLO has achieved top-1 results on 5 popular benchmarks, 3 of them without any fine-tuning on training data.
- Additionally, we have once again demonstrated that a carefully implemented multi-output (multi-task) approach can provide a significant performance boost compared to single-task models.
- We have also shown that multi-input models able to gain generalization ability in the same way as multi-task models.

- The original UTKFace dataset has been restored to include full-sized images.
- The annotations of IMDB-clean, UTK and FairFace datasets have been modified to include annotations of all detectable persons and faces in each image using our models.
- Human-level estimation for the task with a substantial sample size.
- A completely new, very well balanced Layer Age and Gender Dataset that we propose to use as a benchmark for age and gender prediction in the wild.

## 2  Related Works

**Facial Age and Gender Recognition.** With significant advancements in computer vision neural networks, many methods have been developed, some based on face recognition techniques and models [16], suggesting the existence of powerful generic models for faces that can be adapted to downstream tasks. Some papers [22] even employ more general models as encoders, such as VGG16 [23], particularly for ordinal regression approaches in age estimation. Other methods utilize CNN networks for direct classification or regression for age recognition [12,21]. As of the writing of this article, the state-of-the-art model on Adience for age classification used the Attention LSTM Networks approach [27], achieving an accuracy of 67.47. However, they did not employ their model for gender prediction.

**Recognition Using Face and Body Images.** In very few works [4], joint recognition using both face and body pictures has been utilized. The earliest attempt [9] predates the era of neural networks and employed classical image processing techniques to predict age. A more recent study [4] utilized both face and body images together in a single neural network for age and gender prediction. Another paper [10] employed face and body images with a late fusion approach, but solely for gender prediction.

**Visual Transformer Models.** One of the first transformer models applied to computer vision was ViT [6], which achieved great success and inspired the exploration of many other variants [8,24]. VOLO [26] is also a transformer-based model, but it efficiently combines the worlds of CNNs and Transformers and performs extremely well.

**Human Level for Age Estimation.** In [11], a comparison was made between the mean absolute error (MAE) of human and machine age estimation. The study examined the FG-NET dataset and the Pinellas County Sheriff's Office (PCSO) dataset (currently unavailable). The authors found that the human MAE on the FG-NET dataset [1] was 4.7, while on the PCSO dataset it was 7.2. For the machine results, they obtained 4.6 and 5.1, respectively. They also claimed that their algorithm performed worse than humans in the age range $\in [0, 15]$ years. The authors noted that this age range is not present in the FG-NET dataset [1], which caused the observed difference. When excluding this range, the estimated human MAE for FG-NET is also very close - 7.4. Eventually, the authors concluded that their model is more accurate than humans.

# 3  Datasets

## 3.1  IMDB-Clean

We primarily conducted our experiments on the IMDB-Clean dataset, which comprises 183,886 training images, 45,971 validation images, and 56,086 test images. We utilized the original split of the dataset.

For the face-only baseline, we utilized this dataset without making any modifications. For experiments involving body images, we generated face and person bounding boxes for all individuals detectable with our model in each image.

## 3.2  UTKFace

The original dataset only includes bounding boxes for cropped face images, as we performed backward matching to the original full-sized images. This process also involved double-checking by the face encoder. During this process, we encountered 4 faces that did not match back to the original images, so we dropped those images from our annotation. The remaining images maintain the original annotations but with bounding boxes generated by our detector.

The original dataset does not provide any predefined training, validation, and test splits. To align with other works, we utilized the exact same split as described in [3]. In this subset, the ages are in range $\in [21, 60]$, totalling in 13,144 training and 3,287 test images.

## 3.3  FairFace

The FairFace [14] dataset comprises 86,744 training images and 10,954 validation images. The ground truth attributes in this dataset cover race, gender, and age, categorized into nine classes: (0–2, 3–9, 10–19, 20–29, 30–39, 40–49, 50–59, 60–69, 70+). The dataset is also very well balanced by races.

For measuring gender and age classification accuracy, we utilize a validation set. To gather information about bodies, we utilize 'fairface-img-margin125' images and employ our detector model to localize the centered face and its corresponding body region.

## 3.4  Adience

The Adience [7] dataset consists of 26,580 facial images, depicting 2,284 subjects across eight age group classes (0–2, 4–6, 8–13, 15–20, 25–32, 38–43, 48–53, 60–100). The dataset additionally includes labels for the binary gender classification task.

For our analysis, we utilize coarse aligned images and refrain from applying any other aligning methods. To refine the facial localization on the images, we employ our detector model. We deliberately avoid using in-plane aligned versions of the faces to prevent distortion. The validation of our models and computation of classification accuracy are performed using all five-fold sets.

### 3.5  AgeDB

The AgeDB [19] dataset contains 16,488 facial images of celebrities. The annotations for each image include gender and age in range from 1 to 101 years old. Since each image in the dataset is a face crop, we take center crop and use it as input for our models. We use all images for validation purpose.

### 3.6  New Layer Age and Gender Dataset

**LAGENDA Benchmark.** To address issues like bias in datasets with celebrities and professional photos, we introduce a new benchmark called the Layer Age Gender Dataset (LAGENDA) in our paper. This benchmark is specifically designed for age and gender recognition tasks in diverse and challenging conditions. It was created by sampling random person images from the Open Images Dataset [17] (OID), which offers diverse scenes and domains.

The images were annotated using a crowd-sourcing platform, with strict control measures for accurate age estimation. Control tasks (honeypots) were included, with a 7-year age range within ±3 years of the true age. A ban was imposed if the accuracy on these tasks (CS@3 [5.1]) fell below 20%.

Our dataset has a 10-vote overlap for age and gender annotations, and we balanced the age distribution into 5-year groups while ensuring gender distribution within each group. It consists of 67,159 images with 84,192 persons, including 41,457 males and 42,735 females. See Fig. 1 for a visualization of the dataset distribution.

**Votes Ensembling.** After completing the annotation process, we encountered the challenge of determining how to utilize the obtained votes.

Table 2 provides a list of all the methods that were tested. In addition to other statistical methods, we employed a weighted mean strategy. It was implemented as follows:

$$A(v) = \frac{\sum_{i=1}^{N} v_i * e^{(MAE(u_i))^{-1}}}{\sum_{i=1}^{N} e^{(MAE(u_i))^{-1}}}$$

where $A$ is final age prediction for the $v$ vector of user votes, $N$ is size of $v$, amount of users who annotated this sample and $MAE(u_i)$ denotes the individual MAE across all control tasks for the $i$-th user $u$ (Fig. 2).

Gender was aggregated using the simple $mode(v)$, where $v$ is an array of elements $\in male, female$. We discarded all answers where the mode occurred with a frequency of less than 75%. Based on control tasks, the gender accuracy has to be 99.72%.

**LAGENDA Trainset.** In addition to our benchmark data, we gathered training data from other sources that closely resemble the visual domain of OID [17] images. Our training dataset comprises around 500,000 images, annotated in the exact same manner as the LAGENDA benchmark.

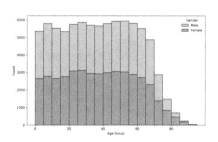

| Method | MAE | CS@5 |
|---|---|---|
| Mean | 4.77 | 62.43 |
| Median | 4.75 | 65.44 |
| Interquartile Mean | 4.74 | 63.80 |
| Mode | 5.70 | 59.28 |
| Maximum Likelihood | 4.81 | 65.86 |
| Winsorized mean (6) | 4.73 | 63.44 |
| Truncated mean (0.3) | 4.75 | 63.62 |
| **Weighted mean** | **3.47** | **74.31** |

**Fig. 1.** Age and gender distributions with bin steps of 5 in the LAGENDA benchmark.

**Fig. 2.** Different statistic methods to aggregate N votes into one age prediction.

In the text, we refer to this training and validation proprietary data as LAGENDA trainset. Although we cannot make this data publicly available, we provide a demo with the model trained on it (the link can be found in the Github repository).

## 4 Method

### 4.1 MiVOLO: Multi-input Age and Gender Model

Our model is depicted in Fig. 3.

For each input pair $(FaceCrop, BodyCrop)$ of size $224 \times 224$, we independently apply the original VOLO [26] patch embedding module, which tokenizes the crops into image patches of size $8 \times 8$.

Two representations are then fed into a feature enhancer module for cross-view feature fusion, which is achieved using cross-attention. The module is illustrated in Fig. 3, block 2. Once the features are enriched with additional information, we perform a simple concatenation, followed by a Multi-Layer Perceptron (MLP) that creates a new fused joint representation and reduces the dimensionality of the features.

This feature fusion allows us to pay attention to important features from both inputs and disregard less significant ones. Additionally, it handles scenarios where one of the inputs is empty, ensuring meaningful information is extracted even from a single view.

The fused features are then processed using the VOLO two-stage architecture [26], which involves a stack of Outlookers, tokens downsampling, a sequence of transformers, and two heads on top.

The last two linear layers update the class embedding into a 3-dimensional vector: two output values for gender and one for the normalized age value. Unlike [2], which uses multiple heads for separate age and gender predictions, MiVOLO produces a single vector for each image containing both outputs.

We use combination of two losses for training:

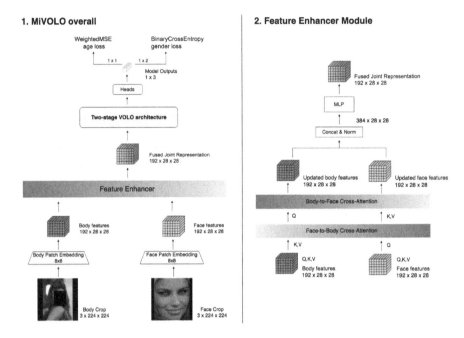

**Fig. 3.** MiVOLO. We present the overall model and a feature enhancer module in block 1 and block 2 respectively.

- WeightedMSE loss function for age prediction with weights from LDS [25]
- BinaryCrossEntropy loss function for gender prediction

Moreover, early feature fusion allows us to maintain almost the same high performance as that of the original VOLO (see Sect. 4.2).

### 4.2   Performance

We take the VOLO-D1 model variation, which has 25.8M parameters, as our baseline. In contrast, the MiVOLO-D1 model has 27.4M parameters. When utilizing $float16$ precision with a batch size of 512 on the NVIDIA V100, the original VOLO-D1 achieves a frame rate of 1200 FPS. On the other hand, MiVOLO-D1, under identical conditions, demonstrates slightly lower performance but still maintains a high frame rate of 971 FPS. Figure 4 demonstrates models performance comparison with different batch sizes.

## 5    Experiments

### 5.1   Evaluation Metrics

In this section, we present the model's performance using various metrics. For gender prediction and age prediction in classification benchmarks, we utilize the classification accuracy metric.

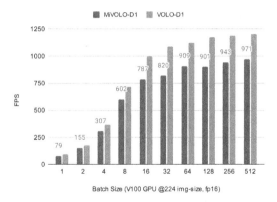

**Fig. 4.** Models performance comparison with a single V100 GPU: MiVOLO-D1 for age and gender estimation by face and body images and single-input VOLO-D1 for age estimation only by face images.

In regression age benchmarks, the model's performance is evaluated based on two metrics: Mean Absolute Error (MAE) and Cumulative Score (CS). MAE is calculated by averaging the absolute differences between the predicted ages and the actual age labels in the testing dataset. CS is computed using the following formula:

$$CS_l = \frac{N_l}{N} \times 100\%$$

Here, $N$ represents the total number of testing examples, while $N_l$ denotes the count of examples for which the absolute error between the estimated age and the true age does not exceed $l$ years.

## 5.2 VOLO Experiments

We conducted experiments on IMDB-clean and UTKFace datasets to establish a good baseline and identify model limitations. In this section original images, annotations and data splits were taken.

For the age estimation task, **our baseline model, VOLO-D1**, was trained using only the face input. We employed the *AdamW* optimizer with an initial learning rate of $1.5e - 5$ and a weight decay of $5e - 5$. The model was trained for 220 epochs individually on both the IMDB-clean and UTKFace datasets. The base learning rate batch size was set to 192. At the start of training, we performed a warmup with $lr = 1e - 6$ for 25 epochs with gradual increase.

The *RandAugment*, *RandomJitter* for bounding box position and size, *Reprob* and *RandomHorizontalFlip* data augmentations were applied during training: Additionally, we incorporated *drop* and *drop-path* with $p = 0.32$.

As shown in Table 1 our results are state-of-the-art without any additional data or advanced techniques on IMDB-clean and UTKFace datasets.

**Table 1.** Comparison of accuracy of VOLO-D1 models and previous SOTA results. **Bold** indicates the best model, trained and evaluated on the same datasets. <u>**Bold**</u> indicates the best model with additional train data. † marks the models that we release to the public domain.

| Model | Output | Train Dataset | Test Dataset | Age MAE | Age CS@5 | Gender Acc |
|---|---|---|---|---|---|---|
| FP-Age [18] | age | IMDB-clean | IMDB-clean | 4.68 | 63.78 | – |
| VOLO-D1 face† | age | IMDB-clean | IMDB-clean | 4.29 | 67.71 | – |
| | | | UTKFace | 5.28 | 56.79 | – |
| | | | Lagenda test | 5.46 | 57.90 | – |
| VOLO-D1 face† | age & gender | IMDB-clean | **IMDB-clean** | **4.22** | **68.68** | **99.38** |
| | | | UTKFace | 5.15 | 56.79 | 97.54 |
| | | | Lagenda test | 5.33 | 59.17 | 90.86 |
| CORAL [5] | age | UTKFace | UTKFace | 5.39 | – | – |
| Randomized Bins [3] | age | UTKFace | UTKFace | 4.55 | – | – |
| MWR [22] | age | UTKFace | UTKFace | 4.37 | – | – |
| VOLO-D1 face† | age | UTKFace | IMDB-clean | 8.59 | 37.96 | – |
| | | | UTKFace | 4.23 | 69.72 | – |
| | | | Lagenda test | 11.16 | 30.51 | – |
| VOLO-D1 face† | age & gender | UTKFace | IMDB-clean | 8.06 | 41.72 | 97.05 |
| | | | **UTKFace** | **4.23** | **69.78** | **97.69** |
| | | | Lagenda test | 11.37 | 30.20 | 83.27 |
| VOLO-D1 face | age | Lagenda train | IMDB-clean | 4.13 | 69.33 | – |
| | | | UTKFace | 3.90 | 72.25 | – |
| | | | Lagenda test | 4.19 | 69.36 | – |
| VOLO-D1 face | age & gender | Lagenda train | IMDB-clean | <u>4.10</u> | <u>69.71</u> | <u>99.57</u> |
| | | | UTKFace | <u>3.82</u> | <u>72.64</u> | <u>98.87</u> |
| | | | **Lagenda test** | 4.11 | 70.11 | 96.89 |

**For the age & gender VOLO-D1 model**, we followed the same training process. To address the discrepancy in the magnitudes of loss between age and gender, we weighted the gender loss with $w = 3e - 2$. We did not change anything else, including the number of epochs.

By adding a second age output to the model, we observed the same effect as reported in the study [15], where a single model performs better than multiple separate models, leveraging the benefits of learning two tasks simultaneously.

We perform identical experiments on our LAGENDA training set. Please refer to Table 1 for the results.

## 5.3   MiVOLO Experiments

The main objective of this paper was to create a model that can perform well even in cases without a visible face, in a joint manner, making it suitable for real-world applications with mixed batches of person and face variations.

We made some minor adjustments to the training process for the MiVOLO model. To reduce training time, we initialized the model from a single-input multi-output VOLO checkpoint. We initialized weights

**Table 2.** Comparison of multi-input MiVOLO-D1 and single-input VOLO-D1 age & gender models accuracy. **Bold** indicates the best model for each benchmark. † marks the model that we release to the public domain.

| Model | Train Set | Test Set | Face | | | Body | | | Face&Body | | |
|---|---|---|---|---|---|---|---|---|---|---|---|
| | | | MAE | CS@5 | Gender Acc | MAE | CS@5 | Gender Acc | MAE | CS@5 | Gender Acc |
| volo-d1 | Lagenda | IMDB | 4.10 | 69.71 | **99.57** | – | – | – | – | – | – |
| | | UTKFace | 3.82 | 72.64 | **98.87** | – | – | – | – | – | – |
| | | Lagenda | 4.11 | 70.11 | 96.89 | – | – | – | – | – | – |
| mivolo-d1† | IMDB | IMDB | 4.35 | 67.18 | 99.39 | 6.87 | 46.32 | 96.48 | 4.24 | 68.32 | 99.46 |
| | | UTKFace | 5.12 | 59.10 | 97.66 | 6.36 | 47.74 | 95.57 | 5.10 | 97.72 | 59.46 |
| | | Lagenda | 5.40 | 58.67 | 91.06 | 10.52 | 31.70 | 87.71 | 5.33 | 59.20 | 91.91 |
| mivolo-d1 | Lagenda | IMDB | 4.15 | 69.20 | 99.52 | 6.66 | 47.53 | 96.74 | **4.09** | **69.72** | 99.55 |
| | | UTKFace | 3.86 | 72.06 | 98.81 | 4.62 | 63.81 | 98.69 | **3.70** | **74.16** | 98.84 |
| | | Lagenda | 4.09 | 70.23 | 96.72 | 7.41 | 49.64 | 93.57 | **3.99** | **71.27** | **97.36** |

of the *body PatchEmbedding* block with the same weights as the *face PatchEmbedding* block. The *Feature Enhancer Module* was initialized with random parameters.

During training, we froze the *face PatchEmbedding* block since it was already trained. We trained the model for an additional 400 epochs, incorporating random dropout of the body input with a probability of 0.1, and random dropout of the face input with a probability of 0.5. Face inputs were only dropped for samples with suitable body crops. If a face input was dropped, the model received an empty (zero tensor) input for *face PatchEmbedding*, and the same for empty body inputs.

These techniques were implemented to adapt the model for various mixed cases and to improve its understanding of input images, resulting in enhanced generalization. We also set the learning rate to $1e-5$. To preserve the structural integrity of the data, all augmentations, excluding jitter, are applied simultaneously.

The remaining parts of the training procedure are unchanged.

We conducted experiments on the IMDB-clean dataset and LAGENDA training set using our MiVOLO. Table 2 shows a comparison between the single-input VOLO and the multi-input MiVOLO. The results indicate that the best performance across all benchmarks is achieved by using both face and body crops. The model trained on our dataset consistently outperforms the one trained on IMDB.

To evaluate the quantitative performance of the MiVOLO when only body images are available, we conducted an experiment where all faces were removed from the data. Additionally, we excluded any images that did not meet our specified requirements mentioned in Sect. 4.1. For IMDB-clean, UTKFace and Lagenda test datasets retained 84%, 99.6% and 89% of images, respectively. Results are displayed in the Table 2.

In Fig. 5, we provide an illustration of a successful recognition result without visible faces in a random picture sourced from the internet. Model generalizes

**Fig. 5.** An illustration of a case where the work is performed without faces on a random picture obtained from the internet.

very well, even though it has never seen images like this with persons shown from the back.

Relationship between MAE and age for the final models is shown in Fig. 6 (a) and (b).

### 5.4    Adience, FairFace, AgeDb Benchmarks

Due to the model's impressive generalization capabilities, we decided to apply MiVOLO to the AgeDb [19] regression benchmark and to popular classification benchmarks such as FairFace [14] and Adience [7]. As our model was not explicitly trained for classification tasks, we applied our final MiVOLO-D1 age & gender model to FairFace and Adience without any modifications. The only change made was mapping the regression output to classification ranges. As shown in Table 3, we achieved SOTA results for the mentioned datasets without any additional changes.

## 6    Human Level Estimation and Votes Ensembling for Age Recognition

### 6.1    Human Level for Age Estimation

As described in Sect. 3, during the annotation of the LAGENDA, control tasks (honeypots) were generated from the IMDB-clean dataset. A total of 3,000 random examples were sampled for this purpose. Users were not aware of which examples were honeypots and annotated them alongside other tasks. This approach provided a reliable source for estimating the human level of performance in the task.

**Table 3.** FairFace, Adience, AgeDB validation results using MiVOLO-D1 trained on LAGENDA train set.

| Method | Test Set | Age Acc | Age MAE | Gender Acc |
|---|---|---|---|---|
| FairFace [14] | FairFace | 59.70 | | 94.20 |
| **MiVOLO-D1 Face&Body** | FairFace | **61.07** | | **95.73** |
| DEX [19,20] | AgeDB | | 13.1 | – |
| **MiVOLO-D1 Face** | AgeDB | | **5.55** | **98.3** |
| MWR [22] | Adience | 62.60 | | – |
| AL-ResNets-34 [27] | Adience | 67.47 | | – |
| Compacting [13] | Adience | – | | 89.66 |
| Gen MLP [16] | Adience | – | | 90.66 |
| **MiVOLO-D1 Face** | Adience | **68.69** | | **96.51** |

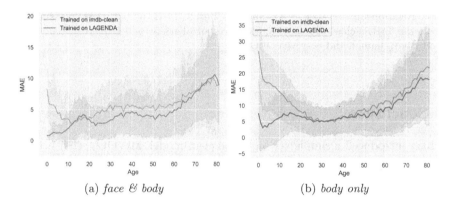

(a) *face & body*          (b) *body only*

**Fig. 6.** Relationship between MAE and age for MiVOLO. Tested on LAGENDA benchmark using: a) face & body; b) body only.

Figure 8 illustrates the distribution of MAE values among the users. The mean of this distribution is 7.22, and the median is 7.05. The averaged maximum error is 28.56, while the minimum mean error for a specific user is 4.54.

We have briefly described paper [11] in Sect. 2. We disagree with the method of excluding certain age ranges as it can potentially lead to incorrect conclusions. To accurately compare human and machine performance, it is crucial to take into account the entire range of ages and images from the wild domain.

As can be seen in Fig. 7, the previous suggestion about low neural network and high human performance in the age range of $[0, 15]$ years no longer holds.

With a human MAE of 7.22 ($\sigma = 1.31$, $n = 2,934$) and a machine MAE of 4.09 ($\sigma = 3.69$, $n = 56,086$), we can state with a high level of confidence that the machine is more accurate than humans.

Furthermore, as shown in Table 2, the model achieved an MAE of 6.66 ($\sigma = 5.66$, $n = 47,185$) on IMDB-clean with body-only images. This demonstrates that, on average, our model outperforms humans even when considering body-only mode.

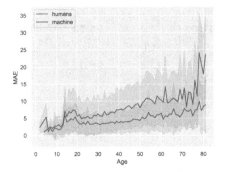

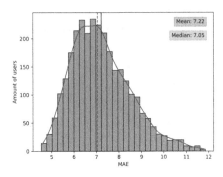

**Fig. 7.** Relationship between MAE and age on the IMDB-clean dataset, annotated by both human and best MiVOLO-D1.

**Fig. 8.** MAE across users, measured on control tasks (n ≥ 20).

# 7    Conclusions

We have introduced a simple yet highly efficient model, MiVOLO, that achieves state-of-the-art results on 5 benchmarks, demonstrating its capability to function robustly even in the absence of a face image.

To contribute to the research community, we are providing the weights of the models, which have been trained on Open Sourced data.

In addition, we have enriched and expanded the annotations of 3 prominent benchmarks, IMDB-clean, UTKFace and FairFace. Furthermore, we have developed our own diverse and unbiased Layer Age and Gender Dataset, which contains challenging real-world images and is publicly available.

For the task of age annotation aggregation, we employed an intuitive yet remarkably accurate method and evaluated its performance.

Our investigation into the comparison of human and machine accuracy in age recognition tasks revealed that our current model consistently outperforms humans across various age ranges, exhibiting superior overall accuracy.

We also need to highlight some limitations of the model:

Despite its high accuracy, we urge caution against drawing hasty conclusions. Like all neural network models, it can produce errors, including hallucinations and significant mistakes in edge cases that are readily apparent to humans.

The model was developed to address practical tasks for our business, so the correct answer for us may not always align with what a human might expect.

Additionally, it's important to note that the current model is not designed to work with altered images. Popular social media filters and effects, such as blurring and others, can significantly impact the results.

# References

1. Fgnet aging database (2004). https://yanweifu.github.io/FG_NET_data/
2. Abdolrashidi, A., Minaei, M., Azimi, E., Minaee, S.: Age and gender prediction from face images using attentional convolutional network (2020)
3. Berg, A., Oskarsson, M., O'Connor, M.: Deep ordinal regression with label diversity (2020). https://doi.org/10.1109/ICPR48806.2021.9412608
4. Bonet Cervera, E.: Age & gender recognition in the wild, September 2022. https://upcommons.upc.edu/handle/2117/373777
5. Cao, W., Mirjalili, V., Raschka, S.: Rank consistent ordinal regression for neural networks with application to age estimation (2019). https://doi.org/10.1016/j.patrec.2020.11.008
6. Dosovitskiy, A., et al.: An image is worth 16 × 16 words: transformers for image recognition at scale (2020)
7. Eidinger, E., Enbar, R., Hassner, T.: Age and gender estimation of unfiltered faces. IEEE Trans. Inf. Forensics Secur. 9(12), 2170–2179 (2014). https://doi.org/10.1109/tifs.2014.2359646
8. El-Nouby, A., et al.: XCiT: cross-covariance image transformers (2021)
9. Ge, Y., Lu, J., Fan, W., Yang, D.: Age estimation from human body images, pp. 2337–2341, October 2013. https://doi.org/10.1109/ICASSP.2013.6638072
10. Gonzalez-Sosa, E., Dantcheva, A., Vera-Rodriguez, R., Dugelay, J.L., Brémond, F., Fierrez, J.: Image-based gender estimation from body and face across distances. In: 2016 23rd International Conference on Pattern Recognition (ICPR), pp. 3061–3066 (2016). https://doi.org/10.1109/ICPR.2016.7900104
11. Han, H., Otto, C., Jain, A.K.: Age estimation from face images: human vs. machine performance. In: 2013 International Conference on Biometrics (ICB), pp. 1–8 (2013). https://doi.org/10.1109/ICB.2013.6613022
12. Hou, L., Samaras, D., Kurc, T.M., Gao, Y., Saltz, J.H.: ConvNets with smooth adaptive activation functions for regression. Proc. Mach. Learn. Res. 54, 430–439 (2017)
13. Hung, C.Y., Tu, C.H., Wu, C.E., Chen, C.H., Chan, Y.M., Chen, C.S.: Compacting, picking and growing for unforgetting continual learning (2019)
14. Karkkainen, K., Joo, J.: Fairface: face attribute dataset for balanced race, gender, and age for bias measurement and mitigation. In: Proceedings of the IEEE/CVF Winter Conference on Applications of Computer Vision, pp. 1548–1558 (2021)
15. Kendall, A., Gal, Y., Cipolla, R.: Multi-task learning using uncertainty to weigh losses for scene geometry and semantics (2017)
16. Kim, T.: Generalizing MLPs with dropouts, batch normalization, and skip connections (2021)
17. Kuznetsova, A., et al.: The open images dataset v4: unified image classification, object detection, and visual relationship detection at scale (2018). https://doi.org/10.1007/s11263-020-01316-z
18. Lin, Y., Shen, J., Wang, Y., Pantic, M.: FP-Age: leveraging face parsing attention for facial age estimation in the wild. arXiv (2021)
19. Moschoglou, S., Papaioannou, A., Sagonas, C., Deng, J., Kotsia, I., Zafeiriou, S.: Agedb: the first manually collected, in-the-wild age database. In: Proceedings of the IEEE Conference on Computer Vision and Pattern Recognition Workshop, vol. 2, p. 5 (2017)
20. Rothe, R., Timofte, R., Gool, L.V.: DEX: deep expectation of apparent age from a single image, December 2015

21. Shaham, U., Zaidman, I., Svirsky, J.: Deep ordinal regression using optimal transport loss and unimodal output probabilities (2020)
22. Shin, N.H., Lee, S.H., Kim, C.S.: Moving window regression: a novel approach to ordinal regression (2022)
23. Simonyan, K., Zisserman, A.: Very deep convolutional networks for large-scale image recognition (2014)
24. Touvron, H., Cord, M., Sablayrolles, A., Synnaeve, G., Jégou, H.: Going deeper with image transformers (2021)
25. Yang, Y., Zha, K., Chen, Y.C., Wang, H., Katabi, D.: Delving into deep imbalanced regression (2021)
26. Yuan, L., Hou, Q., Jiang, Z., Feng, J., Yan, S.: Volo: Vision outlooker for visual recognition (2021)
27. Zhang, K., et al.: Fine-grained age estimation in the wild with attention LSTM networks (2018)
28. Zhang, Z., Song, Y., Qi, H.: Age progression/regression by conditional adversarial autoencoder. In: IEEE Conference on Computer Vision and Pattern Recognition (CVPR). IEEE (2017)

# Handwritten Text Recognition and Browsing in Archive of Prisoners' Letters from Smolensk Convict Prison

Nikita Lomov[1,2]([⊠]) [iD], Dmitry Kropotov[1,3] [iD], Danila Stepochkin[3],
and Anton Laptev[1] [iD]

[1] HSE University, Myasnitskaya str., 20, Moscow 101000, Russia
`nikita-lomov@mail.ru, aklaptev@hse.ru`
[2] Tula State University, Lenina Avenue, 92, Tula 300012, Russia
[3] Lomonosov Moscow State University, GSP-1, Leninskie Gory,
Moscow 119991, Russia

**Abstract.** The task of creating a prototype navigation system for a small archive of historical documents (letters from prisoners of the Smolensk convict prison of the early 20th century) recorded in a single handwriting, is considered. To fit a model for handwritten text recognition, procedures were created for automatic preparation of image collections, including breaking into lines, pen trace segmentation, and deslanting of lines and pages. Experiments have shown that training a modern neural network on about a thousand line samples with the same handwriting allows achieving a decent recognition quality (5.11% CER and 17.55% WER). Further, the automatically recognized text was used for the task of searching by keywords. The text was corrected by dictionaries and prescribed rules, taking into account the peculiarities of Russian pre-reform spelling, recognition errors and the scriptor's own errors. The search engine reached a precision of 97.14% and a recall of 91.35%. Visualization of the results provided highlighting of the found words on the original images. The study conducted demonstrates the possibility of creating a navigation system and its fitting to a specific handwriting with a small number of marked samples and limited human participation.

**Keywords:** Handwritten Text Recognition · Layout Analysis · Vertical Attention · Lexical Analysis · Manuscript Navigation

## 1 Introduction

Deep learning models show considerable progress in solving handwritten text recognition (HTR) problem. Notable approaches here are transformer-based model TR-OCR [12] and different convolutional architectures with attention VAN [7], DAN [8]. This success opens up opportunities for automated intelligent processing of historical archives thus facilitating for many researchers searching and analysis of information from primary sources. However, qualitative processing of real-world historical archives is a challenging problem and requires careful

D. I. Ignatov et al. (Eds.): AIST 2023, LNCS 14486, pp. 227–240, 2024.
https://doi.org/10.1007/978-3-031-54534-4_16

pre- and post-processing procedures [13,28]. The situation here gets more complicated by limited amount of available HTR datasets for a particular language that can be used for training deep learning models.

In the paper, we consider the processing problem of prisoners' letters from Smolensk convict prison of early 20th century written by one scriptor in Russian language. Unfortunately, there are only few available HTR datasets in Russian, e.g., Digital Peter [15] and HKR [14]. Both datasets are not quite suitable for the considered problem since they contain either outdated orthography or too short phrases that are not representative for full paragraph handwritten documents. Hence, we present an approach where neural networks are learned from scratch using only data from the archive. Besides, we propose a series of pre- and postprocessing steps such as automatic text line extraction, their deslanting and denoising, extracted text postprocessing using large language models, etc. In experiments we show that proposed processing procedures are indeed crucial for obtaining good performance. Also we present results for keywords and personal names search that is important for researchers working with the archive.

In sum, our main contribution is two-fold: 1) we present processing results for the historical archive in Russian language and thus expand very limited research works available for this language, 2) we propose a processing pipeline with several pre- and post-processing steps that is crucial for obtaining good performance for handwritten text recognition and keywords search problems, and that can potentially be used for processing other historical archives.

## 2   Data Description

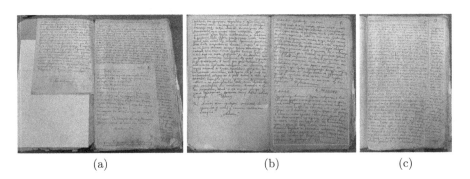

| (a) | (b) | (c) |

**Fig. 1.** Images from the archive of convict letters. Individual text fragments are enclosed in frames.

The input data are 67 photographs of a notebook with letters from prisoners of the Smolensk convict prison dated to period of the First Russian Revolution (1907–1908)[1]. These letters, transmitted inside the prison between political prisoners, were secret and were to be destroyed after reading.

---

[1] State Archive of the Smolensk Oblast (GASO). F. 404. Op. 1. D. 631.

Preserved due to perusal, the letters were copied by the same gendarme responsible for the covert work with the revolutionary elements in the convict prison, what causes a single handwriting for the archive of letters. The key aspect in the work on perlustration was the most accurate copying of the text: the transfer of typos and spelling errors, writing style, spelling and punctuation features (for example: intentional use of capital letters, lack of punctuation marks, etc.). The reason for such accuracy when copying was the use of ciphers by revolutionary prisoners, as well as metaphors and specific vocabulary, understandable only to the recipient. Thus, convict letters are valuable material for historians due to the preservation of both the form of the original letter and its content.

The photographs contained page spreads and sheets of different sizes overlapping each other, in addition, fragments of different letters could be located on one page. While data preparation, 87 image of unique text paragraphs, corresponding to fragments of 61 letters, were manually extracted (see Fig. 1). For each fragment, a transcript was prepared in two versions of spelling: modern and pre-reform, with all the features of writing including line breaking reproduced in the second one.

## 3   Layout Analysis

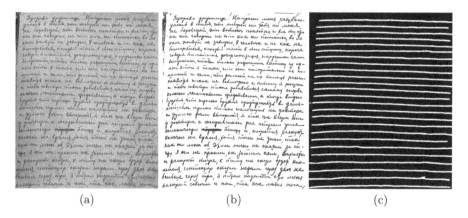

(a)                                 (b)                                 (c)

**Fig. 2.** Neural network preprocessing. (a) Original image, (b) pen trace extraction, (c) initial extraction of baselines.

**Used Tools.** In addition to text fragments extraction, typical layout analysis procedures are foreground text binarization [21] and highlighting lines of text. The former is solved by such methods as local contrast enhancement with SVM classification [27], segmenting neural networks [9] and generative adversarial networks for binarization quality assessment [10]. For the latter one, postprocessing of segmentation minimizing the energy of cutting [1] or analyzing axis of inertia [3] and setting the problem as regression of thinned and smoothed text lines [17] are applied.

We use the output of the existing U-Net based neural networks, shown in Fig. 2, as a basis for preprocessing. For binarization, a neural network prepared by S. Liedes[2] is applied. To create the initial baselines, we utilized one of the networks from the dhSegment package[3] that contains a ResNet-50-based architecture in the convolutional part.

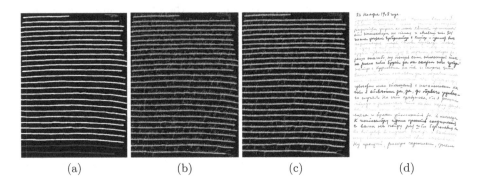

(a)          (b)          (c)          (d)

**Fig. 3.** Postprocessing of baselines. (a) Deletion and linking of components, (b) lines before vertical level correction, (c) lines after vertical level correction, (d) distribution of pen trace components over lines.

**Deleting and Linking Baseline Components.** To overcome degradations specific for this particular dataset (e.g. consideration of decorative elements as pen trace and selection of non-informative marks such as the word <<Копия>> ["Copy"]), the procedure for correcting baselines (see Fig. 3,a) includes the following steps:

- connection of components in the event that the projection of the smaller one onto the $X$ axis intersects with the projection of the larger one by no more than 50%, and there is such a pair of points $p_1 = (x_1, y_1)$, $p_2 = (x_2, y_2)$ such that $|x_2 - x_1| \leq 100$ and $|y_2 - y_1| \leq 20$;
- deletion of a component if its area is less than 500 pixels, or its center of mass is less than 150 pixels away from the upper right corner of the image, or the proportion of pixels of the pen trace on the upper border exceeds 60%.

**Vertical Level Correction.** For further distribution of text components by lines, it is more convenient to have line markings not in the form of baselines, but in the form of mean lines filling the entire area between the upper and lower boundaries of the letters. For each of the $n$ rows, ordered from top to bottom, determine $y_{avg}^{(k)}(x)$, the average vertical position at position $x$, by averaging the

[2] https://github.com/sliedes/binarize.
[3] https://github.com/dhlab-epfl/fdh-tutorials/releases/download/v0.1/line_model. zip.

upper and lower $y$-coordinates of the points of the base lines, $y_{\text{top}}^{(k)}(x)$ and $y_{\text{bot}}^{(k)}(x)$. As a standard line spacing $h$ on a page, take the median of distances between consecutive lines $\{y_{\text{avg}}^{(k+1)}(x) - y_{\text{avg}}^{(k)}(x)\}$ for $k = 1, \ldots, n-1$ and all admissible $x$.

Next, for each row, we define the $p$-th upper level obtained by indenting the component up by $p$ pixels: $T_p^{(k)} = \{(x, y_{\text{top}}^{(k)}(x) - p)\}$, and the $q$-th lower level obtained by offsetting the component down by $q$ pixels: $B_q^{(k)} = \{(x, y_{\text{bot}}^{(k)}(x)+q)\}$. After that, $t_{\text{top}}$ upper levels are added to the line and $t_{\text{bot}}$ lower levels are removed from the line. To determine the values of $t_{\text{top}}$ and $t_{\text{bot}}$, a procedure similar to the Otsu binarization method, that minimizes the intraclass variance for mean pen trace intensities along the levels, is used.

**Assigning Pen Trace Components to Lines.** The connected components of the binarized and morphologically dilated pen trace are analyzed based on the following rules:

- if more than 80% points of a component are at a distance of at least $0.6h$ from all points of the mean lines, the component is considered noise and belongs to the background;
- otherwise, the proportion of pixels that fall into each base line and the number of lines with a share greater than 10% is calculated:
- if there are no such lines, the component is assigned to the line with the largest number of closest pixels for component points;
- if there is only one such line, the item is assigned to that line;
- if there are several such lines, the component is cut between adjacent lines: until the component "falls apart", horizontal sections with the minimum value of the criterion $(l+4)(\max(d_1, d_2)+1)/(\min(d_1, d_2)+1)$, where $l$ is the section length, $d_1$ and $d_2$ are the vertical distances from the section to two lines. Next, the line assignment procedure is called recursively for new components.

The postprocessing procedures described in the previous paragraphs are illustrated by Fig. 3.

**Deslanting Lines and Pages.** For dewarping text documents, the straightening coordinate transformation is usually calculated: using neural networks [26], optimization procedures [11], analytical solutions [4], and reprojection of a two-dimensional distortion grid into space [22].

To straighten the paragraph images it is necessary to shift all lines in a consistent manner. Let us define for each base line its center of mass $(\bar{x}^{(k)}, \bar{y}^{(k)})$. Next, we require that, as a result of straightening, the base line of the row $y_{\text{avg}}^{(k)}(x)$ becomes equal to $\bar{y}^{(k)}$. Thus, for each $x$ we have a function whose value at the points $\{y_{\text{avg}}^{(k)}(x)\}_{k=1}^{n}$ is known, and the other values are determined by linear interpolation. The colors of the pixels that go beyond the boundaries of the original image were replaced by the colors of pixels closest to them. The result of page straightening is shown in Fig. 4.

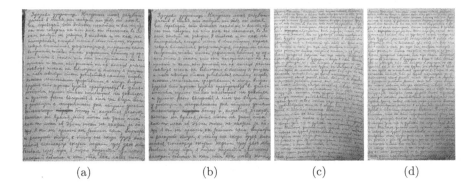

**Fig. 4.** Page deslanting. (a,c) Original images, (b,d) deslanted images.

Individual lines were extracted from the straightened page image according to the bounding boxes of the corresponding pen trace fragment, pixels related to the pen trace of other lines were repainted to the color of the nearest background pixels. The result of straightening and clearing line images is shown in Fig. 5.

**Fig. 5.** Line deslanting. Baselines (top row), pen trace masks (second row), extracted lines with noisy trace (third row) and with denoised trace (bottom row). On the images of the left column, the slant is not eliminated, of the right one — is eliminated.

As a result of the pre-processing procedures, we got a database of 1782 straightened lines without artifacts, and also straightened 87 of images with paragraphs of text.

## 4   Text Recognition and Correction

**Handwritten Text Recognition by Neural Networks.** While a fully convolutional network may be sufficient to recognize a line of text, page-level architectures face the challenge of determining the order in which image fragments are viewed. In [25], the beginnings of the lines are detected by separate module, in [6] horizontal feature blocks are concatenated into one line, in [29] the sequence of horizontal contractions and vertical extensions is performed, and in [8] character-by-character attention is used.

In our task, three models for handwriting recognition were used:

1. A fully convolutional encoder from the VerticalAttentionOCR network [7], designed to recognize lines of text by minimizing CTC-loss. Contains 10 blocks of triplets of regular and depthwise separable convolutions with residual connections if the dimensions match.
2. A complete VerticalAttentionOCR network with no hidden variables in the decoder, designed to recognize pages of text with lines stacked on top of each other At each step, the line representation is calculated as a linear combination of features in different rows, the weights are determined using hybrid attention, which works in a recurrent manner and takes into account the aggregate features for each row, as well as the total weights of the rows in the previous steps.
3. Russian Genetic Handwriting 2 from Transkribus[4]. The model is based on VGG-like model with BLSTMs [16], CTC-loss is minimized during training. The model was trained on 148846 lines of text from various archives, including the National Archives of Estonia and the "Prozhito" project.

**Preparation of the Text for Search.** One of the basic tasks of navigation in the archive of handwritten documents is a search by content, and not only text [2], but also image crops [30] and fragments of pen trace graph [20] can act as a representation.

In the simplest setting search setting, the user specifies individual keywords, which prompts us to represented the text as a sequence of individual lemmas. In addition, for user convenience, it is reasonable to bring the spelling of lemmas to a modern form. Thus, the automatically recognized text goes through the following processing stages provided by various NLP libraries before handling search queries.

1. Text tokenization via `word_tokenize` method from `nltk`.
2. Transition from pre-reform to modern spelling via `russpelling`[5].
3. Correcting errors in selected tokens by searching for the nearest word in the dictionary via `autocorrect` library.
4. Lemmatizing words — converting them to their normal form — via `pymorphy`.

**GPT-based Text Correction.** Text recognition mistakes correction was also performed with the use of large language models (LLM). BPE-tokenizer which comes with LLMs allows to parse any text [19], even in the presence of errors, dividing word sequences into sub-tokens, and pre-trained LLMs generalization ability enables them to comprehend and transform corrupted text into grammatically correct and coherent form.

For this work, an instruction-tuned LLM was chosen. Such models are particularly effective at approaching tasks in zero-shot [23] (only instructions are

---

[4] https://readcoop.eu/model/russian-generic-handwriting/.
[5] https://github.com/ingoboerner/russpelling.

provided) and few-shot [5] (instructions are aided with examples) fashion. Specifically OpenAI's `gpt-3.5-turbo-16k`[6] model was selected, GPT-3.5 models are proficient at comprehension and generation in foreign languages, possess some of the best reasoning capabilities and remain widely available.

An instruction was composed for the task. Main decisions which were incorporated during instruction preparation:

- Instruction is provided in the language of the dataset.
- Model is expected to generate text in modern grammar.
- Model is told to keep to the original text as close as possible.
- Model is given a role and a context (professional philologist working with historical documents).
- Model is provided with examples of correct task completions.

LLM correction does not necessarily lead to better accuracy in all cases. GPT-3.5 is not accustomed to pre-reform Russian language writing, so it often misinterprets and modifies it; furthermore it also changes word-form and order, whereas WER is strict with evaluation in this regard. Steering model to find balance between cautiousness, where it leaves misspellings uncorrected, and boldness, where it actively changes words and order, proves to be a challenge.

There were attempts inspired by Chain-of-Thoughts [24] approach to encourage the model to process input text longer, reason about it and understand it better, by setting auxiliary subtasks, such as Named Entity Search and summarization, which the model had to complete first. Another approach was to make LLM to correct its own output w.r.t. detected text. But those attempts did not lead to accuracy improvement as the model began to interpret input text more. Finding suitable auxiliary tasks and their formulation is an open problem.

**Text Localization in Images.** In order to navigate through the archive, recognized text must be aligned with image areas [18]. To calculate CTC-loss, probability matrix $P \in \mathbb{R}^{w \times m}$, characterizing the presence of characters from the alphabet $A$, $|A| = m$ in $w$ columns of the image, is produced. When a string $\mathbf{l}$ is recognized or given by an expert, it is natural to set the task of finding the alignment of the string that provides the highest probability:

$$\text{maximize} \prod_{i=1}^{w} p_{i,j_i} \text{ subject to } B(a_{j_1}, \ldots, a_{j_w}) = \mathbf{l},$$

where $B$ removes blank characters from the sequence and leaves only one character in a group of identical consecutive ones. This problem is solved using a dynamic programming algorithm. A similar problem arises when restoring line breaks of the text $S^*$ resulting from the correction by the language model of automatic transcription $S$ split on lines $\{\mathbf{l}_i\}_{i=1}^{k}$:

$$\text{minimize} \sum_{i=1}^{k} \text{EditDistance}(\mathbf{l}_i^*, \mathbf{l}_i) \text{ subject to } \text{concat}(\mathbf{l}_1^*, \ldots, \mathbf{l}_k^*) = S^*.$$

---

[6] https://platform.openai.com/docs/models/gpt-3-5.

As a result, we determine the location of the word in the line, and eventually on the page, since the baselines of the lines are known. Positive detection of the query word is considered true if the areas of the image corresponding to the word intersect when restored by the ground truth and by the predicted transcription.

# 5 Experiments

**Handwritten Text Recognition.** The VerticalAttentionOCR (VAOCR) models were trained from scratch for 5000 epochs with standard parameters, except that sign flipping was not used during augmentation and padding color was white, not black. The two-thirds of the pages were assigned to the training set, a quarter — to the test set, and every twelfth to the validation set. Table 1 demonstrates that LLM has a tendency to convert any text into grammatically correct form and it is convenient for reading and search tasks.

**Table 1.** Handwritten text recognition results for various models. The quality scores are for a single page.

| Ground truth | VAOCR (line)<br>CER = 2.63%, WER = 8.38% | VAOCR (line) + GPT<br>CER = 3.98%, WER = 6.91% |
|---|---|---|
| Дорогой дружокъ! Прошу тебя не волнуйся,<br>того который судитъ тебя я лично его не знаю,<br>но помню онъ былъ в околодкѣ и разговоръ проис-<br>ходилъ въ клѣткѣ откуда насъ выгналъ ментъ.<br>Клянусь тебѣ честью не знаю его фамилiи,<br>но изъ словъ понялъ, что негодяй. Далѣе ты<br>совѣтуешь проситься отсюда я нахожу это<br>лишнимъ т.к. это приведетъ не к лучшему,<br>они и сами меня спровадятъ, если будетъ<br>отправка, но что пока я не вѣрю. А вотъ сооб-<br>щу тебѣ новость: либералъ к новому году<br>уходитъ это он сам говорилъ мнѣ,- пере-<br>ходитъ в какое то управление чиновникомъ,<br>что я охотно вѣрю, а пока прошу не от-<br>чаивайся и не прибѣгай къ крайнимъ мѣрамъ,<br>это отъ насъ никогда не уйдетъ. О адресахъ и<br>ходахъ пока говоритъ не время, то покажетъ<br>будущее, пока конечно хода у меня нѣтъ, но<br>удочку уже закинулъ и уже клюетъ. Сегодня<br>узналъ: пришелъ большой этапъ, пришли и<br>вѣчники, может среди их и герои, которые<br>разбудятъ сонное царство и тамъ образумятся | Дорогой дружокъ! Прошу тебя не волнуцйіся,<br>того который судилъ тебя я лично его не знаю,<br>но помню онъ былъ в околодкѣ и разговоръ пром<br>ходятъ въ клозетъ откуда насъ выгналъ ментъ.<br>Клянусь тебѣ честью не знаю его фамити,<br>но словъ понякъ, что негодял. Залѣе ты<br>совѣтуешь, проситься отсюда я нахожу это<br>лишнимъ т.к. это приведетъ не к лучшему,<br>они и сами меня спровадятъ, если будетъ<br>отправка, но что пока я не вѣрю. А вомъ сооб-<br>щу тебѣ новость либералъ к новому году<br>уходитъ это он сам говорилъ мнѣ, - пере-<br>ходитъ в какое то управление чиновникомъ,<br>что я охотно вѣрю, а пока прошу не от-<br>чатвайся и не прибѣгай къ крайнимъ мѣрамъ,<br>это отъ насъ никогда не уйдетъ. О адресахъ и<br>ходахъ пока говоритъ не время, то покажетъ<br>будущее, пока конечно хода у меня нѣтъ, но<br>удочку уже закинулъ и уже клюетъ. Сегодня<br>узналъ: пришелъ большой этапъ, причили и<br>вѣчники, может среди их и чероя, которые<br>разбудятъ сонное царство и тамъ образумятся | Дорогой дружок! Прошу тебя не волноваться,<br>того, который судил тебя, я лично его не знаю,<br>но помню он был в околодке и разговор про-<br>ходил в клозете, откуда нас выгнал мент.<br>Клянусь тебе честью, не знаю его фамилии,<br>но из слов понял, что негодяй. Зачем ты<br>советуешь проситься отсюда, я нахожу это<br>лишним, так как это приведет не к лучшему,<br>они и сами меня спровадят, если будет<br>отправка, но что пока я не верю. А вот сооб-<br>щу тебе новость: Либерал к новому году<br>уходит, это он сам говорил мне, пере-<br>ходит в какое-то управление чиновником,<br>что я охотно верю, а пока прошу не от-<br>чаивайся и не прибегай к крайним мерам,<br>это от нас никогда не уйдет. О адресах и<br>ходах пока говорить не время, то покажет<br>будущее, пока конечно хода у меня нет, но<br>удочку уже закинул и уже клюет. Сегодня<br>узнал: пришел большой этап, пришли и<br>вечники, может, среди них и герои, которые<br>разбудят сонное царство и там образумятся |

| Rus. Gen. Handwriting (Transkribus)<br>CER = 6.67%, WER = 20.99% | VAOCR (page)<br>CER = 3.84%, WER = 14.37% | VAOCR (page) + GPT<br>CER = 4.66%, WER = 9.43% |
|---|---|---|
| Дорогой друпюок. Прошу тебя не волнизисся,<br>того, который судил тебя я лично его не знаю<br>но помню онъ был в околодке и разговор проис<br>ходин в клохете откуда нас выгнал мент.<br>клянусь тебе честью не знаю его фамимии<br>но из слов понял, что негодяй. Далее ть<br>сквечуешь проситься отсюда я нахожу это<br>лишнимъ т. к. это приведеть не к лучшему<br>они и сами меня спровадить если будет<br>отправка, но что пока я не верю. А вот<br>щу тебе новость: либерал к новому год<br>уходи это он сам говорил мне, - пере<br>ходит в какое то управление чиновником,<br>что я охотно верю, а пока прошу не от<br>чайвался и не прибѣгал к крайним мером<br>это от нас никогда не уйдет. О адресах и<br>ходах пока говорить не время, то покажет<br>будущее, пока конечно хода у меня нет, но<br>удочну уже закинла и уже клюст. Сегодня<br>утяй: пришел большой этоп причили и<br>вечники, может среди их и герои, которые<br>позбудым сонное царство и там образумя | Дорогой дружокъ! Прошу тебя не воинуйся9<br>того который судил тебя я мично его не знаю,<br>но помню онъ былъ в околодкѣ и разговор проис<br>ходилъ въ клозетъ откуда насъ выгналъ ментъ.<br>клянусь тебѣ честью не знаю его дФрамии,<br>но изъ словъ понялъ, что негодяй. Залѣе ты<br>совѣтуешь, проситься отсюда я нахожу это<br>лишнимъ т.к. это приведетъ не к лучшему,<br>они и сами меня спровадятъ, если будетъ<br>отправка, но что пока я не вѣрю. А вотъ соб-<br>щу тебѣ новость. либералъ к новому годду<br>уходитъ это он сам говорилъ мнѣ, - пере-<br>ходитъ в какое то управление чиновникомъ,<br>что я охотно вѣрю, а пока прошу не от-<br>чайвайся и не прибѣгай къ крайнимъ мѣрамъ,<br>это отъ насъ никогда не уйдетъ. О адресахъ и<br>ходахъ пока говоритъ не время, то покажетъ<br>будущее, пока конечно хода у меня нѣтъ, но<br>удочку уже закинулъ и уже клюетъ. Сегодня<br>узналъ! пришелъ большой этапъ, пришли и<br>вѣчники, может среди их и чероя, комором<br>разбудятъ сонное царство и тамъ образумятя | Дорогой дружок! Прошу тебя не волнуйся,<br>того который судил тебя я лично его не знаю,<br>но помню он был в околодке и разговор проис<br>ходил в клозете, откуда нас выгнал мент.<br>Клянусь тебе честью, не знаю его фамилии,<br>но из слов понял, что негодяй. Залей ты<br>советуешь проситься отсюда, я нахожу это<br>лишним, так как это приведет не к лучшему,<br>они и сами меня спровадят, если будет<br>отправка, но что пока я не верю. А вот сооб-<br>щу тебе новость. Либерал к новому году<br>уходит, это он сам говорил мне, пере-<br>ходит в какое-то управление чиновником,<br>что я охотно верю, а пока прошу не от-<br>чаивайся и не прибегай к крайним мерам,<br>это от нас никогда не уйдет. О адресах и<br>ходах пока говорить, то покажет<br>будущее, пока конечно хода у меня нет, но<br>удочку уже закинул и уже клюет. Сегодня<br>узнали! Пришел большой этап, пришли и<br>вечнини, может среди них и чероя, которые<br>разбудят сонное царство и там образумятся |

Table 2 shows that the recognition quality of the line-level model for the entire sample (5, 11%) is comparable to the results of VerticalAttentionOCR on the standard handwritten datasets, such as IAM (4, 97%) and READ2016 (4, 10%). The weaker performance of the page model can be explained by smaller line spacing compared to these datasets and an very small number of page images for training. These results also show the importance of image preprocessing. Although noise removal provides the main gain in quality, deslanting is also useful because it reduces data volume and processing time by about 25%.

**Table 2.** Handwriting recognition quality scores for various models on the full dataset.

| Model | CER | WER |
|---|---|---|
| VAOCR (line) | 5.11% | 17.55% |
| VAOCR (page) | 6.48% | 20.82% |
| Transkribus | 12.23% | 34.88% |
| VAOCR (line, not deslanted) | 5.20% | 17.98% |
| VAOCR (line, not denoised) | 5.87% | 19.70% |
| VAOCR (line, not deslanted, not denoised) | 6.87% | 21.60% |
| VAOCR (page, not deslanted) | 9.18% | 28.52% |

Table 3 shows that formulating role and finding good examples for the task requires some manual work but it does provide an improvement and is thus recommended. Since GPT changes word-form and order, quality was assessed by WER as more consistent with human perception in this case. WER was measured without considering punctuation and capitalization, which can also be affected by GPT.

**Search by Keywords.** The experts identified four topics systematically present in the letters, defined by their keywords: criminal, medical (corporal), emotional and temporal (topic of the future). The search for keywords is illustrated by Fig. 6, the selected words are underlined in the color corresponding to the topic: blue, red, green and yellow in the order of mention. The token is included in the search results if its lemma fully matches the query word. In preparation for search in GPT-corrected text, only the missing tokenization and lemmatization steps were performed.

The difficulties of GPT with improving initially high quality of the search reflected in Table 4 can be caused by too loose correction with the inversion of words and the addition of affixes.

**Search for Personal Names.** To assess the quality of the search for personal names, two approaches were considered: with the selection of all words written with a capital letter and with the selection of such words that are not at the beginning of the sentence. The recognition results in Fig. 7 show that some errors are caused by the peculiarities of the handwriting, which does not allow

**Table 3.** Quality of GPT correction of VAOCR output measured by WER w/o punctuation and capitalization.

| Instruction | Line-level | Page-level |
|---|---|---|
| Task only | 16.47% | 18.85% |
| Task and role | 14.94% | 19.90% |
| Task, role and examples | 14.18% | 16.91% |
| Dictionary-based correction | 15.91% | 18.71% |

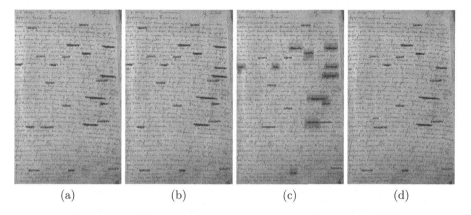

| (a) | (b) | (c) | (d) |

**Fig. 6.** An example of keyword extraction: (a) by ground truth transcription, (b) and (c) by automatic transcription (VAOCR, models for lines and pages, respectively), (d) by Transkribus transcription.

**Table 4.** Keyword search quality scores for various models on the full dataset.

| Model | Precision | Recall | F-score |
|---|---|---|---|
| VAOCR (line) | 97.14% | 91.35% | 94.15% |
| VAOCR (page) | 94.58% | 88.70% | 91.55% |
| VAOCR (line) + GPT | 95.04% | 87.49% | 91.11% |
| Transkribus | 94.04% | 73.58% | 82.56% |

to strictly distinguish between uppercase and lowercase writing of some letters. In terms of recognition quality (Table 5), the GPT model proves its ability to deal effectively with punctuation and correct obvious errors in case. Implementation of more advanced analysis methods, such as case checking by several pages and extraction of grammatical and semantic categories, remains a future work.

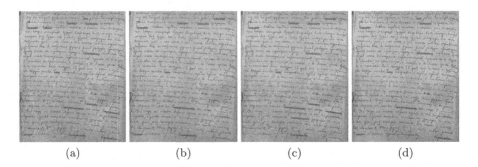

Fig. 7. An example of highlighting personal names by transcription: (a) from ground truth, (b) from line-level VAOCR (c) from line-level VAOCR + GPT, (d) from Transkribus. Magenta indicates recognized personal names, cyan indicates all words starting with capital letters.

Table 5. Quality scores for finding personal names for various models on the full dataset.

| Model | Precision | Recall | F-score |
|---|---|---|---|
| VAOCR (line, w/o punctuation) | 45.86% | **87.36%** | 60.14% |
| VAOCR + GPT (line, w/o punctuation) | 36.28% | 86.91% | 51.19% |
| Transkribus (w/o punctuation) | 43.86% | 78.95% | 56.39% |
| VAOCR (line, with punctuation) | 74.03% | 60.00% | 66.28% |
| VAOCR + GPT (line, with punctuation) | **89.81%** | 66.01% | **76.09%** |
| Transkribus (with punctuation) | 60.22% | 58.95% | 59.57% |

# 6   Conclusion

The results of the work demonstrate the possibility of creating from scratch a navigation system for an archive of historical handwritten documents with a single handwriting. The system integrates elements related to a wide class of computer vision and text processing tasks, such as document markup analysis, handwriting recognition, spelling correction, personal names extraction, and text-image alignment. It is demonstrated that several dozens of recognized pages of text may be enough to train a model tailored to a specific handwriting. The prospects of using modern large language models to enhance navigation are outlined. In future work, it is planned to scale the experience gained to create a navigation system for larger personal archives containing hundreds of pages.

**Acknowledgements.** This work was supported by the Russian Science Foundation, grant no. 22-68-00066, https://rscf.ru/en/project/22-68-00066/.

# References

1. Alberti, M., Vogtlin, L., Pondenkandath, V., Seuret, M., Ingold, R., Liwicki, M.: Labeling, cutting, grouping: an efficient text line segmentation method for medieval manuscripts. In: 2019 International Conference on Document Analysis and Recognition (ICDAR), pp. 1200–1206 (2019)
2. Andrés, J., Toselli, A.H., Vidal, E.: Approximate search for keywords in handwritten text images. In: Uchida, S., Barney, E., Eglin, V. (eds.) DAS 2022. LCS, vol. 13237, pp. 367–381. Springer, Cham (2022). https://doi.org/10.1007/978-3-031-06555-2_25
3. Barakat, B., Droby, A., Kassis, M., El-Sana, J.: Text line segmentation for challenging handwritten document images using fully convolutional network. In: 2018 16th International Conference on Frontiers in Handwriting Recognition (ICFHR), pp. 374–379 (2018)
4. Bolelli, F.: Indexing of historical document images: ad hoc dewarping technique for handwritten text. In: Digital Libraries and Archives. IRCDL 2017. Communications in Computer and Information Science, vol. 733, pp. 45–55 (2017)
5. Brown, T., et. al.: Language models are few-shot learners. In: Larochelle, H., Ranzato, M., Hadsell, R., Balcan, M., Lin, H. (eds.) Advances in Neural Information Processing Systems, vol. 33, pp. 1877–1901. Curran Associates, Inc. (2020)
6. Coquenet, D., Chatelain, C., Paquet, T.: SPAN: a simple predict & align network for handwritten paragraph recognition. In: Lladós, J., Lopresti, D., Uchida, S. (eds.) ICDAR 2021. LNCS, vol. 12823, pp. 70–84. Springer, Cham (2021). https://doi.org/10.1007/978-3-030-86334-0_5
7. Coquenet, D., Chatelain, C., Paquet, T.: End-to-end handwritten paragraph text recognition using a vertical attention network. IEEE Trans. Pattern Anal. Mach. Intell. **45**(1), 508–524 (2023)
8. Coquenet, D., Chatelain, C., Paquet, T.: Dan: a segmentation-free document attention network for handwritten document recognition. IEEE Trans. Pattern Anal. Mach. Intell. **45**(7), 8227–8243 (2023)
9. Kang, S., Iwana, B.K., Uchida, S.: Complex image processing with less data-Document image binarization by integrating multiple pre-trained U-Net modules. Pattern Recogn. **109**, 107577 (2021)
10. Khamekhem Jemni, S., Souibgui, M.A., Kessentini, Y., Fornés, A.: Enhance to read better: a multi-task adversarial network for handwritten document image enhancement. Pattern Recogn. **123**, 108370 (2022)
11. Kim, B.S., Koo, H.I., Cho, N.I.: Document dewarping via text-line based optimization. Pattern Recogn. **48**(11), 3600–3614 (2015)
12. Li, M., et al.: TrOCR: transformer-based optical character recognition with pre-trained models (2022)
13. Nockels, J., Gooding, P., Ames, S., Terras, M.: Understanding the application of handwritten text recognition technology in heritage contexts: a systematic review of transkribus in published research. Arch. Sci. **22**(3), 367–392 (2022)
14. Nurseitov, D., Bostanbekov, K., Kurmankhojayev, D., Alimova, A., Abdallah, A., Tolegenov, R.: Handwritten Kazakh and Russian (HKR) database for text recognition. Multimedia Tools Appl. **80**(21–23), 33075–33097 (2021)
15. Potanin, M., Dimitrov, D., Shonenkov, A., Bataev, V., Karachev, D., Novopoltsev, M.: Digital peter: dataset, competition and handwriting recognition methods. CoRR abs/2103.09354 (2021)

16. Puigcerver, J.: Are multidimensional recurrent layers really necessary for handwritten text recognition? In: 2017 14th IAPR International Conference on Document Analysis and Recognition (ICDAR), vol. 01, pp. 67–72 (2017)
17. Renton, G., Chatelain, C., Adam, S., Kermorvant, C., Paquet, T.: Handwritten text line segmentation using fully convolutional network. In: 2017 14th IAPR International Conference on Document Analysis and Recognition (ICDAR), vol. 05, pp. 5–9 (2017)
18. Romero-Gomez, V., Toselli, A., Bosch, V., Sánchez, J.A., Vidal, E.: Automatic alignment of handwritten images and transcripts for training handwritten text recognition systems. In: 2018 13th IAPR International Workshop on Document Analysis Systems (DAS), pp. 328–333 (2018)
19. Sennrich, R., Haddow, B., Birch, A.: Neural machine translation of rare words with subword units. In: Proceedings of the 54th Annual Meeting of the Association for Computational Linguistics (Volume 1: Long Papers), pp. 1715–1725. Association for Computational Linguistics, Berlin, Germany (2016)
20. Stauffer, M., Fischer, A., Riesen, K.: Keyword spotting in historical handwritten documents based on graph matching. Pattern Recogn. **81**, 240–253 (2018)
21. Sulaiman, A., Omar, K., Nasrudin, M.F.: Document binarization: a review on issues, challenges, techniques, and future directions. J. Imaging **5**(4), 48 (2019)
22. Tian, Y., Narasimhan, S.G.: Rectification and 3D reconstruction of curved document images. In: Proceedings of the 2011 IEEE Conference on Computer Vision and Pattern Recognition, pp. 377–384. CVPR 2011, IEEE Computer Society, USA (2011)
23. Wei, J., et al.: Finetuned language models are zero-shot learners. In: The Tenth International Conference on Learning Representations, ICLR 2022 (2022)
24. Wei, J., et al.: Chain of thought prompting elicits reasoning in large language models. CoRR abs/2201.11903 (2022)
25. Wigington, C., Tensmeyer, C., Davis, B., Barrett, W., Price, B., Cohen, S.: Start, follow, read: end-to-end full-page handwriting recognition. In: Ferrari, V., Hebert, M., Sminchisescu, C., Weiss, Y. (eds.) ECCV 2018. LNCS, vol. 11210, pp. 372–388. Springer, Cham (2018). https://doi.org/10.1007/978-3-030-01231-1_23
26. Xie, G.W., Yin, F., Zhang, X.Y., Liu, C.L.: Dewarping document image by displacement flow estimation with fully convolutional network. In: International Workshop on Document Analysis Systems, pp. 131–144 (2020)
27. Xiong, W., Xu, J., Xiong, Z., Wang, J., Liu, M.: Degraded historical document image binarization using local features and support vector machine (SVM). Optik **164**, 218–223 (2018)
28. Yandex service "Search in archives". https://yandex.ru/archive
29. Yousef, M., Bishop, T.E.: OrigamiNet: weakly-supervised, segmentation-free, one-step, full page text recognition by learning to unfold. In: 2020 IEEE/CVF Conference on Computer Vision and Pattern Recognition (CVPR), pp. 14698–14707 (2020)
30. Zhang, X., Sugumaran, V.: Content based search engine for historical calligraphy images. Int. J. Intell. Inf. Technol. **10**, 1–18 (2014)

# Greedy Algorithm for Fast Finding Curvilinear Symmetry of Binary Raster Images

Oleg Seredin$^{(\boxtimes)}$ (iD), Daniil Liakhov (iD), Nikita Lomov (iD), Olesia Kushnir (iD), and Andrei Kopylov (iD)

Tula State University, Tula, Russia
`oseredin@yandex.ru`

**Abstract.** The article proposes a fast method for detecting curved symmetry for binary images by greedily searching for locally symmetric nodes of a polyline inside a figure, starting from a user-specified point. The advantage is that the procedure is virtually devoid of any manually adjusted inputs. The key procedure inputs (the increment and angular range used at the next step) are estimated by an adaptive procedure. Further, the obtained fragments of the image corresponding to the found segments of the polyline are sequentially superimposed on a single axis by rotation, forming a "straightened" figure, which makes it possible to calculate the Jaccard measure of symmetry for the entire figure. The experimental results demonstrate the successful operation of the method even on images with significant curvature at a perfect calculation speed.

**Keywords:** Reflection symmetry · Curved symmetry · Jaccard index · Binary Image

## 1 Introduction

The estimation of the reflective symmetry measure is studied for many years, and an extensive range of approaches is now available. The most recent works [1–5] propose a range of mathematical models and algorithms. There are several problems where an initially symmetric (or nearly symmetric) object is articulated (see Fig. 1). Obviously, a human easily comprehends the concept of symmetry in such shapes and can detect such an articulation. Some real-world applications of curvilinear symmetry axis detection tools are the analysis of biomedical images [6], scanned plant leaves, lab observation of animal movements, and symmetrization of 2D polygonal shapes [7].

Detection of reflective symmetry with respect to a straight line using some index (e.g., Jaccard index of similarity between the original and reflected image) is a suitable solution for such images (see Fig. 1)). To solve the problem in terms

This study was supported by the Russian Science Foundation, Grant No. 22-21-00575. https://rscf.ru/project/22-21-00575/.

of the computer vision theory, we need to define the "curvilinear axis" either as a parametric analytical representation or by explicitly specifying the pixels or articulated segments of the axis. Another issue is the formalization of the symmetry index (e.g., the Jaccard index proved to be quite good). Lomov N. et al. in their most recent work [8] present some existing approaches to curvilinear symmetry detection of planar figures [6,9–12] and proposes a dynamic programming method applicable to segmented images. The experiments demonstrated acceptable results of curvilinear symmetry axis detection and symmetry index evaluation. The disadvantages of the global optimum search methods are high computational costs and serious limitations in terms of the curvature of the symmetry axis, so significantly articulated images cannot be processed. The curvature constraint will also be decisive for the method based on the Hough transform [5].

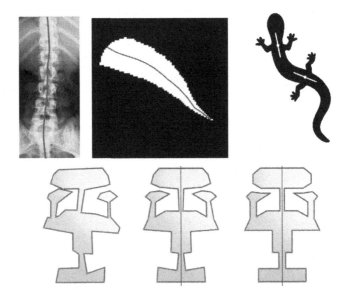

**Fig. 1.** Test images and human-marked curvilinear symmetry axes or their segments (top row); symmetrization of 2D polygonal shapes (bottom row).

In this study, we propose a fast curvilinear symmetry detection method applicable to binary images. It uses a greedy incremental search for locally symmetric broken line nodes within the shape, starting from a user-specified point. The symmetry axis in this case is a chain of connected segments. The image fragments containing the segments found are successively rotated and superimposed on a single axis resulting in a "straightened" shape. Such an approach makes it possible to estimate the symmetry index for the entire shape.

## 2   A Greedy Algorithm for Curvilinear Symmetry Axis Detection in Binary Raster Images

Let us denote the original binary image as $I$. The shape $I^1$ within the image is represented with 1 (black color). The background is represented with 0 (white color): $I = \{I^0 \cup I^1\}$.

Kushnir O. et al. [1] proposed to use the Jaccard index [13] to estimate the degree of reflective symmetry of the shape $I^1$. The index is a measure of similarity between the original shape and the shape reflected with respect to some axis $\bar{I}$:

$$J = \frac{|I \cap \bar{I}|}{|I \cup \bar{I}|} = \frac{|I \cap \bar{I}|}{2|I| - |I \cap \bar{I}|}. \tag{1}$$

Papers [4,14] offer a formal definition of the reflection with respect to an axis defined either in polar coordinates or as two points on the boundary.

To give a formal definition of the proposed procedure, we should first define a symmetry index of a segment that lies within the shape with respect to a point on this segment. The following lemma defines such an index.

**Lemma 1.** *For a line that crosses the shape boundary at two points* $(\mathbf{c}_l, \mathbf{c}_r)$, *and a given point* $\mathbf{p}$ *on this line (see Fig. 2), the measure of symmetry (Jaccard index) with respect to this point is defined as*

$$J(\mathbf{p}|\mathbf{c}_l, \mathbf{c}_r) = \frac{\min\left(d(\mathbf{p}, \mathbf{c}_l), d(\mathbf{p}, \mathbf{c}_r)\right)}{\max\left(d(\mathbf{p}, \mathbf{c}_l), d(\mathbf{p}, \mathbf{c}_r)\right)}. \tag{2}$$

*Proof.* Let the segment $\mathbf{c}_l, \mathbf{c}_r$ is given with the point $\mathbf{p}$ on it. The segment $\mathbf{c}'_l, \mathbf{c}'_r$ is the segment $\mathbf{c}_l, \mathbf{c}_r$ mirrored with respect to the point $\mathbf{p}$. Obviously, $d(\mathbf{p}, \mathbf{c}'_l) = d(\mathbf{p}, \mathbf{c}_r)$ and $d(\mathbf{p}, \mathbf{c}'_r) = d(\mathbf{p}, \mathbf{c}_l)$. Let $d(\mathbf{p}, \mathbf{c}_l) > d(\mathbf{p}, \mathbf{c}_r)$). Then as the two segments overlap, the total length is $2d(\mathbf{p}, \mathbf{c}_l)$. The length of their intersection is $2d(\mathbf{p}, \mathbf{c}_r)$. Then Jaccard index is estimated as:

$$J(\mathbf{p}|\mathbf{c}_l, \mathbf{c}_r) = \frac{2d(\mathbf{p}, \mathbf{c}_r)}{2d(\mathbf{p}, \mathbf{c}_l)} = \frac{d(\mathbf{p}, \mathbf{c}_r)}{d(\mathbf{p}, \mathbf{c}_l)}.$$

In general, the intersection length is equal to the shortest length of the two segments $d(\mathbf{p}, \mathbf{c}_l), d(\mathbf{p}, \mathbf{c}_r)$ multiplied by 2: $2\min\left(d(\mathbf{p}, \mathbf{c}_l), d(\mathbf{p}, \mathbf{c}_r)\right)$. The length of the union of the two segments is then equal to $2\max\left(d(\mathbf{p}, \mathbf{c}_l), d(\mathbf{p}, \mathbf{c}_r)\right)$.

Then the Jaccard index estimation is reduced to finding the ratio of the smaller part of the segment to its larger part:

$$J(\mathbf{p}|\mathbf{c}_l, \mathbf{c}_r) = \frac{2\min\left(d(\mathbf{p}, \mathbf{c}_l), d(\mathbf{p}, \mathbf{c}_r)\right)}{2\max\left(d(\mathbf{p}, \mathbf{c}_l), d(\mathbf{p}, \mathbf{c}_r)\right)} = \frac{\min\left(d(\mathbf{p}, \mathbf{c}_l), d(\mathbf{p}, \mathbf{c}_r)\right)}{\max\left(d(\mathbf{p}, \mathbf{c}_l), d(\mathbf{p}, \mathbf{c}_r)\right)}. \qquad \blacksquare$$

The concept behind the algorithm for finding a broken line composed of consecutive segments locally symmetric at the nodes is as follows (refer to Figs. 3–4 for visual explanations). A sector is built from some initial point within the

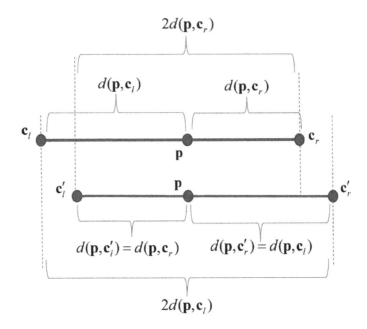

**Fig. 2.** Definition of the symmetry index for a segment with respect to a point on the segment.

shape, The sector defines the directions to search for the next trial point at a given distance from the current point. The symmetry index of the segment orthogonal to the direction from the initial to the trial point (a transverse segment) is calculated at each trial point.

To select the next point, we both maximize the symmetry index of the transverse segment and consider certain assumptions about the resulting broken line. These assumptions are expressed as penalties for drastic changes in the orientation of subsequent segments. and in the length of transverse segments. The algorithm is quite similar to the method proposed by Lomov N. et al. [12] which constructs the key branch of a binary shape skeleton through the medial representation of circles inscribed in the shape to identify the planaria flatworms. The use of skeletons for image analysis is limited due to their excessive responsiveness to boundary noise. Therefore, special regularization procedures are used (pruning).

The algorithm inputs are:

User-defined initial point position: $\hat{\mathbf{p}}_0 = (x, y)^T$;

initial direction: $\hat{\mathbf{v}}_0$ (produced by an algorithm described below);

$h_i$: $i$ th increment (produced by an algorithm described below);

$\alpha_i^{\max}$ is the max angular range in which the best direction is sought at the $i$ th step (produced by an algorithm described below);

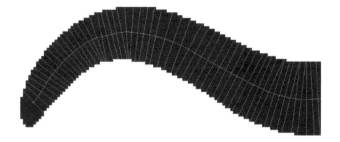

**Fig. 3.** Example: an articulated curvilinear axis of symmetry (yellow) that maximizes the symmetry index for the transverse segments (green) at its nodes. (Color figure online)

$\lambda \in (0, ..., 1)$ is the factor (user-defined) representing the contribution of the direction selected in the previous step.

Algorithm:

$$i = 1, 2, ...,$$

$$\hat{\mathbf{p}}_i = \hat{\mathbf{p}}_{i-1} + \hat{\mathbf{v}}_i,$$

$$\hat{\mathbf{v}}_i = \min\left(h_i, S\left(\mathbf{p}_i^\alpha | \hat{\mathbf{p}}_{i-1}\right)\right) \frac{\mathbf{v}_i}{|\mathbf{v}_i|},$$

$$\mathbf{v}_i = \lambda q_i \frac{\hat{\mathbf{v}}_{i-1}}{|\hat{\mathbf{v}}_{i-1}|} + (1-\lambda)(1-q_i) \frac{\mathbf{v}_i^{\tilde{\alpha}}}{|\mathbf{v}_i^{\tilde{\alpha}}|}, q_i = \frac{\min\left(J\left(\mathbf{p}_i | \hat{\mathbf{p}}_{i-1}\right), J\left(\mathbf{p}_i^{\tilde{\alpha}} | \hat{\mathbf{p}}_{i-1}\right)\right)}{\max\left(J\left(\mathbf{p}_i | \hat{\mathbf{p}}_{i-1}\right), J\left(\mathbf{p}_i^{\tilde{\alpha}} | \hat{\mathbf{p}}_{i-1}\right)\right)},$$

$$\mathbf{v}_i^\alpha = \min\left(h_i, S\left(\mathbf{p}_i^\alpha | \hat{\mathbf{p}}_{i-1}\right)\right) \begin{pmatrix} \cos\alpha & \sin\alpha \\ -\sin\alpha & \cos\alpha \end{pmatrix} \frac{\hat{\mathbf{v}}_{i-1}}{|\hat{\mathbf{v}}_{i-1}|},$$

$$\mathbf{p}_i^\alpha = \hat{\mathbf{p}}_{i-1} + \mathbf{v}_i^\alpha,$$

$$\tilde{\alpha} = \arg\max_\alpha \left(\frac{J\left(\mathbf{p}_i^\alpha | \hat{\mathbf{p}}_{i-1}\right) \cdot \sqrt{\cos(\alpha)}}{S\left(\mathbf{p}_i^\alpha | \hat{\mathbf{p}}_{i-1}\right)}\right), \alpha \in [\hat{\mathbf{v}}_{i-1} \mp \alpha_i^{\max}],$$

$$\underset{\sim}{\alpha} = \arg\min_\alpha S\left(\mathbf{p}_i^\alpha | \hat{\mathbf{p}}_{i-1}\right), \alpha \in [\hat{\mathbf{v}}_{i-1} \mp \alpha_i^{\max}],$$

$$S\left(\mathbf{p}_i^\alpha | \hat{\mathbf{p}}_{i-1}\right) = d\left(\mathbf{c}_l^\alpha, \mathbf{c}_r^\alpha\right).$$

Here $d(\mathbf{c}_l^\alpha, \mathbf{c}_r^\alpha)$ is the Euclidean distance between $\mathbf{c}_l^\alpha$ and $\mathbf{c}_r^\alpha$, the extreme points of the shape located on the line orthogonal to the vector $\hat{\mathbf{v}}_{i-1}$ at the point $\mathbf{p}_i^\alpha$ (refer to Fig. 4). The $\mathbf{c}_l^\alpha$ and $\mathbf{c}_r^\alpha$ point coordinates are calculated with Bresenham's algorithm. When the points $\mathbf{c}_l^\alpha$, $\mathbf{c}_r^\alpha$ and $\mathbf{p}_i^\alpha$ are available, we can calculate the Jaccard index for the $\mathbf{c}_l^\alpha$. $\mathbf{c}_r^\alpha$ segment mirrored with respect to the point $\mathbf{p}_i^\alpha$ from (2):

$$J\left(\mathbf{p}_i^\alpha | \mathbf{p}_{i-1}\right) = \frac{\min\left(d(\mathbf{p}_i^\alpha, \mathbf{c}_l^\alpha), d(\mathbf{p}_i^\alpha, \mathbf{c}_r^\alpha)\right)}{\max\left(d(\mathbf{p}_i^\alpha, \mathbf{c}_l^\alpha), d(\mathbf{p}_i^\alpha, \mathbf{c}_r^\alpha)\right)}.$$

Stop rule:

$$stop = if \begin{cases} \mathbf{p}_i \notin I, \\ \mathbf{p}_i \notin I^1, \\ h_i \leq 1. \end{cases}$$

So, the optimal step direction depends on the Jaccard measure $J(\mathbf{p}_i^\alpha|\hat{\mathbf{p}}_{i-1})$ of the orthogonal line and the local width of the figure $S(\mathbf{p}_i^\alpha|\hat{\mathbf{p}}_{i-1})$. The regularizer $\sqrt{\cos(\alpha)}$ and coefficient of the procedure $q_i$ provides priority to the previous direction in the case when the Jaccard measure in the trial direction will not differ much.

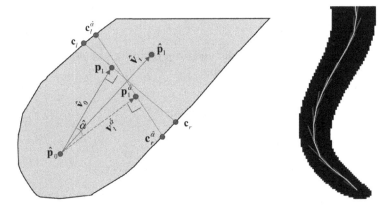

**Fig. 4.** Algorithm explanation: constructing of the first segment of the curvilinear axis of symmetry (left) and the algorithm steps. Red segment: previous direction; blue segment: best greedy choice; yellow segment: weighted compromise. (Color figure online)

## 3  Selecting the Initial Direction

Let some set of initial directions be given, e.g., a set of eight elements defined on a plane with the $\pi/4$ angle between them:

$$\mathbf{V}_0 = \left\{ (0, -1)^T, (1, -1)^T, (1, 0)^T, ..., (-1, -1)^T \right\}.$$

As the initial point $\mathbf{p}_0$ is given, the distances to the points $\mathbf{c}_{\mathbf{v}_k}^{\mathbf{P}_0}$, $k = 1, ..., 8$ on the shape boundary along the directions contained in the set $\mathbf{V}_0$ are found. Then the initial direction is the one in which the distance to the boundary is maximized:

$$\hat{\mathbf{v}}_0 = \arg \max_{\mathbf{v}_k \in \mathbf{V}_0} \left( d\left( \mathbf{p}_0, \mathbf{c}_{\mathbf{v}_k}^{\mathbf{P}_0} \right) \right).$$

## 4   Adaptive Increment Selection

The greedy algorithm for curvilinear symmetry axis detection needs an increment value for the $h_i$ procedure to find the next point. Using an empirical, small, and fixed increment may lead to too many steps and slow down the procedure. On the contrary, a large value will deprive the procedure of its flexibility and may result in going beyond the shape boundaries. We propose an automatic selection of the increment when finding the next point of the curvilinear axis of symmetry.

A new point $\mathbf{p}_i$ is sought at some step with the given direction $\hat{\mathbf{v}}_{i-1}$. Let us denote the ray from the point $\hat{\mathbf{p}}_{i-1}$ and figure's boundary intersection point as $\mathbf{c}_{\hat{\mathbf{v}}_{i-1}}^{\hat{\mathbf{p}}_{i-1}}$. Then the increment can be calculated by a rule of thumb from the distance to the shape boundary along the ray direction:

$$h_i = \sqrt{d\left(\hat{\mathbf{p}}_{i-1}, \mathbf{c}_{\hat{\mathbf{v}}_{i-1}}^{\hat{\mathbf{p}}_{i-1}}\right)}.$$

Note that if $h_i \leq 1$, the procedure stops because we reach the shape boundary. The square root in the equation for the increment is intended to correctly handle sharp curves.

It is proposed to define the first increment as $h_1 = 2$.

## 5   Adaptive Adjustment of the Search Angular Range

A fixed angular range relative to the direction from the previous point (e.g., $(-\pi/3, \pi/3)$) is inefficient and results in too many lookups of potential directions. We propose an adaptive adjustment of the angular range depending on the local width and distance to the shape boundary in the tested direction. At each step, the angular range is selected from the calculated increment according to the rule of thumb:

$$\alpha_i^{\max} = \frac{\pi}{2 \cdot \max\left(\ln\left(h_i\right), 2\right)}.$$

When the procedure reaches a narrow, long section within the shape, the angular range is narrowed. If the procedure approaches a turn or "dead end", the increment is reduced while the angular range expands. As the equation indicates, the range will not exceed $\pi/4$. Changes to the increment significantly affect the angular range, so a logarithm function is added for smoothing.

## 6   Determining the Degree of Symmetry by "Shape Straightening"

The segments orthogonal to the segments of the curvilinear axis of symmetry generated by the proposed algorithm divide the shape into several fragments (Fig. 5(a)). To estimate the general figure symmetry, each segment of the broken line and the respective fragment are successively aligned with the ordinate axis by rotating the shape in such a way that the segment gets vertical and its beginning coincides with the end of the previous segment. This operation produces a "straightened" shape (Fig. 5(b)). Therefore, the problem is reduced to the estimation of the reflective symmetry with respect to the vertical axis found (Fig. 5(c)).

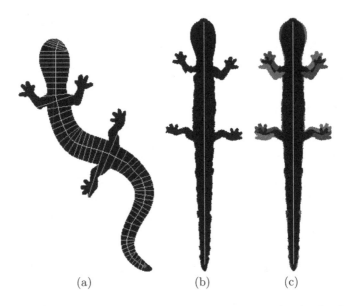

(a)              (b)              (c)

**Fig. 5.** Shape "straightening" and image reflection. The Jaccard index for the straightened figure is 0.777.

The procedure of shape reflection relative to the axis becomes trivial, and the Jaccard index can be easily found from (1). This method is rough: the "sliced" shape fragments being vertically "stacked" on each other may overlap or have gaps. Note the jagged edge of the straightened figure which affects the estimation of the symmetry index.

## 7    Experiments

We implemented the algorithm in C++ using the OpenCV library. The angular increment within the angular range is one degree. We did not use any parallelization. The computer specifications were as follows: Intel® Core™ i7-1260P, 2.10 GHz, RAM: 16 Gb DDR5, SSD: 1,024 Gb. Table 1 presents examples of processing some images from the MPEG-7 CE Shape-1 Part B database [15] and such values as the Jaccard index after straightening, number of segments found, total runtime and its fraction spent on the shape straightening. The red markers are the starting points.

**Table 1.** Curvilinear symmetry detection examples and numerical values produced by the proposed procedure.

| Image | Numerical values |
|---|---|
| | $J = 0.886878$<br>Number of segments is 74<br>Time: 0.010691 s.<br>Straightening time: 0.000391 s. |
| | $J = 0.921821$<br>Number of segments is 85<br>Time: 0.023221 s.<br>Straightening time: 0.000872 s. |
| | $J = 0.777386$<br>Number of segments is 61<br>Time: 0.013486 s.<br>Straightening time: 0.000746 s. |
| | $J = 0.782661$<br>Number of segments is 125<br>Time: 0.032127 s.<br>Straightening time: 0.001331 s. |
| | $J = 0.693615$<br>Number of segments is 137<br>Time: 0.054227 s.<br>Straightening time: 0.002106 s. |
| | $J = 0.860022$<br>Number of segments is 160<br>Time: 0.038669 s.<br>Straightening time: 0.001954 s. |
| | $J = 0.863597$<br>Number of segments is 86<br>Time: 0.030487 s.<br>Straightening time: 0.001469 s. |

Note the excellent performance: the processing takes 10...50 milliseconds. 3...6% of the time is spent on the shape straightening and calculating the symmetry index. The path instability at the final steps in the wide shape parts (the lizard image in the fifth row and the stingray image) should be noted as an obvious shortcoming.

## 8    Conclusion

The advantages of the proposed method are better performance and applicability to shapes with significant curvature. Another advantage is that the procedure is virtually devoid of any manually adjusted inputs. The key procedure inputs (the increment and angular range used at the next step) are estimated by an adaptive procedure. The algorithm is suitable for parallelization.

The key drawback is the inability to find an optimal solution (which is a feature of greedy search) and the need to specify the initial point. The algorithm should be applied to prolonged, narrow shapes. The algorithm has no protection from infinite looping which may occur when processing shapes with holes.

We plan a large experimental study to compare the proposed procedure with the dynamic programming-based global optimization [8], It is also of interest to compare the results for low-articulated images with the reference values obtained by brute force search for reflective symmetry (we have access to such a database).

**Acknowledgements.** This study was supported by the Russian Science Foundation, Grant No. 22-21- 00575. https://rscf.ru/project/22-21-00575/.

## References

1. Kushnir, O., Fedotova, S., Seredin, O., Karkishchenko, A.: Reflection symmetry of shapes based on skeleton primitive chains. In: Ignatov, D., et al. (eds.) Analysis of Images, Social Networks and Texts. AIST 2016. CCIS, vol. 661, pp. 293–304. Springer, Cham (2017). https://doi.org/10.1007/978-3-319-52920-2_27
2. Mestetskiy, L., Zhuravskaya, A.: Method for assessing the symmetry of objects on digital binary images based on Fourier descriptor. Int. Arch. Photogramm. Remote Sens. Spat. Inf. Sci. **XLII-2/W12**, 143–148 (2019)
3. Nguyen, T.P., Truong, H.P., Nguyen, T.T., Kim, Y.-G.: Reflection symmetry detection of shapes based on shape signatures. Pattern Recognit. **128**, 108667 (2022)
4. Lomov, N., Seredin, O., Kushnir, O.: Detection of the optimal reflection symmetry axis with the Jaccard index and the radon transform. In: 2022 International Russian Automation Conference (RusAutoCon), Sochi, Russian Federation, 2022, pp. 489–498 (2022). https://doi.org/10.1109/RusAutoCon54946.2022.9896373
5. Ricca, G., Beltrametti, M.C., Massone, A.M.: Detecting curves of symmetry in images via hough transform. Math. Comput. Sci. **10**, 179–205 (2016). https://doi.org/10.1007/s11786-016-0245-5
6. Liu, J., Liu, Y.: Curved reflection symmetry detection with self-validation. In: Kimmel, R., Klette, R., Sugimoto, A. (eds.) Computer Vision – ACCV 2010. ACCV 2010. LNCS, vol. 6495, pp. 102–114. Springer, Berlin, Heidelberg (2011). https://doi.org/10.1007/978-3-642-19282-1_9

7. Huang, J., Stoter, J., Nan, L.: Symmetrization of 2D polygonal shapes using mixed-integer programming. Comput.-Aided Des. **163**, 103572 (2023). ISSN 0010–4485. https://doi.org/10.1016/j.cad.2023.103572

8. Lomov, N., Seredin, O.: Dynamic programming for curved reflection symmetry detection in segmented images. Int. Arch. Photogramm. Remote. Sens. Spat. Inf. Sci. **48**, 157–163 (2023)

9. Peng, H., Long, F., Liu, X., Kim, S., Myers, E.: Straightening Caenorhabditis elegans images. Bioinformatics (Oxford, England) **24**, 234–242 (2008)

10. Lee, S., Liu, Y.: Curved glide-reflection symmetry detection. In: IEEE Conference on Computer Vision and Pattern Recognition, 2009, pp. 1046–1053 (2009)

11. Teo, C.L., Fermuller, C., Aloimonos, Y.: Detection and segmentation of 2D curved reflection symmetric structures. In: IEEE International Conference on Computer Vision (ICCV), pp. 1644–1652 (2015)

12. Lomov, N., Tiras, K., Mestetskiy, L.M.: Identification of planarian individuals by spot patterns in texture. In: Farinella, G.M., Radeva, P., Bouatouch, K. (eds.), Proceedings of the 17th International Joint Conference on Computer Vision, Imaging and Computer Graphics Theory and Applications, VISIGRAPP 2022, Volume 4: VISAPP, Online Streaming, 6–8 February 2022, SCITEPRESS, pp. 87–96 (2022)

13. Jaccard, P.: Étude comparative de la distribution florale dans une portion des Alpes et des Jura. Bull. Soc. Vaudoise Sci. Nat. **37**, 547–579 (1901)

14. Seredin, O.S., Kushnir, O.A., Fedotova, S.A.: Comparative analysis of reflection symmetry detection methods in binary raster images with skeletal and contour representations. Comput. Opt. **46**(6), 921–928 (2022). https://doi.org/10.18287/2412-6179-CO-1115

15. Latecki, L.J., Lakamper, R.: Shape similarity measure based on correspondence of visual parts. IEEE Trans. Pattern Anal. Mach. Intell. **22**(10), 1185–1190 (2000)

# Data Analysis and Machine Learning

# Ensemble Clustering with Heterogeneous Transfer Learning

Vladimir Berikov[✉][iD]

Sobolev Institute of mathematics, Novosibirsk, Russia
`berikov@math.nsc.ru`

**Abstract.** This work introduces a novel approach to ensemble clustering by incorporating transfer learning. We address a clustering problem where, in addition to the data being analyzed, we have access to "similar" labeled data. The datasets may have different feature descriptions. Our method revolves around constructing meta-features that capture the structural characteristics of the data and transferring them from the source domain to the target domain. To define meta-features, we use the cluster ensemble method. Through extensive Monte Carlo modeling experiments, we have demonstrated the effectiveness of our proposed method. Notably, compared to other similar approaches, our method exhibits the capability to handle arbitrary feature descriptions in both the source and target domains. Additionally, it offers a reduced computational complexity, making it more efficient in practice.

**Keywords:** Ensemble Clustering · Transfer Learning · Co-association matrix

## 1 Introduction

In machine learning, there exist various models, methods, and algorithms based on different approaches. Two important areas of research in this field are transfer learning and ensemble clustering. Transfer learning aims to improve decision making by using additional information from related fields, while ensemble clustering focuses on improving the quality and stability of clustering results.

Transfer learning, closely related to domain adaptation and knowledge transfer, involves using additional data known as source data that is similar, to some extent, to the target data of interest. For example, this could involve using digital images of the same landscape captured at different moments in time or utilizing data and results of text document classification in one language to classify documents written in another language.

There are different ways to set up the transfer learning problem. It can be formulated with or without information about class labels in the target and source

---

This work was supported by the State Contract of Sobolev Institute of Mathematics, Project No. FWNF-2022-0015.

D. I. Ignatov et al. (Eds.): AIST 2023, LNCS 14486, pp. 255–266, 2024.
https://doi.org/10.1007/978-3-031-54534-4_18

domains, and the feature descriptions can be the same (homogeneous domain adaptation) or different (heterogeneous domain adaptation). Typically, transfer learning assumes that the probability distributions of the target and source domains do not coincide. Various transfer learning methods have been developed for pattern recognition problems, regression, and cluster analysis [5, 9]. These methods incorporate different approaches to knowledge transfer, such as using source examples with adjustable weights for constructing decision functions, seeking common feature descriptions, or making use of assumptions about the distribution of certain hyperparameters.

Ensemble clustering focuses on finding a consensus decision considering multiple ways to partition the analyzed data [3, 4, 6, 11]. This methodology often provides robust and effective solutions, especially when the data structure is uncertain. Even if we combine "weak" algorithms, a well-organized ensemble can significantly improve the overall quality of clustering.

We consider a clustering problem in which, along with the target dataset, we have an option to utilize an already classified dataset described by different features and sampled from another statistical population. A practical illustration of this scenario is the segmentation of color images by utilizing pre-segmented gray-scale images that share similar characteristics.

The problem of transfer learning with the use of cluster ensembles was considered in [1]. The authors assume that the source and target domains share common feature space and class labels. The authors of [10] propose a versatile framework that accommodates arbitrary feature descriptions. However, their suggested algorithm has a cubic time complexity, which means that its computational requirements increase rapidly with the maximum size of both the source and target datasets.

In this paper, we propose a method that combines ensemble clustering with transfer learning. The method operates in the following stages.

- *Independent analysis.* Both target and source data are analyzed independently using a cluster ensemble to identify stable structural patterns. This analysis results in the extraction of meta-features that describe the structural characteristics of data.
- *Relationship extraction.* The relationships between the elements of the co-association matrix of the source data and their corresponding meta-features are established using supervised classification. This step aims to capture the dependencies between the data and its structural characteristics.
- *Transfer and prediction.* The identified dependencies from the source domain are transferred to the target domain, enabling the prediction of the co-association matrix for the target data. This step leverages the learned relationships to estimate the similarity of elements in the target data.
- *Final clustering.* Based on the predicted co-association matrix, we obtain the final clustering partition for the target data using some of the known techniques (hierarchical agglomerative clustering, spectral algorithm, etc.).

The rest of the paper is organized as follows. Section 2 presents methodological issues that need to be addressed when discussing cluster ensemble and knowledge transfer. Section 3 gives the details of the proposed method. Section 4 describes numerical experiments and reports on the results obtained. In the concluding part we summarize our work and outline future extensions.

## 2  Methodology

This section presents the notations used in this paper, explains the problem considered, and gives a brief description of the cluster ensemble methodology.

### 2.1  Basic Notation and Problem Statement

Let us consider a set $T = \{a_1, \ldots, a_{N_T}\}$ of objects from some statistical population. Each object is described by a set of real-valued features $X_1, \ldots, X_{d_T}$. Through $x = x(a) = (x_1, \ldots, x_{d_T})$ we denote a feature vector for an object $a$, where $x_j = X_j(a)$, $j = 1, \ldots, d_T$, and through $X_T$ we denote data matrix $X_T = (x_{ij})$, $j = 1, \ldots, d_T$, $i = 1, \ldots, N_T$. It is required to obtain a partition $P = \{C_1, \ldots, C_{K_T}\}$ of $X_T$ on some number of $K_T$ groups (clusters) in according to a given quality criterion. The number of clusters can be set in advance or not; in this paper, we assume that the required number is a given parameter.

Suppose that there is an additional dataset $S = \{b_1, \ldots, b_{N_S}\}$ where each object $b$ is described by real-valued features $X'_1, \ldots, X'_{d_S}$, and the data matrix $X_S$ is given. A categorical attribute $Y$ is specified, denoting a class to which an object $b \in S$ belongs: $Y(b) \in \{1, \ldots, K_S\}$, where $K_S$ is the total number of classes. By classification vector we understand a vector $Y = \{y_1, \ldots, y_{N_S}\}$, where $y_i = Y(b_i)$, $i = 1, \ldots, N_S$.

The set $T$ is called target data, and the set $S$ source data. It is assumed that $X_T$ and $X_S$ share some common regularities in their structure, which can be detected by cluster analysis and used as additional information when setting up the desired partition of $X_T$.

### 2.2  Cluster Ensemble

Suppose that we are able to create variants of the partitioning of $X_T$ into clusters using some clustering algorithm $\mu$. The algorithm may work under different parameter settings, or, more generally, "learning conditions" such as initial centroids locations, subsets of selected features, number of clusters, or random subsamples. On the $l$th run it gives a partition of $X_T$ on $K_l$ clusters, $l = 1, \ldots, L$, where $L$ is the total number of runs.

For each pair of points $a_i, a_j \in X_T$ we define the value $h_l^T(i,j) = I[\mu_l(x_i) = \mu_l(x_j)]$, where $\mathbb{I}[\cdot]$ is an indicator function: $\mathbb{I}[true] = 1$; $\mathbb{I}[false] = 0$, $\mu_l(x)$ is an index of the cluster assigned to a point $x \in X_T$ by the algorithm $\mu$ in the $l$-th run. Let $H_l^T = (h_l^T(i,j))$ be the co-association matrix for $l$th partition. We then calculate the averaged co-association matrix $H^T = (\bar{h}^T(i,j))$ with elements $\bar{h}^T(i,j) = \frac{1}{L} \sum_{l=1}^{L} h_l^T(i,j)$, $i, j = 1, \ldots, N_T$.

The next stage is aimed at constructing the final partition of $X_T$. Elements of the matrix $H^T$ are considered as measures of similarity between pairs of objects. To form the partition, any algorithm that uses these measures as input information can be used. In this paper, we apply the ensemble spectral clustering algorithm [2] based on a low-rank decomposition of the averaged co-association matrix, which has near-linear time and storage complexity.

The basic steps of the ensemble clustering algorithm EC used are described below.

---

**Algorithm 1. EC.**

---

**Input:**
$X_T$: target data;
$L$: number of runs of the ensemble clustering algorithm;
$\Omega$: set of algorithm parameters;
$K_T$: required number of clusters in the partition of $X_T$.
**Output:**
$P = \{C_1, ..., C_{K_T}\}$: partition of $X_T$.
**Steps:**
1. Get $L$ variants of the cluster partition for objects from $X_T$ by randomly choosing the algorithm parameters from $\Omega$;
2. Calculate the averaged co-association matrix $H^T$ in low-rank representation;
3. Using spectral clustering with input matrix $H^T$, find the final partition $P$.
**end.**

---

### 2.3  Probabilistic Properties of Cluster Ensemble

Suppose that we have the *i.i.d.* sample $X = \{x_1, \ldots, x_N\}$ generated from a mixture of $K$ distributions (classes). Suppose that there also exists a ground truth (latent, directly unobserved) variable $Y$ that determines the class to which an element $x_i$ belongs: $Y_i \in \{1, \ldots, K\}$. Let $Z(i,j) = I[Y_i \neq Y_j]$, where $i, j = 1, \ldots, N$. Let the algorithm $\mu$ be randomized, that is, it depends on a random set of parameters $\Omega \in \mathbf{\Omega}$, and the sample partitions are obtained using independently selected statistical copies of $\Omega$.

Each ensemble algorithm contributes to the overall collective decision. Denote by $v_1(i,j)$ the number of votes for the union of $x_i$, $x_j$ into same cluster; and by $v_0(i,j)$ the number of votes for their separation. The value $c(i,j) = I[v_1(i,j) > v_0(i,j)]$ is called the ensemble solution for $x_i$ and $x_j$ obtained according to the voting procedure. The conditional probability of classification error for each pair is defined as $P_{err}(i,j) = P[c(i,j) \neq Z(i,j)|X]$. The following property was proved in [3].

**Theorem.** *Let us suppose that for any $i, j$ $(i \neq j)$ the symmetry condition is satisfied:* $P[h(i,j) = 1|Z(i,j) = 1] = P[h(i,j) = 0|Z(i,j) = 0] = q(i,j)$ *where* $q(i,j)$ *is the conditional probability of the correct decision. If the condition of weak learnability $0.5 < q(i,j) \leq 1$ holds, $P_{err}(i,j) \to 0$ as $L$ approaches infinity.*

Therefore, under certain regularity conditions, the quality of ensemble decisions is improved with an increase in ensemble size. However, in case of the violation of the assumptions, as well as with a small number of ensemble elements, the quality of the decisions can turn into a degenerate. To improve the quality of the ensemble, it is possible to use information contained in the additional (source) data.

## 3 Proposed Method: Usage of Source Data

For source data $X_S$ the classification vector $Y$ is known. Therefore it is possible to calculate the coincidence matrix $Z^S = (z^S(i,j))$, where $z^S(i,j) = \mathbb{I}\,[y_i = y_j]$, $i,j = 1, \ldots, N_S$. Despite the fact $X_T$ and $X_S$ belong to different statistical populations, we may assume that some general structural regularities characterizing both populations exist. These regularities can be found using cluster analysis. To obtain more robust results, we apply the cluster ensemble algorithm independently for the source and target data (for simplicity, we assume the same number $L$ of runs in the ensemble for each dataset).

As the analysis proceeds, the averaged co-association matrix $H^S$ is determined for $X_S$.

For specifying the regularities under interest, one may use different approaches. In this paper, we consider characteristics of mutual positions of data points with respect to the found clusters and use them as meta-features describing common properties of source and target domains.

As the first type of meta-feature, we use frequencies of the assignment of object pairs to the same clusters (elements of matrix $H^S$). These values belong to the interval $[0,1]$ and do not explicitly depend on the dimensionality of the initial feature space.

Another meta-feature used in this work is based on Silhouette index which is defined for each data point and reflects its similarity to other points in the same cluster and dissimilarity to points from different clusters. Let $Sil(x_i)$, $Sil(x_j)$ denote Silhouette indices, averaged over ensemble partitions, respectively for points $x_i, x_j \in X_S$. Denote $P^S(i,j) = \frac{1}{2}(Sil(x_i) + Sil(x_j))$; matrix $P^S$ is determined for all pairs of points of the source data. Similarly, matrix $P^T$ is defined for the target data.

Let $U^S$ denote a set of meta-features used, for example $U^S = (H^S, P^S)$. Consider a problem of finding a decision function

$$f : U^S \mapsto Z^S \tag{1}$$

for predicting elements of $Z^S$ viewed as new class labels (0 or 1). The classifier can be found by using existing machine learning models such as Support Vector Machine (SVM), Random Forest (RF) or Artificial Neural Net (ANN).

By employing the existing validation techniques such as hold-out or cross-validation, it is also possible to assess the accuracy of forecasting, determine the optimal prediction model, and select the most important meta-features.

The found classifier is then transferred to target domain for predicting $\hat{Z}^T = f(U^T)$. The resulting coincidence matrix $\hat{Z}^T$ cannot be used directly for clustering (for example, by finding connected components), since it can cause violation of the metric properties in the feature space. For example, if for some $i, j, k$, $\hat{Z}^T(i, j) = 1$ and $\hat{Z}^T(i, k) = 1$, there is no guarantee that $\hat{Z}^T(j, k) \neq 0$. For that reason, we search for a partition by solving the following optimization problem:

$$\text{find } \hat{Z}^* \in \Psi_{K_T} : \ \hat{Z}^* = \arg\min (\hat{Z}^* - \hat{Z}^T)_2 \tag{2}$$

where $\Psi_{K_T}$ is a set of Boolean coincidence matrices corresponding to all possible partitions of $X_T$ into $K_T$ clusters, $(\cdot)_2$ is the Frobenius norm of a matrix.

To find an approximate solution, we apply the following procedure:

1: Choose an initial partition of $X_T$;
2: Correct the partition by finding such point which gives the best improvement of functional in (2) when migrating to another cluster;
3: Continue iterations until the relative improvement of the optimized functional becomes less than a given parameter $Q_{\min}$, or the number of iterations (migrated points) exceed the preset value $\text{It}_{\max}$.

In our implementation, we start from the initial partition found with EC. The main steps of the proposed algorithm TrEC (Transfer Ensemble Clustering) are described below.

---

**Algorithm 2.** TrEC.

---

**Input:**
$X_T$: target data;
$X_S$: source data;
$Y$: source data class labels;
$L$: number of runs for the clustering algorithm $\mu$;
$\Omega_S$, $\Omega_T$: parameters of algorithm $\mu$ working on $X_S$ and $X_T$, respectively;
$K_T$: required number of clusters in $X_T$.
**Output:**
$P = \{C_1, \ldots, C_{K_T}\}$: clustering partition of $X_T$.
**Steps:**
1. Generate $L$ clustering variants for $X_S$ and $L$ clustering variants for $X_T$ by random choice of parameters from $\Omega_S$, $\Omega_T$, respectively;
2. Calculate co-association matrices $H^T$, $H^S$ and matrices $P^T$, $P^S$, $Z^S$;
3. Find a decision function $f(P^S, H^S)$ for predicting elements of $Z^S$;
4. Calculate matrix $\hat{Z}^T = f(P^T, H^T)$;
5. Find matrix $\hat{Z}^*$ and corresponding partition $P$, using the above-mentioned approximate procedure of solving (2);
**end.**

---

The complexity of the algorithm with respect to the sample size can be estimated as $O(\max(N_T, N_S)^2)$.

# 4    Numerical Experiments

To verify the applicability of the suggested approach, we have designed Monte Carlo experiments with artificial and real datasets.

## 4.1    Details of Experiments

**Synthetic Data.** To generate target data, we use the following distribution model. In 12-dimensional feature space, the two modeled classes have a spherical form, and the other two have a strip-like form. The first and second classes are of Gauss distribution $N(\nu_i, \sigma_1 \mathbf{I})$, where $\nu_1 = (0, \ldots, 0)^T$, $\nu_2 = (8, \ldots, 8)^T$, $\sigma_1 = 1.5$, and $\mathbf{I}$ denotes the unit covariance matrix. The coordinates of objects of the other two classes are determined recursively: $x_{k_{i+1}} = x_{k_i} + \gamma_1 \cdot \mathbf{1} + \varepsilon$, where $\mathbf{1} = (1, \ldots, 1)^T$, $\gamma_1 = 0.2$, $\varepsilon$ is the Gaussian random vector $N(0, \gamma_2 \cdot \mathbf{I})$, $\gamma_2 = 0.25$, $k = 3, 4$. For class 3, $x_{3_1} = (-5, 5, \ldots, -6, 6)^T + \varepsilon$; for class 4, $x_{4_1} = (5, -5, \ldots, 6, -6)^T + \varepsilon$. The number of objects for each class is 20.

Source data set has a similar structure, with the following differences: feature space dimensionality equals 8; the first and second classes follow Gauss distribution $N(\lambda_i, \sigma_2 \mathbf{I})$, $i = 1, 2$, where $\lambda_1 = (0, \ldots, 0)^T$, $\lambda_2 = (6, \ldots, 6)^T$, $\sigma_2 = 1.6$; the third and fourth classes are determined as follows: $x'_{k_{i+1}} = x'_{k_i} + \gamma'_1 \cdot \mathbf{1} + \varepsilon'$, where $\gamma'_1 = 0.2$, $\varepsilon'$ is normally distributed vector $N(0, \gamma'_2 \cdot \mathbf{I})$, $\gamma'_2 = 0.2$, $k = 3, 4$. For the third class $x'_{3_1} = (-5, 5, \ldots, -5, 5)^T + \varepsilon'$; for the fourth class $x'_{4_1} = (5, -5, \ldots, 5, -5)^T + \varepsilon'$. The sample size for each class is equal to 25.

Examples of generated data are shown in Fig. 1.

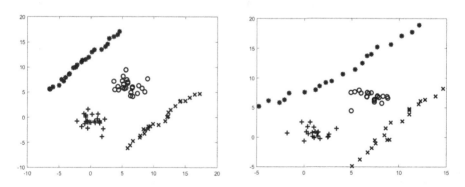

**Fig. 1.** Examples of sampled data (left: target data, right: source data); projection on first two coordinate axes.

In the implementation of TrEC, we use $K$-means as the base algorithm in the ensemble. To design the ensemble, we use clustering with a different number of clusters taking its value from the interval $[2, L+1]$, where $L$ is the size of the ensemble ($L = 10$ in our experiments).

For training on $(P^S, H^S)$, $\hat{Z}^S$ and predicting the elements of $\hat{Z}^T$ using $(P^T, H^T)$, we apply SVM and RF with parameters set by default. Parameter $Q_{\min} = 0.01$, $It_{\max} = 40$.

In the Monte Carlo modeling process, synthetic datasets are repeatedly generated according to the specified distribution model. To increase the reliability of the results, the accuracy estimates are averaged over 250 Monte Carlo simulations.

**Real Data.** In the second example we analyze the MNIST database of handwritten digits [8]. The database includes 70,000 normalized grayscale images in the $28 \times 28$ pixel format. We used 2000 randomly chosen images with reduced size $14 \times 14$ as source data and 100 independent images with original resolution as target data. Figure 2 exemplifies the images. Pixel values from images are utilized as the analyzed features. As in the previous example, $K$-means is the base ensemble algorithm (the number of clusters varies from 10 to 35). In addition to the introduced meta-features, we apply the following characteristics describing object pairs: normalized distance and averaged distance to closest centroids.

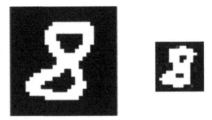

**Fig. 2.** Examples of images (left: original size, right: reduced size)

To train the model on source data, a feedforward ANN is used. The ANN consists of three fully connected hidden layers with 80, 80 and 20 neurons correspondingly; batch normalization and ReLU activation function are implemented. The network is trained using the following settings: batch size of 1024, stochastic gradient descent with momentum (SGDM) optimizer, initial learning rate of $10^{-4}$, maximum number of epochs is 4. The algorithm was implemented in Python 3.

All other settings are the same as in the previous experiment.

Due to the limited computational resources, we perform only 25 Monte Carlo simulations with randomly chosen subsets of data.

**Evaluation Metric.** To evaluate the quality of ensemble clustering, we use the Adjusted Rand Index $(ARI)$ [7]. This index estimates the degree of correspondence between two partitions $P_1 = \{C_{1,1}, \ldots, C_{k,1}, \ldots, C_{K_1,1}\}$ and $P_2 = \{C_{1,2}, \ldots, C_{l,2}, \ldots, C_{K_2,2}\}$. Let $N_{k,1} = ||C_{k,1}||$, $N_{l,2} = ||C_{l,2}||$ denote the sizes of clusters, and $N^{(k,l)} = \|C_{k,1} \cap C_{l,2}\|$, $k = 1, \ldots, K_1$, $l = 1, \ldots, K_2$. The index is defined with the following expression:

$$ARI = \frac{\sum\limits_{k,l} \binom{N^{(k,l)}}{2} - Q_1 Q_2 / G_0}{\frac{1}{2}(Q_1 + Q_2) - Q_1 Q_2 / G_0},$$

where $Q_1 = \sum\limits_{k} \binom{N_{k,1}}{2}$, $Q_2 = \sum\limits_{l} \binom{N_{l,2}}{2}$, $G_0 = \binom{N}{2}$, $N$ is sample size. The maximum value of $ARI$ is 1; the index can take negative values. $ARI$ close to 1 means a high degree of matching between the found partition and the true one; $ARI$ close to zero indicates nearly random correspondence.

## 4.2   Results

**Synthetic Data.** The following data processing strategies are compared: a, b) with the use of the suggested TrEC algorithm (combined either with SVM or RF); c) with the EC algorithm, in which transfer learning is not utilized; d) $K$-means without ensemble usage.

Statistical analysis of the significance of the differences between the estimates is carried out using a paired Student's t-test. The results of the experiments are shown in Table 1.

**Table 1.** Averaged $ARI$ over Monte Carlo simulations for different methods.

| TrEC+SVM | TrEC+RF | EC | $K$-means |
|----------|---------|-------|-------|
| 0.831 | 0.789 | 0.788 | 0.658 |

The Student test shows significant differences between the TrEC + SVM estimates and the EC (p-value $< 0.05$). Therefore, despite the fact that the data distribution is quite difficult for $K$ means (which is oriented on spherical clusters), the suggested method demonstrates a statistically significant increase in the quality of decisions.

Figure 2 presents an example of the SVM decision in the coordinate plane with axes determined by values of $P^S$ (horizontal axis) and $H^S$ (vertical axis). From this example, it can be seen that Silhouette index-based meta-feature provides some additional information for correct discrimination of object pairs into same or different clusters.

We also estimate the significance of meta-features with respect to clustering quality. To this end, we try to exclude one of the features from the analysis and repeat the experiment. The performance of TrEC+SVM is evaluated then the co-association matrix-based meta-feature is used alone, and then only the Silhouette-based meta-feature is utilized. Table 2 shows the obtained results. One can conclude from the experiment that the co-association matrix-based meta-feature is more important than the Silhouette-based; however, both of them are useful in combination.

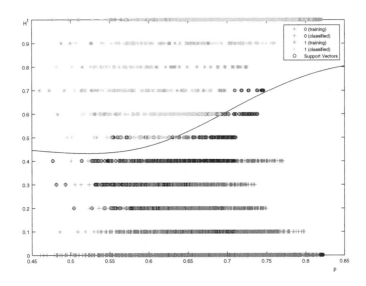

**Fig. 3.** An example of the decision boundary.

**Real Data.** Table 3 presents the results of experiments with subsets of the MNIST database. For comparison purposes, the table demonstrates the averaged quality metrics over Monte Carlo experiments, as well as the best and worst results (with respect to the *ARI* values).

Evidently, the clustering performance is rather low in this example, however, transfer learning allows us to increase it to a certain extent.

**Table 2.** Averaged *ARI* for used meta-features.

| Meta-feature used | Estimated quality |
|---|---|
| Co-association matrix-based | 0.776 |
| Silhouette-based | 0.578 |

**Table 3.** *ARI* for different methods.

| | TrEC | EC | $K$-means |
|---|---|---|---|
| averaged | 0.260 | 0.239 | 0.244 |
| best | 0.454 | 0.321 | 0.218 |
| worst | 0.156 | 0.116 | 0.139 |

# 5   Conclusion

This work has introduced an ensemble clustering method using the transfer learning methodology. The method is based on finding meta-features describing structural characteristics of data and their transfer from source to target domain. The proposed method allows one to consider different feature sets describing the domains, as well as a different number of classes for both of them. The complexity of the method is of quadratic order and is less than the complexity of analogous algorithms.

An experimental study of the method using Monte Carlo modeling has confirmed its efficiency.

Compared to similar methods, our proposed method has the following advantages.

- *Compatibility with arbitrary feature descriptions.* The proposed method can handle arbitrary feature descriptions in both the source and the target domains. This flexibility allows it to adapt to diverse data representations.
- *Reduced complexity.* Our method has a lower computational complexity compared to existing approaches. This reduction in complexity makes it more efficient for large-scale datasets.

While our proposed method holds promise in improving clustering quality and stability through ensemble clustering with transfer learning, there are certain limitations that should be acknowledged:

- *Data suitability.* The effectiveness of our method heavily relies on the availability of "similar" labeled data in addition to the target dataset. If such data is not available or not adequately representative, the performance of the method could be compromised.
- *Computational demands.* Though our method offers reduced complexity compared to other approaches, the process of constructing meta-features and transferring them between domains can still demand substantial computational resources, especially for large datasets.

In the future, we plan to continue studying theoretical properties of the proposed method and its further development aimed at faster processing speed. Determining of useful types of meta-features, in addition to such as those proposed in the present work, is another important problem. A detailed comparison with existing combined ensemble clustering and transfer learning methods is our next objective. Application of the method in various fields is also planned.

# References

1. Acharya, A., Hruschka, E.R., Ghosh, J., Acharyya, S.: Transfer learning with cluster ensembles. In: Proceedings of the 2011 International Conference on Unsupervised and Transfer Learning Workshop, vol. 28, pp. 123–133 (2011)
2. Berikov, V.: Autoencoder-based low-rank spectral ensemble clustering of biological data. In: 2020 Cognitive Sciences, Genomics and Bioinformatics (CSGB), pp. 43–46 (2020)
3. Berikov, V., Pestunov, I.: Ensemble clustering based on weighted co-association matrices: error bound and convergence properties. Pattern Recognit. **63**, 427–436 (2017)
4. Boongoen, T., Iam-On, N.: Cluster ensembles: a survey of approaches with recent extensions and applications. Comput. Sci. Rev. **28**, 1–25 (2018)
5. Fang, Z., Lu, J., Liu, F., Zhang, G.: Semi-supervised heterogeneous domain adaptation: theory and algorithms. IEEE Trans. Pattern Anal. Mach. Intell. **45**(1), 1087–1105 (2022)
6. Ghosh, J., Acharya, A.: Cluster ensembles. Wiley Interdiscip. Rev. Data Min. Knowl. Discov. **1**(5), 305–315 (2011)
7. Hubert, L., Arabie, P.: Comparing partitions. J. Classif. **2**(1), 1193–218 (1985)
8. LeCun, Y.: MNIST handwritten digit database (2013). http://yann.lecun.com/exdb/mnist
9. Pan, S., Yang, Q.: A survey on transfer learning. IEEE Trans. Knowl. Data Eng. **22**(10), 1345–1359 (2020)
10. Shi, Y., et al.: Transfer clustering ensemble selection. IEEE Trans. Cybern. **50**(6), 2872–2885 (2018)
11. Shi, Y., Yu, Z., Chen, C.P., Zeng, H.: Consensus clustering with co-association matrix optimization. IEEE Transactions on Neural Networks and Learning Systems (2022)

# Detecting Design Patterns in Android Applications with CodeBERT Embeddings and CK Metrics

Gcinizwe Dlamini📖, Usman Ahmad📖, Lionel Randall Kharkrang📖, and Vladimir Ivanov(✉)📖

Innopolis University, Innopolis, Russian Federation
{g.dlamini,us.ahmad,l.kharkrang,v.ivanov}@innopolis.university

**Abstract.** As the software codebase increases and the complexity of software increases software developers tend to use simplified tested solutions such as design patterns. These design patterns in most cases tend to be transformed into clients' requirements documents. Having large software increases the difficulty of understanding the overall architectural design of the overall application. With a speedily growing application of natural language processing (NLP) to source code understanding, in this paper, we propose an approach for detecting design patterns in the Android development domain. Our approach combines source code embeddings together with source code metrics to detect Android applications' architectural design patterns namely model view controller (MVC), model view presenter (MVP), and model-view-view model (MVVM). We also detect if a design pattern is missing from a project. Our proposed approach was evaluated on a standard publicly available benchmark dataset retrieved from GitHub. The results showed that incorporating CodeBERT embeddings together with CK metrics improves classification performance. In addition, we compare our proposed approach results with the current state-of-the-art.

**Keywords:** Design Patterns Detection · CodeBERT · CK metrics

## 1 Introduction

Software design and maintenance is a challenging task [12], especially in modern large Android applications due to the lack of conformity to proper coding practices. The implementation of design patterns is considered one of the most vital components of a codebase as it can improve the understandability, reusability, and maintenance of a system [5]. However, the design pattern implementation may be overlooked and their manual detection in a codebase is an inefficient and challenging task due to the diverse and complex structure of the Android applications. The primary motivation behind detecting design patterns lies in two aspects. Firstly, it empowers program managers to assess the conformity

© The Author(s), under exclusive license to Springer Nature Switzerland AG 2024
D. I. Ignatov et al. (Eds.): AIST 2023, LNCS 14486, pp. 267–280, 2024.
https://doi.org/10.1007/978-3-031-54534-4_19

of the delivered implementation with the contracted design in large-scale system acquisitions. Secondly, it helps software developers in crafting maintainable source code during continuous integration and delivery, thereby minimizing the accumulation of technical debt.

With the increasing size of codebases pattern detection is a challenging task and is not very straightforward due to the complexity and variations of the patterns that could be implemented in a software project [16]. To counter the aforementioned problem, machine and deep learning approaches have been proposed over the years [13,16,18]. However, data scarcity is a major problem and it poses a huge challenge, especially for design pattern detection in Android applications [7,10].

Previous works employed static code features for detecting design patterns. Later, the detection was treated as a classification-based problem by employing Machine Learning (ML) approaches over hand-crafted code features. Recent techniques also explore Graph Neural Networks and vector-based representations using word2vec with limited success [18]. Most of the existing approaches, including the current state-of-the-art, either rely on manual feature engineering or perform two-phase classification that is (i) extracting features to form a pattern definition and (ii) match an input with the definition to detect a pattern. On the contrary, we propose an end-to-end approach which leverages transfer learning paradigm, that is, relying on features provided from a pretrained transformer-based model, CodeBERT [11]. Our approach is inspired by the recent advancement of transformer-based models in code-related tasks [19].

Modern NLP approaches, especially transformer-based architectures, have been increasingly applied for source code understanding. For example, Incheon Paik and Jun-Wei Wang [19] investigated the use of code graphs in code generation tasks using the generative pre-trained transformer (GPT) architecture. Specifically, the researchers [19] focused on GPT-2 model [21] and its ability to learn the effects of the data stream using additional code graph features. The study divided the experimental phase into fine-tuning an existing GPT-2 model and pre-training from scratch using code data. The authors suggest that incorporating code graphs into GPT-2 can improve performance in generating code, and their findings demonstrate the potential of this approach.

Empirical investigation of the effectiveness of using the Transformer model for the source code summarization task was presented in [2]. The study found that the Transformer model with relative position representations and copy attention significantly outperforms the performance of prevailing approaches. The authors suggest that incorporating code structure into the Transformer model may further improve its performance in source code summarization and other software engineering sequence generation tasks, such as commit message generation for source code changes. Wei Hua and Guangzhong Liu [14] proposed a novel transformer-based neural network called Transformer-Based Code Classifier (TBCC) for code classification and clone detection tasks in software engineering. TBCC can avoid gradient vanishing problems and capture long-distance dependencies in code statements by splitting deep abstract syntax trees into smaller subtrees. The proposed approach leads in two program comprehension tasks and outperformed baseline methods in terms of accuracy, recall, and F1 score.

Typically, pre-trained models used as feature extractors helping to avoid manual feature engineering. However, one of the classical approaches to source code analysis is based on a set of well-established measures of code, CK-metrics. These metrics can be used as input features for classifier (design pattern detector), and therefore constitute and alternative to code features provided by a pre-trained transformer model. However, the relative impact of the CK-metrics and deep neural features is still an open question. In the present paper we consider two research questions which potentially have an impact on software engineering:

**RQ1** What are relative importance of CK metrics for design patterns detection?
**RQ2** Is it possible to improve design pattern detection ML models performance by incorporating source code embeddings features?

The main contributions of the paper are:

– An approach for architectural design patterns detection in Android applications
– Analysis of the level to which source code embeddings together with CK metrics expose architectural design patterns information in source code

The structure of this paper is as follows: Sect. 2 presents the related work and the background of existing approaches. Section 3, our proposed model for design patterns is presented in detail together with the experimental setup. The results are presented in Sect. 4 followed by the achieved results discussions. Section 7 concludes our paper by presenting the limitations of our approach and possible future directions.

## 2   Related Work

A number of approaches have been proposed for detecting design patterns using source code metrics, static code features, ML classifiers, and graph-based methods. In particular, pattern detection in Android-based applications is a recent area of interest; however, numerous studies have been conducted on the general GoF (Gang of Four) design pattern detection [3,15,17]. In this section, we present an overview of design pattern detection specifically in Android applications, followed by a discussion on pattern detection in general.

### 2.1   Patterns Detection in Android Apps

The area of Android design pattern detection has recently seen a rise in the development of new approaches. Aymen et al. [10] proposed a system called RIMAZ that aims to identify MVC-based architectural patterns in Android apps using heuristics. RIMAZ leverages the SOOT framework and analyze Dalvik bytecode contained within apk files. They validate their approach on a sample of open-source apps from F-DROID apps. The downside of RIMAZ is that their approach is unable to detect any MVVM pattern and relies completely on heuristics.

COACH is another recent ML-based classification approach that focuses on the detection of design patterns from Android applications [7]. The researchers proposed a two-fold approach: training and detection. In the training phase, COACH builds a classifier for assigning roles to classes and recognizing MVC-based patterns. The detection phase consists of using trained classifiers to predict MVC-based patterns. They also contribute a training set comprising 69 Android applications in Java language. COACH leads the current state-of-the-art for pattern detection in Android apps.

## 2.2    Patterns Detection (GoF)

While pattern detection in Android applications is relatively newer, general design pattern detection (for GoF) has been researched well over the past decade. Dabain et al. [9] proposed a system called FINDER where they extracted static relations from Java class files using Javex and QL. They further extracted additional relations among variables and finally integrate all the information into a factbook. The factbook is then used to filter candidates that satisfy the static definition of design patterns that need to be detected. Their approach can detect twenty-two out of twenty-three GoF design patterns, however, it requires manual thresholding for candidate selection and is inefficient due to the large number of steps involved.

Kouli et al. [16] also proposed a similar two-phase approach that utilizes structural features in addition to static behavioral features. They first extract structural and behavioral features that represent the main roles of a design pattern from its signature. These features are then organized into a feature-based textual design pattern definition form and then added to a reusable vocabulary. In the second phase, the design pattern definition is matched with the input system to detect patterns having a matching design pattern definition. They also propose additional object-oriented features to improve accuracy over the previous approaches. However, their approach is not directly scalable over other design patterns as it might require adding additional features to the vocabulary.

Another group of researchers has also investigated Machine Learning based approaches in combination with code metrics. Uchiyama et al. [13] proposed pattern detection that defines code metrics through Goal-Question Metrics (GQM) method and assigns roles to every pattern. These roles are then fed as input features to machine learning models in order to detect patterns. Their approach is, however, limited to detecting five GoF patterns only. Chihada et al. [8] also proposed a very similar solution with additional features. The proposed method also extended the SVM-PHGS classifier to achieve better accuracy.

A few researchers have also explored NLP-based features for pattern detection. The vector-based representation of natural language through word embeddings has proven to be a useful feature in many NLP-related tasks. Nazar et al.

[18] experimented with the utilization of code semantics as features by utilizing word2vec embeddings. The intuition behind the approach was that code can also be treated as natural language [18]. Instead of only relying on static code features, they also employ lexical and syntactic features. They introduce a novel approach called Feature-based design pattern detector that uses fifteen source code features and combine them with lexical and semantic representation called Java Embedded Model (JEM). These JEM, computed through word2vec embeddings, are then fed as input features to a supervised ML classifier.

Almost all the aforementioned design pattern detection techniques focus only on Java-based systems and their solutions cannot be generalized to other programming languages. In some cases, manual feature extraction also inhibits the generalization capability of the proposed systems. Moreover, recent techniques (like graph and semantic-based approaches) still require thorough experimentation before they can be compared with the SOTA techniques.

## 3   Methodology

This section presents our proposed approach which is visualized in Fig. 1. The following sections outline the main parts in detail.

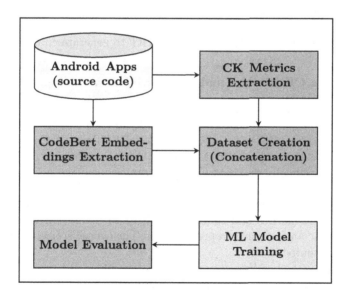

**Fig. 1.** The proposed model pipeline

### 3.1   Dataset: Source Code of Android Apps

To effectively benchmark the proposed methodology, we employ the same dataset utilized by the current state-of-the-art in design pattern detection, COACH [7].

The dataset consists of 69 open-source Android applications, which were gathered from the F-DROID repository [1]. Each directory represents a complete Java-based Android project, containing Java, JSON, and XML files. Table 1 delineates the distribution of samples for each class within our dataset.

**Table 1.** Dataset distribution

| Pattern | MVC | MVVM | MVP | None | Total |
|---------|-----|------|-----|------|-------|
| Count   | 26  | 9    | 12  | 22   | 69    |

### 3.2  Data Preprocessing

During the preprocessing stage, we filter out non-Java files from the source directory of each project to ensure that only Java code files are included for the embeddings. We then remove all comments and unnecessary spaces from each code file to maximize the input of relevant code for pattern detection. By doing this, we prioritize the most important and relevant parts of the code, while ignoring any unnecessary details that may not contribute much to the pattern detection process.

Due to the limitations of CodeBERT, we truncate the code for each file to a maximum length of 512 tokens. This ensures that the input size of each code file is within the acceptable limit for CodeBERT to generate embeddings.

### 3.3  Features: CK Metrics and CodeBERT Embeddings

Once we are done preprocessing, we are left with clean Java code against each file truncated to a max sequence length of 512, we then convert each file to an embedding vector by passing it through the CodeBERT model. CodeBERT is an NLP model recently proposed by Microsoft [11]. It is a transformer-based language model specifically designed for programming language understanding. It is trained on code-related data from GitHub, Stack Overflow and other sources. In addition to the features from CodeBERT, we also incorporate static code features from CK metrics and concatenate them with CodeBERT embeddings for rich syntactic and semantic representation. The overall steps of feature engineering are presented in Fig. 2.

We obtain one embedding vector for each Java file in the project. To generate a single feature representation for the complete project, we stack the feature vectors and apply sum pooling strategy. Following the application of sum pooling, we are left with a single feature vector for each Android app. To avoid the curse of dimensionality and enable more effective training, we reduce the size of the CodeBERT embeddings to 5 using the isomap, a nonlinear dimensionality reduction technique [22]. We apply this technique because our initial experiments, without dimensionality reduction, have shown that the training is less effective and leads to inferior performance, likely due to the difficulty of learning

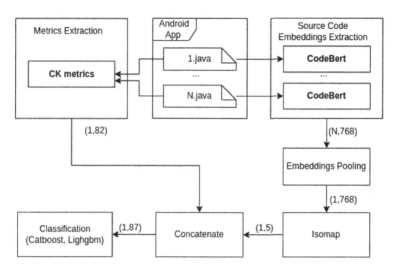

**Fig. 2.** Feature extraction with CodeBERT pipeline

from high-dimensional features and lack of data samples. By applying isomap dimensionality reduction, we enable more robust learning while still retaining the semantic information encoded by CodeBERT. We then concatenate these reduced CodeBERT embeddings with CK metrics to create the final feature vector input for model training.

The extraction of CK metrics from each an every Android project is done using a publicly available tool [4]. The CK metrics are extracted at different levels of the project namely class and method level. For each an every file CK metrics are calculated and to get the overall metrics for a given project aggregation is done using mean. The full list of the CK metrics is presented in [4].

### 3.4  Modeling Design Pattern Detection

After feature engineering, the resulting data is used as input in the training of machine learning model which will later be used to predict design pattern. We use a gradient boosting based model named CatBoost [20]. The optimized CatBoost training parameters are as follows:

- Learning Rate: 0.1
- Iterations: 15
- Tree Depth: 16

### 3.5  Model Performance Evaluation

Since design pattern detection is regarded as a classification problem, we assess the performance of our proposed method using class-wise recall, precision and F1-Score. These metrics are also favored by existing approaches, such as COACH, enabling direct comparisons with contemporary methods. Computing the class-wise F1-score is essential for tackling the label imbalance problem. Due to the absence of design patterns detection benchmark datasets, we use a 5-fold cross-validation strategy for our proposed approach model assessment (see Table 3).

## 4   Implementation Details and Results

Our approach implementation is in Python programming language. CodeBERT [11] python implementation is from transformers library [23] (version 4.28.1). The pooling techniques that we experimented with are sum pooling, mean pooling and max-pooling. The best results were obtained using sum pooling. Full implementation code is publicly available for the research community[1]

Table 2. F1-measure of our approach performance results with different experimental settings

| Input Features | MVC | MVP | MVVM | NONE |
|---|---|---|---|---|
| CK metrics | 0.67±0.18 | 0.62±0.24 | **0.69±0.19** | 0.45±0.28 |
| CodeBERT Embeddings and CK metrics | **0.68 ±0.12** | 0.61 ±0.21 | 0.69±0.37 | **0.63±0.15** |

Table 3. F1-measure comparison of our proposed approach models with the state-of-the-art

| | MVC | MVP | MVVM | NONE |
|---|---|---|---|---|
| RIMAZ | **0.68** | 0.22 | 0 | 0.62 |
| COACH | 0.68 | 0.56 | 0.67 | 0.61 |
| CK ONLY | 0.67±0.18 | 0.62±0.24 | **0.69±0.19** | 0.45±0.28 |
| OUR MODEL | **0.68 ±0.12** | 0.61 ±0.21 | 0.69±0.36 | **0.63±0.15** |

Table 4. Classification performance on Ck-metrics (CK) and **CodeBERT** Embeddings (CBEM) joined with Ck-metrics

| Pattern | Input Features | Precision | Recall | F1-score |
|---|---|---|---|---|
| MVC | CK | 0.62 | 0.74 | 0.67 |
| | CBEM and CK | **0.62** | **0.77** | **0.68** |
| MVP | CK | 0.66 | 0.60 | **0.62** |
| | CBEM and CK | **0.70** | **0.60** | 0.61 |
| MVVM | CK | 0.73 | 0.70 | 0.69 |
| | CBEM and CK | 0.73 | 0.70 | 0.69 |
| NONE | CK | 0.53 | 0.41 | 0.45 |
| | CBEM and CK | **0.76** | **0.56** | **0.63** |

[1] https://github.com/Gci04/design-patterns-detection.

# 5  Discussion

Based on the dataset used statistics presented in Table 1 it is evident that there is still a lack of benchmark data in the domain of design patterns detection. As much as deep learning approaches are showing outstanding performances in different fields (i.e. health care [25], cyber security [24], finance [6] etc.), these deep learning approaches are data hungry, requiring carefully labelled data for performance evaluation. To minimize this issue in this research we resorted for k-fold validation to evaluate our proposed approach. However there are some other available data generative (i.e. SMOTE [26], adversarial neural networks [24]) approaches that could be used to minimize the data scarcity issue (see Table 4).

To investigate the effectiveness of our proposed approach and the inspect what the employed machine learning model is learning from the data, we answer are aforementioned research questions (**RQ1** and **RQ2**).

## 5.1  RQ1: What Are Relative Importance of CK Metrics for Design Patterns Detection?

To address this research question, we employed multiple classifiers trained on CK metrics and subsequently selected the optimal model, in our case, the CatBoost classifier, to investigate feature explainability. We utilized CatBoost weights to represent relative feature importance. Our analysis unveiled the following top-5 contributing CK metrics:

- The number of references from other classes to a particular class
- The number of declared modifiers in a method
- The number of private fields in a class
- Response for a Class - the number of unique method executions in a method
- The number of methods with default modifier in a class

Despite these findings, it should be noted that the CK metrics alone may not provide a comprehensive understanding of the structural patterns within Android applications. This limitation is further substantiated by our classification tuning results, presented in Table 2, which demonstrate the performance of the classifier when fine-tuned solely on CK metrics.

In conclusion, while CK metrics do offer some insights into design pattern detection, their relative importance may be limited in capturing the complete structural intricacies of Android applications. Further research will explore the integration of additional features (e.g., features based on source code graph) to enhance the effectiveness of design pattern detection.

## 5.2  RQ2: Is it Possible to Improve ML Models Performance in Design Pattern Detection by Incorporating CodeBERT Features?

Building upon the findings from 1, we observed that while CK metrics were able to capture high-level source code features, they were insufficient in surpassing the current state-of-the-art approaches for design pattern detection. To address this limitation, we explored the potential of leveraging CodeBERT embeddings with CK metrics in order to improve classification performance.

Our experimental results demonstrated moderate enhancement in design pattern detection when combining CK metrics with CodeBERT embeddings, surpassing current leading approaches. Furthermore, the feature importance analysis of this combined approach indicated that the features derived from Code-BERT embeddings contributed, although not as much as initially expected, to the classification task. Hence, the CK metrics alone are not sufficient for the detection of design patterns in Android applications. However, by incorporating CodeBERT embeddings, we were able to improve the performance of CK metrics and achieve a more effective design pattern detection method.

This finding highlights the importance of considering complementary feature extraction techniques from NLP, such as CodeBERT embeddings, to supplement CK metrics and improve design pattern detection outcomes. What is more, we observed a drammatic change in CK metrics importance after introducing the CodeBERT embeddings as features to the model. The following list shows changes of metric position in the ranked list relative to the model with CK metrics only.

- class_fanin (-64)
- method_modifiers (-16)
- class_privateFieldsQty (-36)
- method_rfc (-9)
- class_defaultMethodsQty (+2)

A more detailed analysis of the contributions from CodeBERT embeddings can be observed in the plots in the Fig. 3 and 4.

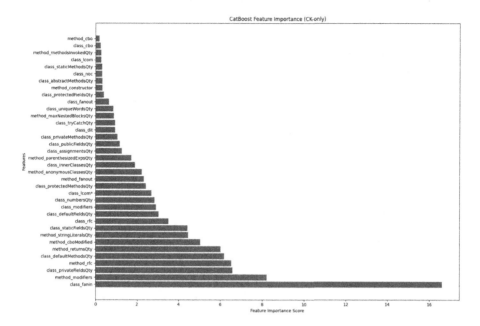

**Fig. 3.** Importances of CK metrics only

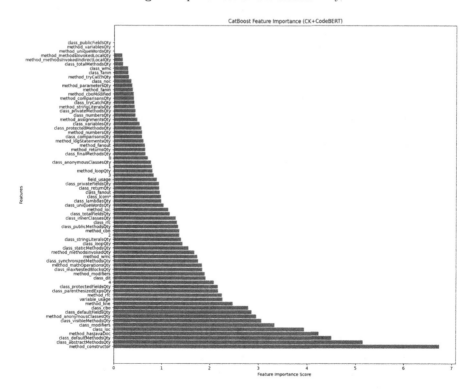

**Fig. 4.** Importances of CK metrics + CodeBERT features. Features with numerical names correspond to CodeBERT metrics after dimensionality reduction.

# 6    Limitations

The major limitation of our study is the limited size of our training and evaluation dataset. This could impact the reliability of our results by introducing errors or biases due to the small sample size. Another limitation is the lack of interpretability of embeddings, which is compounded by the application of dimensionality reduction. This makes it difficult to interpret the information contained in the embeddings, which may limit the usefulness of our approach for certain applications. Additionally, we were unable to fine-tune CodeBERT on our training data, which may have limited the performance of our model.

To address these limitations, future research could explore strategies for obtaining more data to provide a more robust benchmark for our approach. In addition, different techniques such as ablation study without dimensionality reduction could be used to gain a more granular understanding of the impact of embeddings. Finally, future research could explore the feasibility of fine-tuning the CodeBERT model to improve its performance for our specific task.

# 7    Conclusion

In this paper, we propose a transfer learning-based approach for detecting architectural design patterns in android application. In our proposed approach we use CK metrics joined source code embeddings extracted using CodeBert. We use joined ck metrics and source code embeddings as input to gradient boosting model to detect architectural design pattern. Our approach detects 3 patterns namely MVP, MVC and MVVM. We compare our achieved results with those of the state-of-the-art namely COACH [7] and RIMAZ [10], and our approach outperforms. In addition we analyse the extent to which CodeBert generated source code embeddings enhances machine learning models performance in design patterns detection. Our work serves as a major milestone in the application of natural language processing techniques to design pattern detection and provides a path for future research in this field. In the future we plan to fine-tuning the CodeBERT. Secondly we plan to increase the size of the benchmark dataset using statistical or deep learning approaches.

**Acknowledgment.** The study was funded by a Russian Science Foundation grant number 22-21-00493 https://rscf.ru/en/project/22-21-00493/.

# References

1. F droid website. f-droid (2019). https://fdroid.org/en/about/. Accessed 17 Nov 17 2019
2. Ahmad, W.U., Chakraborty, S., Ray, B., Chang, K.W.: A transformer-based approach for source code summarization. arXiv preprint: arXiv:2005.00653 (2020)
3. Ampatzoglou, A., Charalampidou, S., Stamelos, I.: Research state of the art on GoF design patterns: a mapping study. J. Syst. Softw. **86**(7), 1945–1964 (2013)

4. Aniche, M.: Java code metrics calculator (CK) (2015). https://github.com/mauricioaniche/ck/
5. Arcos-Medina, G., Menéndez, J., Vallejo, J.: Comparative study of performance and productivity of MVC and MVVM design patterns. KnE Eng., 241–252 (2018)
6. Ba, H.: Improving detection of credit card fraudulent transactions using generative adversarial networks. arXiv preprint: arXiv:1907.03355 (2019)
7. Chekhaba, C., Rebatchi, H., ElBoussaidi, G., Moha, N., Kpodjedo, S.: Coach: classification-based architectural patterns detection in android apps. In: Proceedings of the 36th Annual ACM Symposium on Applied Computing, pp. 1429–1438 (2021)
8. Chihada, A., Jalili, S., Hasheminejad, S.M.H., Zangooei, M.H.: Source code and design conformance, design pattern detection from source code by classification approach. Appl. Soft Comput. **26**, 357–367 (2015)
9. Dabain, H., Manzer, A., Tzerpos, V.: Design pattern detection using finder. In: Proceedings of the 30th Annual ACM Symposium on Applied Computing, pp. 1586–1593 (2015)
10. Daoudi, A., ElBoussaidi, G., Moha, N., Kpodjedo, S.: An exploratory study of MVC-based architectural patterns in android apps. In: Proceedings of the 34th ACM/SIGAPP Symposium on Applied Computing, pp. 1711–1720 (2019)
11. Feng, Z., et al.: CodeBERT: a pre-trained model for programming and natural languages. arXiv preprint: arXiv:2002.08155 (2020)
12. Gupta, A., Sharma, S.: Software maintenance: challenges and issues. Issues **1**(1), 23–25 (2015)
13. Heidenreich, F., et al.: Model-driven modernisation of java programs with JaMoPP. In: Joint Proceedings of the First International Workshop on Model-Driven Software Migration (MDSM 2011) and the Fifth International Workshop on System Quality and Maintainability (SQM 2011), pp. 8–11 (2011)
14. Hua, W., Liu, G.: Transformer-based networks over tree structures for code classification. Appl. Intell., 1–15 (2022)
15. Hunt, J., Hunt, J.: Gang of four design patterns. Scala Des. Patterns: Patterns Pract. Reuse Des., 135–136 (2013)
16. Kouli, M., Rasoolzadegan, A.: A feature-based method for detecting design patterns in source code. Symmetry **14**(7), 1491 (2022)
17. Mayvan, B.B., Rasoolzadegan, A., Yazdi, Z.G.: The state of the art on design patterns: a systematic mapping of the literature. J. Syst. Softw. **125**, 93–118 (2017)
18. Nazar, N., Aleti, A., Zheng, Y.: Feature-based software design pattern detection. J. Syst. Softw. **185**, 111179 (2022)
19. Paik, I., Wang, J.W.: Improving text-to-code generation with features of code graph on GPT-2. Electronics **10**(21), 2706 (2021)
20. Prokhorenkova, L., Gusev, G., Vorobev, A., Dorogush, A.V., Gulin, A.: CatBoost: unbiased boosting with categorical features. In: Advances in Neural Information Processing Systems, vol. 31 (2018)
21. Radford, A., Wu, J., Child, R., Luan, D., Amodei, D., Sutskever, I.: Language models are unsupervised multitask learners (2019)
22. Tenenbaum, J.B., Silva, V.D., Langford, J.C.: A global geometric framework for nonlinear dimensionality reduction. Science **290**(5500), 2319–2323 (2000)
23. Wolf, T., et al.: Transformers: state-of-the-art natural language processing. In: Proceedings of the 2020 Conference on Empirical Methods in Natural Language Processing: System Demonstrations, pp. 38–45. Association for Computational Linguistics (2020). https://www.aclweb.org/anthology/2020.emnlp-demos.6

24. Yuan, L., Yu, S., Yang, Z., Duan, M., Li, K.: A data balancing approach based on generative adversarial network. Futur. Gener. Comput. Syst. **141**, 768–776 (2023)
25. Zhang, H., Zhang, H., Pirbhulal, S., Wu, W., Albuquerque, V.H.C.D.: Active balancing mechanism for imbalanced medical data in deep learning-based classification models. ACM Trans. Multimedia Comput., Commun. Appl. (TOMM) **16**(1s), 1–15 (2020)
26. Zhu, T., Lin, Y., Liu, Y.: Synthetic minority oversampling technique for multiclass imbalance problems. Pattern Recogn. **72**, 327–340 (2017)

# Metamorphic Testing for Recommender Systems

Sofia Iakusheva[1(✉)] and Anton Khritankov[1,2]

[1] Moscow Institute of Physics and Technology, Institutsky dr. 9, Dolgoprudny,
Moscow Region, Russia
yakusheva.sf@phystech.edu
[2] HSE University, Pokrovskiy blvd., 11, Moscow, Russia

**Abstract.** Recommender systems are commonly based on a multi-armed bandit model. This model should be carefully tested because it affects the users, but it is technically complicated because of the test oracle problem and the stochastical nature of multi-armed bandit algorithms. Metamorphic testing is a testing method for problems without test oracles. In this paper, we propose a novel approach that applies metamorphic testing to the verification of the requirements for stochastic models. We propose a stochastic metamorphic relation (SMR) which is a composition of a sampling procedure and a determination function. We propose several relations for multi-armed bandit models and algorithms. Then, we implement those relations and test algorithms. Our experiment demonstrates that the proposed method can identify errors and that stochasticity of relations is essential.

**Keywords:** metamorphic testing · recommendation system ·
multi-armed bandit · stochastic metamorphic relation

Nowadays, recommender systems play an important role in our daily life. Online services like marketplaces, cinemas, music platforms, and news portals use automatic recommendation technologies. Such systems can consist of a large number of components and take years to develop. The process of creating a software product consists of many stages: planning, collecting and formalizing of requirements, design, implementation, debugging, testing using various levels of abstraction, analysis, improvements, and reporting. These stages can go in one sequence or another due to the specific development model [11]. The stage of collecting the requirements is very important because the requirements completely determine the functionality of the program. Verification of the program takes place during the testing stage. It ensures that the implemented code satisfies all the requirements. This also applies to recommendation algorithms: it is necessary to verify that the implementation has the same internal specific features as the theoretical model.

Recommender systems may harm their users, thus thorough functional and safety tests must be carried out. Any operating service should satisfy technical requirements and technological and ethical quality standards. But the test oracle

© The Author(s), under exclusive license to Springer Nature Switzerland AG 2024
D. I. Ignatov et al. (Eds.): AIST 2023, LNCS 14486, pp. 281–293, 2024.
https://doi.org/10.1007/978-3-031-54534-4_20

problem can make verifying compliance with the stated requirements very complicated. This problem states that sometimes it is difficult or impossible to compute the right test answer [2]. In the case of recommender systems, an even more general question arises: what is the answer? One can only evaluate the overall behavior of the system in various situations.

Many recommender systems are based on the multi-armed bandit model [16]. The multi-armed bandit problem is briefly formulated as follows. A bandit is a model of slot machine with many arms, each arm generates a reward according to some distribution. On each iteration, a recommender algorithm chooses some arms and gains the reward as a sample of the selected arms distributions. The algorithm does not know the distributions, thus its objective is to find an optimal recommendation that maximizes the reward. In this formulation of the problem, it is hard or even impossible to obtain the test oracle, which will check if the selected recommendation is an optimal one. Moreover, the verification of the algorithm is complicated by the stochastic nature of the bandit or the algorithm itself: it is allowed to behave suboptimally in a certain percentage of cases.

According to recent studies, the behavior of multi-armed bandits is not yet fully understood. For example, hidden feedback loops result in degradation of bandit recommendations to a small subset and loss of coverage and novelty [8]. So additional actions to verify the algorithms of multi-armed bandits are required.

Let us consider the verification procedure that the multi-armed bandit algorithm satisfies a given set of requirements. Some special testing methods, such as metamorphic testing, can be used to avoid the test oracle problem. In this study, we propose a new stochastic metamorphic testing procedure. We then apply it to the verification of multi-armed bandit algorithms used in recommender systems. Our contributions to the field are as follows:

– We consider stochastic metamorphic relations (SMRs) which are combinations of sampling procedures and following determination functions.
– We propose a set of SMRs for multi-armed bandit algorithms.

We conduct a controlled experiment to demonstrate that the proposed SMRs can be implemented and find errors.

The article has the following structure. Section 1 provides the background on the problems of testing algorithms for multi-armed bandits. Section 2 proposes the method of stochastic metamorphic relations. Section 3 gives stochastic metamorphic relations for bandits. Then Sect. 4 checks the applicability and realizability of the proposed method as well as the significance of stochasticity. Section 5 concludes our work.

# 1    Related Work

## 1.1    Multi-armed Bandits

A multi-armed bandit is a model of a slot machine with several arms. Each arm has some reward distribution, which can be either deterministic or stochastic. One can select a certain number of arms per turn and receive a reward. There are many variations of multi-armed bandits, including classic (stationary) and non-stationary and contextual ones. The task of the recommender algorithm is to choose arms in such a way as to maximize the reward. Often the loss (a lack of the reward) is considered instead of the reward. The loss can be estimated by the regret function [16]. Recommender systems use this model to solve the problem of choosing one or more recommendations from a limited set.

## 1.2   The Problem of Verification for Recommender Systems

The main approaches for recommender systems verification are *offline experiments*, *online experiments*, and *user studies* [14].

*Offline testing* requires a pre-labeled dataset, on which the recommender system is trained and tested. Standard quality metrics evaluate the performance of the model on a test dataset. Unfortunately, these metrics only evaluate the model quality on a specific dataset. Also, the sample size is often insufficient for correct model training and evaluation, and obtaining new datasets is associated with expenses. Another method of offline testing is simulating user behavior [5], for example, using bots. However, it may also be expensive.

*User research* involves hiring a number of people, who interact with the system and then describe their experience. Such studies are expensive and difficult to perform.

*Online testing* of the recommender system is carried out during the exploitation of the system [9]. Some users are redirected to alternative recommender systems for comparison. This method is quite risky because alternative systems may give low-quality recommendations.

Thus, offline testing seems to be more secure for the users and accessible for the researchers. However, the test oracle problem is essential for recommender systems because evaluation of recommendations is subjective. Thus, a large number of developed methods allow for the estimation of the quality of the recommender system, including systems based on multi-armed bandit algorithms. However, many of those methods require labeled data or are expensive, while others carry risks for the user.

## 1.3   Metamorphic Testing

Metamorphic testing is one of the popular methods for solving the test oracle problem [4]. Its difference from the usual oracle testing is as follows.

When testing with an oracle, there is a direct comparison of the system response to the correct reference response. When testing with a metamorphic relation, there is no need for evaluating each specific system response for correctness. The system is run on several test inputs with some relation between them. Then, we check the presence of the corresponding relation between the test outputs. Thus, we consider the evolution in the system response when we alter the input data in a specific way.

Generally, metamorphic relation is a function

$$\mathcal{R}(x_1, x_2, ..., x_n, f(x_1), f(x_2), \ldots, f(x_n)) \longrightarrow \{0, 1\}, \tag{1}$$

where $f(x_i)$ is a program's output on the $i$-th test input $x_i$.

Metamorphic testing is useful in many areas such as machine learning, self-driving cars, QA systems, bioinformatics, graph models, compilers, databases, distributed systems. Mao et al. [10] successfully apply metamorphic testing to recommender systems that do not use the multi-armed bandit model.

In the previous work [6], we have shown that strict deterministic metamorphic relations can be not suitable for systems that allow for some errors during the operating process. Some authors [12,13,18] use different statistical criteria to check if metamorphic relations are fulfilled. In this study, we consider stochastic metamorphic relations for stochastic systems which are more complex and informative.

## 2    Proposed Method

### 2.1    Stochastic Metamorphic Relations

Classic metamorphic relation can be formulated as a function

$$\mathcal{R}(X, f(X)) \longrightarrow \{0, 1\}. \tag{2}$$

where $X = (x_1, x_2, \ldots, x_n)$ is the input matrix, $x_i$ is a single input.

We consider $X$ as a matrix of numeric continuous random values. Thus, the stochastic metamorphic relation (SMR) is a sampling procedure $\mathcal{S}(X, f(X))$ and the function of determination $\mathcal{T}(\mathcal{S})$:

$$\mathcal{R}(X, f(X)) = \mathcal{T}(\mathcal{S}(X, f(X))). \tag{3}$$

For example, we can consider a sample of mean rewards got on $i$-th step for different runs of the algorithm as $x_i$. Or $x_i$ can be a result for a whole run if we want to check the presence of a trend and then collect information for different runs.

The stochasticity of the relation is important for at least two reasons: 1) the opportunity to consider $\mathcal{S}$ as sampled from unknown distribution and study this distribution (in time, for example). And 2) the possibility of using compositions of stochastic relations for system testing.

The definition of stochastic relations is in no way tied to a specific subject area, so the method is not limited to the recommender systems.

### 2.2    Method

The procedure $\mathcal{S}$ provide information about some random values, so we can analyze them and determine distribution properties. Also, SMRs can be more useful for the composition of relations than classic MRs because they provide more information about the system. Nevertheless, test answer should be only yes or no, so we can not directly use data sampled with $\mathcal{S}$ as the test results.

In Sect. 2.1 we denote the function of determination as $\mathcal{T}$. We can calculate the final test result $\mathcal{T}(\mathcal{S}(\mathcal{X}))$ if we consider $\mathcal{S}$ in terms of statistical hypotheses.

In this case, $\mathcal{T}$ would be a proper statistical criterion. As a result of testing, we either reject or do not reject the hypothesis with a given significance level.

So, we can briefly describe the approach for using stochastic relations as follows.

1. Consider the system's mathematical model properties or functional requirements to it.
2. Formalize requirements as stochastic relations:
   - choose experiment's parameters to change and output data to analyze,
   - determine $\mathcal{S}$,
   - formulate the hypotheses,
   - choose an appropriate statistic criterion $\mathcal{T}$ and a level of significance.
3. Conduct a controlled computational experiment:
   - choose a test bench and provide the necessary environments for the test object,
   - implement data generator and validation procedure for SMRs,
   - put the test object into the test bench,
   - perform tests,
   - collect data.
4. Analyze the obtained results; study the data sampled with $\mathcal{S}$; study the distribution of failed cases;
5. Report failures if SMRs are violated.

Data generation and validation of SMRs can be automated in the same way as for classic metamorphic relations. So, one can implement an automated generator and use generated data to test units or entire systems. However, while simple systems only require to the generation of a few parameters, more complex ones may require large context matrices or special user models.

# 3    Stochastic Metamorphic Relations for Multi-armed Bandits

Metamorphic relations can be obtained from different sources: mathematical models, external technical requirements, etc. Tables 1 and 2 represent different types of relations.

The main criterion for the correct behavior of the recommendation algorithm is its ability to learn, (i.e. to reduce regret over time). This feature is deduced from the formulation of the problem. We can test the algorithm's ability to learn using smooth or abrupt changes in the bandit's reward distributions. As technical relations, we can consider checking the constancy of metric distributions from run to run with the same initial parameters. Based on the individual features of the algorithms, it is possible to evaluate the metric curves for non-negativity, the presence of a non-zero trend, linearity, and convexity.

We propose several SMRs. We use technical requirements and features of the multi-armed bandits that can be deduced from the mathematical model. We use these three Propositions as requirements.

We denote mathematical expectation of bandit's arm $i$ reward as $a_i$, and probability to choose the arm $i$ as $p_i$.

**Table 1.** Stochastic metamorphic relations for multi-armed bandit models

| Group | Requirement | Bugs to detect |
|---|---|---|
| A. Technical reproducibility | Bandit's behavior is stable when parameters are fixed | Technical irreproducibility, major bugs |
| B. Changing of parameters | Depending on the parameters change, the bandit's behavior alters (or does not alter) in an expected way | Major bugs |
| C. Model's individual features | Expected values are equal to the given ones, sampled values are from a pre-determined distribution | Absence of the specific features |

**Proposition 1.** *The algorithm should increase the reward in time, which means that more profitable arms should be chosen more often than less profitable ones. After the bandit's preference suddenly changes due to swapping the arms with the highest and lowest mathematical expectations of the reward, there should be a statistically significant decrease in the mean reward.*

**Proof:** Let us denote the probability to choose the arm with the highest mean reward $a_i$ as $p_i$, and denote the probability to choose the arm with the lowest mean reward $a_j$ as $p_j$. Then if the algorithm learned the distributions correctly, $p_i > p_j$ and
$$p_i a_i + p_j a_j - p_i a_j - p_j a_i = (p_i - p_j)(a_i - a_j) \geq 0.$$
That means that if the algorithm chose a more profitable arm more often, its payoff should decrease after the permutation.

**Proposition 2.** *If we train several algorithms with the same initial parameters on the bandits with the same rewards distributions, but with permuted arms, the distributions of the average rewards should be the same.*

**Proof:** Let us consider two bandits. The first has arms numbered 1, 2, ..., n. The second has arms numbered $\sigma(1)$, $\sigma(2)$, ..., $\sigma(n)$, where $\sigma$ is a permutation. The optimal probability of choosing arm number i is $p_i$ for the first bandit. Probability $p_{\sigma(i)}$ will be optimal for arm $\sigma(i)$ of the second. Then $p_i a_i = q_{\sigma(i)} b_{\sigma(i)}$, where $a_i$ and $b_i$ are the expected rewards i for the arm of the first and second bandits, respectively. That means that the formula of the expectation of the reward of the first bandit can be transformed into the formula of the expectation of the reward of the second.

**Proposition 3.** *If we compare two runs of the algorithm with the same initial parameters on the bandits with the distributions of the parameters p and $\phi \cdot p$, where $\phi$ is a positive value greater than 1, then the distributions of the average reward of the second run should be greater than the average reward of the first run.*

**Table 2.** Stochastic metamorphic relations for multi-armed bandit algorithms

| Group | Requirement | Bugs to detect |
|---|---|---|
| A. Technical reproducibility | Algorithm's behavior is stable when parameters are fixed | Irreproducibility, major failures |
| B. Ability to learn | Metrics are getting better | Major bugs |
| C. Response to changes | The metrics decrease in comparison to the metrics on the bandit without changes, then improve | Major failures, algorithm's poor quality |
| D. Homogeneity of parameters | Permutations of isomorphic parameters do not change the final quality | Implementation failures, dependency on initialization |
| E. Comparison on different bandits | Results are similar for similar bandits and different for significantly different bandits | Poor sensibility and universality |
| F. Asymptotics, bounds | Linear dependence of regret and asymptotic values, check of upper and lower bounds | Incorrect implementation or theoretical estimates, poor effectiveness of the algorithm |
| G. Linear transformations of bandit distribution parameters | Rewards are also expected to be linearly transformed | Accuracy errors, arithmetic errors |
| H. Algorithms's individual features | Convexity, specific shapes of plots, presence or absence of a correlation between performance and parameters | Absence of the specific features |

**Proof:** This follows from the linearity of the mathematical expectation.

We can rewrite Proposition 3 in more general and strict way for the optimal algorithm.

**Proposition 4.** *If we compare two runs of the optimal algorithm with the same initial parameters on the bandits with the distributions of the parameters $p$ and $\phi(p)$, where $\phi$ is a linear transformation, the distributions of the average reward of the second run should be obtained by the same linear transformation $\phi$ of first run reward.*

**Proof:** This follows from the linearity of the mathematical expectation.

Now, we formalize technical and mathematical requirements as SMRs.

1. **Requirement** (2 A): requirement of the stable behaviour.
   **SMR0**: during repeated runs of the algorithm with the same initial parameters on the same bandit, the algorithm's reward (or another metric) $\psi(t)$ at each step $t$ should be the same. $\psi_1(t) = \psi_2(t)$.
   $\mathcal{S} = (\psi_1(t), \psi_2(t))$, $\mathcal{T}(\mathcal{S})$: $\mathbb{E}\psi_1 = \mathbb{E}\psi_2$. We use a two-sample criterion to compare expectations.

2. **Requirement** (2 B): algorithm should increase the reward in time.
   **SMR1**: the mean rewards $\psi$ of the algorithm at the end of training on the stationary bandit should not be statistically lower than the mean rewards at the beginning, $\mathbb{E}(\psi(n)) \geq \mathbb{E}(\psi(1))$.
   $\mathcal{S} = (\psi(n), \psi(1))$, $\mathcal{T}(\mathcal{S})$: $\mathbb{E}\psi(n) \geq \mathbb{E}\psi(1)$. We a the two-sample criterion to compare expectations.

3. **Requirement** (2 B): algorithm should increase the reward in time.
   **SMR2**: a non-zero trend in the algorithm's reward (and other metrics) should be statistically significant.
   $\mathcal{S} = \psi(t)$, $\mathcal{T}(\mathcal{S})$: $\mathbb{E}\mathcal{S}(t)$ has a trend. We use the Mann-Kendall test to find the trend.

4. **Requirement** (2 C, Proposition 1): algorithm should response to changes.
   **SMR3**: If on step $i$ bandit's distributions change according to Proposition 1, then the reward's expectation changes accordingly: $\mathbb{E}(\psi(i-1)) \geq \mathbb{E}(\psi(i))$, $\mathbb{E}(\psi(i)) \leq \mathbb{E}(\psi(n))$ for the algorithm's reward $\psi$ and step $i$.
   $\mathcal{S} = (\psi(0), \psi(i), \psi(i+1), \psi(n))$, $\mathcal{T}(\mathcal{S})$: $\mathbb{E}\psi(i) \geq \mathbb{E}\psi(i+1) \cap \mathbb{E}\psi(n) \geq \mathbb{E}\psi(i+1)$. We use a two-sample criterion to compare expectations.

5. **Requirement** (2 D, Proposition 2): assumption about homogeneity of parameters.
   **SMR4**: if we permute bandit's arms according to Proposition 2, then the reward should remain the same, $\psi_{p_1} = \psi_{p_2}$ for the algorithm's reward $\psi$.
   $\mathcal{S} = (\psi_1(t), \psi_2(t))$, $\mathcal{T}(\mathcal{S})$: $\mathbb{E}\psi_1 = \mathbb{E}\psi_2$. We use a two-sample criterion to compare expectations.

6. **Requirement** (2 E, Proposition 3): comparison of less and more profitable bandits.
   **SMR5**: if the bandit's rewards changes according to Proposition 3, then the reward also changes, $\mathbb{E}(\psi(t)) \leq \mathbb{E}(\psi(\phi \cdot p, t))$ for the algorithm's reward $\psi$, step $t$ and coefficient $\phi > 1$.
   $\mathcal{S} = (\psi_1(t), \psi_2(t))$, $\mathcal{T}(\mathcal{S})$: $\mathbb{E}\psi_1(t) \leq \mathbb{E}\psi_2(t)$. We use a two-sample criterion to compare expectations.

# 4    Experiment

In addition to the theoretical description, we implement and check the proposed method. We formulate the following research questions:

**RQ1.** Is the SMR method realizable and applicable to the multi-armed bandit problem?

**RQ2.** Can SMRs detect errors?

**RQ3.** Is stochasticity of SMRs essential?

## 4.1    Experiment Setup

We conduct a controlled computational experiment to answer the research questions. We perform all experiments using the MLDev (specialized machine learning experiment automation tool [7]) to provide automation and ensure reproducibility.

**System Model and Test Objects.** We consider a two-component system consists of a multi-armed bandit model and an arm selection algorithm. We choose the $\varepsilon$-greedy [14], EXP3 and EXP3M [1], Thompson sampling [15], F-DSW-TS [3], random sampling algorithm and optimal algorithm as test objects. We use multi-armed bandits with Bernoulli distributions, the expectations of which are randomly sampled from uniform distribution $U[0, 1]$. The total number of arms varies from 3 to 10, the number of selected arms varies from 1 to 3. The duration of the experiment is from 1500 to 3000 steps.

**Metamorphic Relations.** We use the proposed SMR0-5. We use Wilcoxon test for SMRs 0, 4, 5, Mann-Whithey test for SMRs1-2, Mann-Kendall test for SMR3. We chose standard p-value=0.05.

**Test Bench and Requirement.** We use test bases on MLDev research template and use automated experiment runs and data collection. It is provided by the "Code-AI" project named MLDev Recommender Systems (link to the source code: https://gitlab.com/mlrep/mldev-recommender-systems).

### 4.2  Results

Table 3 represents the results of testing. We run the experiment with different parameters to study the parameter space and find parameter combinations that cause SMRs failures. We determine SMRs to get the final test results.

**Table 3.** Proportion of fulfilled SMRs in the studied parameter space for multi-armed bandit algorithms.

| Algorithm | SMR0 | SMR1 | SMR2 | SMR3 | SMR4 | SMR5 |
|---|---|---|---|---|---|---|
| Random | 1.0 | 0.03 | 0.0 | 0.1 | 1.0 | 1.0 |
| $\varepsilon$-greedy | 1.0 | 0.97 | 1.0 | 0.9 | 0.43 | 1.0 |
| EXP3 ($\gamma = 0.05$) | 1.0 | 0.97 | 1.0 | 0.37 | 0.73 | 1.0 |
| EXP3 ($\gamma = 0.1$) | 1.0 | 1.0 | 1.0 | 1.0 | 0.6 | 1.0 |
| EXP3M | 1.0 | 1.0 | 1.0 | 0.05 | 0.45 | 1.0 |
| Thompson sampling | 1.0 | 1.0 | 1.0 | 1.0 | 0.97 | 1.0 |
| F-DSW-TS | 1.0 | 1.0 | 1.0 | 0.97 | 0.43 | 1.0 |
| Optimal | 1.0 | 0.05 | 0.0 | 0.0 | 1.0 | 1.0 |

SMR0 is based on the technical requirement of reproducibility. SMR0 is completely true for all the algorithms, so MLDev provides correct automatization and reproducibility. SMR1-3 fail for the random algorithm because this algorithm does not learn anything. Also SMR1-3 fail for the optimal algorithm because this algorithm already knows the distributions of rewards. So these relations are useful for checking the ability to learn.

Relation SMR4 fails for five algorithms. We study failed cases and notice that SMRs tend to fail on permutations with big position difference between the most profitable arms. Thus it seems that the amount of reward depends in time on arms positions. This is expected for $\varepsilon$-greedy because this algorithm is greedy and mostly chooses the best option among the known ones.

Also, SMR3 unexpectedly fails for EXP3 algorithm with the parameter $\gamma = 0.05$. We assume that it can be caused by the low learning speed of the particular EXP3 configuration that we have tested (algorithm's parameter $\gamma = 0.05$ means low exploration rate, which causes slow increasing of the reward). SMR3 is fine for EXP3 with the parameter $\gamma = 0.1$. So, we can use SMRs for studying parameters spaces. Also SMR3 fails for the implementation of EXP3M algorithm with $\gamma = 0.1$. We detect that this algorithm does not change the choices after arms swapping.

SMR5 is completely true for all the algorithms. So, we use Proposition 4 and $S$ for more detailed analysis. We use known $\phi$ and study the value $\Delta = \psi_1 - \phi(\psi_2)$. Figure 2 shows that the mean reward of the F-DSW-TS algorithm was not multiplied by the same coefficient as the rewards of the rest of the algorithms. For example, for one test run $\varepsilon$-greedy's sample has deviation $\Delta \sim \mathcal{N}(0.0175, 0.05)$ which is low in comparison to the reward expectation 0.31. For F-DSW-TS $\Delta \sim \mathcal{N}(0.05, 0.074)$, and for the random algorithm $\Delta \sim \mathcal{N}(-0.0003, 0.04)$. So, the mathematical expectation of $\Delta$ for F-DSW-TS is greater than for $\varepsilon$-greedy and random. Figure 1 represents samples of reward deviation $\Delta$ for three algorithms for this test run. Also we use B-spline approximation [17] to analyze $\Delta(t)$ (red lines on Fig. 1). We also can see that expectation of $\Delta$ for EXP3 tends to increase in time which can be caused by different learning speeds on different bandits. Also, it looks like $\Delta(t)$ tends to decrease in time for $\varepsilon$-greedy and Thompson sampling.

The proposed SMRs allow us to identify errors in the experiment source code. Figure 3 shows the case when the bandit's preferences changes only for the first algorithm and do not for the others.

SMRs allow us to detect different types of errors: wrong paths (a few dozens of cases), errors in the experiment configurations (about 5 cases), abnormalities in bandits' behavior (EXP3, EXP3M, F-DSW-TS), defects in validation procedures (3–4 uncaught exceptions). Mostly these errors are caused by the lack of developers' attention. SMR3-4 and additional analysis of SMR5 detect anomalies that can not be found without metamorphic testing.

### 4.3    Analysis

The experiment shows that SMRs are applicable to multi-armed bandits and implementable in practice. Also, examples of founded failures demonstrate that SMRs are useful for error detection. Implemented relations helped us to identify and correct many errors in the source code. So, the answers to RQ1 and RQ2 are yes.

**Fig. 1.** Distribution of $\psi_1 - \phi(\psi_2)$ for SMR5 for one test run. Red lines show the approximation. (Color figure online)

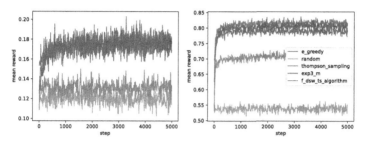

**Fig. 2.** An example of SMR5. Despite SMR5 is true for all the algorithms, the mean reward of F-DSW-TS is not multiplied by the same coefficient as the rewards of the rest of the algorithms.

We provide an example of SMR analysis for Proposition 4. SMRs allow us to check normality of the deviation, presence of trend, stationarity of the deviation. Thus we can detect the presence of systematic errors and anomalies. The stochasticity of the relations is important since the studied probabilistic models allow non-optimal or unstable behavior in some cases, and exact deterministic relations do not take into account such natural features. One can use similar comparative statistical analysis, but our analysis was based on model requirements. It proves that SMRs can be useful for detailed analysis, so the answer for RQ3 is also yes.

Internal threats to the validity of the results may be unreasonable requirements for the analyzed models, unidentified errors in the source code, incomplete analysis of the results, and poorly chosen parameters of the experiments. An external threat to validity is accidental results because of pseudo-random generators and the stochastic nature of the studied models.

We faced difficulties when we tried to strictly formalize the behavior of the system and the requirements for it. This procedure is crucial for an experiment. Also choosing the appropriate p-value is a big question. Another question is the appropriate number of trials for every SMR. It is possible that SMR3 failed on EXP3 because of a small number of trials in each test. Thus, formalizing the choice of the appropriate parameters for the SMR testing procedure may be a direction for future work.

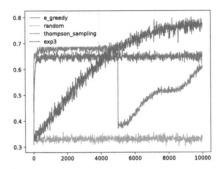

**Fig. 3.** A failed SMR4 preliminary test. This helps to find an error in the experiment setup: bandit's preferences changes only for the first algorithm in the experiment.

## 5  Conclusion

In this paper, we propose a novel approach to applying metamorphic relations to stochastic algorithms. We study a problem of multi-armed bandits verification, reveal the necessity of specific verification procedures, and propose a solution. We propose SMR as composition of sampling procedure and a determination function, propose testing method, and develop a collection of SMRs for multi-armed bandit models and algorithms. We implement SMRs in practice, use them to test several algorithms, and show that SMRs can be useful for analysis and error detection. So, our approach can be useful for testing stochastic models.

Proposed SMRs helped us identify and correct errors in the experiment source code. The SMRs are not limited to any particular area, so our method can be useful for testing other stochastic models with the test oracle problem. Thus, our method potentially can solve many new problems and help to improve the quality of verification in practical applications.

Whereas in this paper, we focus on testing single models. It will be useful to combine SMRs to test complex systems, which contain many modules. The problem is that it can be easy to propose SMR for a module and hard to do the same for the whole system. An extension of the proposed approach to system testing could be a direction of future research.

## References

1. Auer, P., Cesa-Bianchi, N., Freund, Y., Schapire, R.: The nonstochastic multiarmed bandit problem. SIAM J. Comput. **32**, 48–77 (2002)
2. Barr, E., Harman, M., McMinn, P., Shahbaz, M., Yoo, S.: The oracle problem in software testing: a survey. IEEE Trans. Software Eng. **41**(5), 507–525 (2015). https://doi.org/10.1109/TSE.2014.2372785
3. Cavenaghi, E., Sottocornola, G., Stella, F., Zanker, M.: Non stationary multi-armed bandit: empirical evaluation of a new concept drift-aware algorithm. Entropy **23**(3), 380 (2021)

4. Chen, T.Y., et al.: Metamorphic testing: a review of challenges and opportunities. ACM Comput. Surv. (CSUR) **51**(1), 1–27 (2018). https://doi.org/10.1145/3143561
5. Fischer, G.: User modeling in human-computer interaction. User Model. User-Adap. Inter. **11**, 65–86 (2001)
6. Iakusheva, S., Khritankov, A.: Composite metamorphic relations for integration testing. In: 2022 8th International Conference on Computer Technology Applications, May 12–14, Vienna, Austria (2022). https://doi.org/10.1145/3543712.3543725
7. Khritankov, A., Pershin, N., Ukhov, N., Ukhov., A.: MLDev: data science experiment automation and reproducibility software. In: Pozanenko, A., Stupnikov, S., Thalheim, B., Mendez, E., Kiselyova, N. (eds.) Data Analytics and Management in Data Intensive Domains. Communications in Computer and Information Science, vol. 1620, pp. 3–18. Springer, Cham (2021). https://doi.org/10.1007/978-3-031-12285-9_1
8. Khritankov, A., Pilkevich, A.: Existence conditions for hidden feedback loops in online recommender systems. In: Zhang, W., Zou, L., Maamar, Z., Chen, L. (eds.) Web Information Systems Engineering - WISE 2021. Lecture Notes in Computer Science(), vol. 13081, pp. 267–274. Springer, Cham (2021). https://doi.org/10.1007/978-3-030-91560-5_19
9. Kohavi, R., Longbotham, R., Sommerfield, D., Henne, R.M.: Controlled experiments on the web: survey and practical guide. Data Min. Knowl. Disc. **18**, 140–181 (2009)
10. Mao, C., Yi, X., Chen, T. Y.: Metamorphic robustness testing for recommender systems: a case study. In: 2020 7th International Conference on Dependable Systems and Their Applications (DSA), pp. 331–336. IEEE (2020)
11. Matković, P., Tumbas, P.: A comparative overview of the evolution of software development models. Int. J. Ind. Eng. Manag. **1**(4), 163 (2010)
12. Pesu, D., Zhou, Z. Q., Zhen, J., Towey, D.: A monte Carlo method for metamorphic testing of machine translation services. In: Proceedings of the 3rd International Workshop on Metamorphic Testing, pp. 38–45 (2018)
13. ur Rehman, F., Izurieta, C.: Statistical metamorphic testing of neural network based intrusion detection systems. In: 2021 IEEE International Conference on Cyber Security and Resilience (CSR), pp. 20–26. IEEE(2021)
14. Ricci, F., Rokach, L., Shapira, B., Kantor, P.B.: Recommender Systems Handbook, 1st edn. Springer-Verlag, Berlin (2010)
15. Russo, D.J., Roy, B.V., Kazerouni, A., Osband, I., Wen, Z.: A Tutorial on Thompson Sampling. Found. Trends R Mach. Learn. **11**(1), 1–96 (2018)
16. Slivkins, A.: Introduction to multi-armed bandits. Found. Trends Mach. Learn. **12**(1–2), 1–286 (2019). https://doi.org/10.1561/2200000068
17. Wang, J.C., Meyer, M.C.: Testing the monotonicity or convexity of a function using regression splines. Can. J. Stat. **39**(1), 89–107 (2011)
18. Zhou, Z.Q., Tse, T.H., Witheridge, M.: Metamorphic robustness testing: exposing hidden defects in citation statistics and journal impact factors. IEEE Trans. Softw. Eng. **47**(6), 1164–1183 (2019)

# Application of Dynamic Graph CNN* and FICP for Detection and Research Archaeology Sites

Aleksandr Vokhmintcev[1,2]($\boxtimes$), Olga Khristodulo[3], Andrey Melnikov[2], and Matvei Romanov[1,4]

[1] Chelyabinsk State University, Chelyabinsk, Russia
vav@csu.ru
[2] Ugra State University, Khanty-Mansiysk, Russia
melnikovav@uriit.ru
[3] Ufa University of Science and Technology, Ufa, Russia
hristodulo.oi@ugatu.su
[4] South Ural State University, Chelyabinsk, Russia

**Abstract.** The paper proposes a methodology for solving the task of accurate semantic classification of 3D data using a combination of 2D and 3D methods based on the YOLO detector and the modified DGCNN network. The methodology is tested on the example of the problem of classification of large-scale geospatial objects, such as digital relief models of archaeological sites. A method for accurate registration of objects (FCIP) in the class of affine transformations using geometric and color features was proposed. The results of computer modeling of the proposed methodology based on FICP+DGCNN*+YOLO were presented and discussed. The methodology has theoretical and applied significance not only for the decryption and research of archaeological sites, but also for many applications of digital information processing and robotics in general.

**Keywords:** 3D semantic segmentation and classification methods · object detector · DTM · DGCNN · ICP

## 1 Introduction

Currently, remote research methods are becoming more important in archaeology, which allow obtaining information about archaeological sites without resorting to excavation methods. Remote research methods are very diverse, for example, scanning of the archaeology sites using magnetometers (SQUID, MOKE, Torque) and georadars is often used, low-altitude aeromagnetic photography using UAVs is used, recently more and more attention has been paid to satellite remote sensing, which allows obtaining satellite images with high spatial resolution. Remote research methods make it possible to create extensive sets of archaeological data in the form of a digital terrain model (DTM) or a digital surface model (DSM), with the help of which it is possible to study the spatial organization of various types of archaeological sites. To date, the following approaches for solving the problem of semantic classifying objects are known: the first approach uses feature points and other of object properties; the second approach uses statistical methods; the third approach is based on the use of machine learning methods and, above all, neural networks. Methods based on the search for object characteristics [1,2] have high computational complexity, are poorly formally described and don't control the detection quality. These methods show poor results when detecting objects in an image with a complex spatially inhomogeneous background distorted by interference of various nature, for example, uneven illumination [3,4]. For statistical pattern recognition, object recognition methods based on ordinal statistics using linear or nonlinear filtering are used [5,6]. The main disadvantage of these methods is the erroneous identification of the object of interest with objects in the background for contextually complex scenes. To overcome this disadvantage, various methods and algorithms have been proposed, which are based on spatial differentiation, distortion of the shape of the object in the image, adaptive and locally adaptive linear and nonlinear filters [7]. Recently, when deciphering archaeological sites, researchers often use 3D mathematical models and machine learning methods to analyze geospatial data. To solve the problem of high computational resource intensity in the analysis of archaeological data, neural network models are used on sparse data structures [8] and variational autoencoders [9]. Moreover, there are certain achievements in reducing the size of the training sample: for example, neural networks are known that are robust to changing the spatial position of 3D objects [10]; models trained with partial (semi-supervised) markup [7].

Modern methods for classification and segmentation of 3D data use semantic signals and sequences in data to increase accuracy, and the development of appropriate algorithms for processing point clouds is an active area of scientific research. Methods for classifying and segmenting 3D objects in the form of a point cloud can be divided into indirect and direct. Indirect methods use a sequence of 2D images of the original point cloud obtained from different viewpoints in 3D space. Then this sequence is sent to the CNN input, and after all, the CNN output, which is a pixel-by-pixel semantic markup map, is projected back into 3D space. Indirect methods include MVCNN (Multi-view Convolutional Neural Networks), SnapNet, SnapNet-R methods [11], which use RGB data and depth data, and VoxNET, SEGCloud, PointGrid methods [12], which use voxel representation. The main disadvantage of indirect methods is that they work well only with polygonal models, for processing contextually complex models, including those containing non-convex point clouds, their use is extremely limited due to the poor quality of segmentation. Also, the process of data registration in multiview methods requires large computational costs when taking pictures from the required angles, and application using voxel models require a significant amount of memory to store the results. Recently, when developing methods for semantic processing of 3D data, researchers have focused on direct methods that use various neural network architectures and extract features automatically, for example, based on Kohonen maps SO-Net [13], RNN (Recurrent Neural Networks),

ResNet CNN (Residual Neural Networks) [14]. The disadvantages of these methods are the complexity of the model and low quality values in terms of accuracy and completeness when processing local features of the object, in our case, details of the microrelief. To overcome this disadvantage, methods have been proposed that can be divided into two groups: the PointNet [15] and PointNet++ group [16] and a group of methods based on GCNN (Graph Convolutional Neural Networks) [17] and its variants DGCNN (Dynamic Graph CNN) [18], RGGNN (Regularized Graph CNN) [19]. The methods of the first group in the 3D semantic processing of the point cloud use local objects properties and don't use data about a complex geometric relationships between points. These methods are invariant to permutations. The methods of the second group use information about the surface of a 3D object and are based on the use of convolution operations on spatial graphs when solving the problem of semantic 3D classification and segmentation. As an example, 3D classification method based on RGCNN uses the Laplace matrix to construct an undirected graph, which causes one of its main disadvantages - low performance, the computational complexity of graph construction in the method can be estimated as $O(n)^3$. To improve the performance of methods of this group, spectral filtering approximation methods (Cayley and Chebyshev polynomials, the Lanczos method, etc.) are often used. Despite numerous modifications, the GCNN method and its variants are not local, which is due to the fact that these methods usually use only one modality - geometric relationships between points in the cloud. The most promising method of the second group is DGCNN, which has a special EdgeConv layer that allows to achieve invariance to rotation, parallel transfer or scaling. This paper proposes a new method of semantic data processing based on a dynamic weighted graph, which combines the advantages of the well-known architectures of DGCNN and RGCNN, but is devoid of their known disadvantages, such as:

- the dependence of the accuracy of the method on the shape of the point cloud and the method of obtaining the analyzed point cloud;
- limiting the dimension of the point cloud when performing the 3D segmentation and classification procedure;
- the use of one modality when performing the convolution operation on spatial graphs and, as a consequence, the insufficient quality of semantic processing of 3D data.

In our research we use following data sources: materials of aerial photography (from the 50 s and the 60 s-80 s last century for the purposes of agriculture and geodesy respectively, the aerial frames were taken at a scale of 1:14 000 with high resolution for the entire territory of the Kizilsky district of the Chelyabinsk region; results of remote sensing of the Resource-P (from 2013 to 2021), Canopus-B (from 2013 to 2023) satellites; total station survey data obtained using the Trimble 3300 (Elta R55); DTM and DSM created as a result of archeology expeditions to the settlements of Stepnoye and Levoberezhnoye from 2006 to the present. These materials form a data set on archaeological sites and objects of the Bronze Age on the territory along the river Sintashta. Data sources were divided into two groups: 2D in the form of RGB frames $I - RGB = \{F_1, \ldots, F_n\}$ and 3D data in the form of point clouds $I - D = \{d_1, \ldots, d_k\}$. Let's consider the algorithms for processing these two groups of data: Algorithm 1 for 2D data and Algorithm 2 for 3D data. Algorithm 1 is represented as the following sequence of steps:

- Step 1. Trimming the edges (bottom = 1.5%, top = 1.5%, left (right) sides = 1%);
- Step 2. Unification of the direction of the images;
- Step 3. Image restoration and noise removal;
- Step 4. Increase the contrast of images;
- Step 5. Normalization of the snapshot size.

Algorithm 2 is represented as the following sequence of steps:

- Step 1. Division of a 3D model into semantic blocks (step = 0.01);
- Step 2. Sampling of point clouds (Down Sampling);
- Step 3. Extraction of singular points;
- Step 4. Calculations of normals in the point cloud.

This article is organized as follows: in the second chapter an accurate algorithm FICP for registering 3D data and the results of a comparative analysis of the proposed algorithm with known solutions are proposed, in the third chapter, a multimodal modified neural network architecture based on DGCNN (DGCNN*) for classifying archaeological objects is proposed, the fourth chapter presents and discusses the results of computer simulation for the proposed methodology of 3D semantic classification of archaeological objects in comparison with known modern approaches of solving this problem.

## 2    Fusion 3D Registration Algorithm (FICP)

It is known that the convergence and accuracy of the ICP (Iterative Closest Point) algorithm proposed in the works [20] can be significantly improved. Known ICP methods are characterized by the following disadvantages: firstly, classical ICP methods do not take into account the local shape of the surface around each point in a 3D point cloud when analyzing and processing information, and secondly, known ICP-based data logging methods have great computational complexity, while the most expensive operation is the search for the nearest points; thirdly, the result of solving the variational problem depends on the correctness of the choice of the initial approximation. The last drawback of the ICP method can be eliminated, for this purpose special points are used in the study, which, as is known, match data frames without specifying initialization parameters. Horn proposed a solution to the conditional variational problem in closed form for affine and orthogonal transformations, in this paper a closed form solution for affine transformations is obtained, which allows: to register non-convex objects on a 3D data; to obtain a solution of the ICP variational problem in a closed form for various degenerate cases associated with the location of points in a 3D data on the same straight line (plane). The data matching algorithm proposed in the articles is used to process characteristic points on an image in an RGB-D frame. The process of processing feature points is represented as the following sequence of steps (Algorithm 3):

- Step 1. Determination of feature points in an RGB-D frame using DHNG (descriptor based on histograms of directional gradients) [21];
- Step 2. Matching feature points in two consecutive RGB-D frames;
- Step 3. Elimination of outliers in the data by estimating the parameters of a 2D image model based on random samples (RANSAC);
- Step 4. Solving the variational problem of 2D data registration with respect to singular points in frames.

Let's define the values of the normalized centered DHNG

$$\overline{HOG_i^R}(\alpha) = (HOG_i^R(\alpha) - Mean^R)/\sqrt{(Var^R)}, \tag{1}$$

where $HOG_i^R(\alpha)$ is GNG value in each position of the i-th round sliding window, $Mean^R$ is the mean HNG value, $Var^R$ is the HNG variance. Let's take a closer look at the step. 4. The solution of the variational problem with respect to singular points in the frame is represented as

$$J(R^F) = \frac{1}{|A_f|} \sum_{i \in A_f}^{n} w_i \parallel M(R^F Hog_x^i) - M(Hog_y^i) \parallel^2, \tag{2}$$

where $R^F$ is the affine transformation matrix for feature points for color data; $w_i$ are the weight characteristics; $Hog_x^i$ and $Hog_y^i$ are singular points in two consecutive frames, respectively: $Hog_x^i = (x_{1f}^i, x_{2f}^i, x_{3f}^i)^T$, $Hog_y^i = (y_{1f}^i, y_{2f}^i, y_{3f}^i)^T$, where $M$ is the function of converting the coordinates of points of a 3D scene $Hog_x^i$ and $Hog_y^i \in R^3$ into the coordinate system of the camera $C^i = (C_x^i, C_y^i, D^i) \in R^3$, where $C_x^i$, $C_y^i$ are the corresponding coordinates of points in pixel space, $D^i$ is the depth value in pixel space.

$$C_x^i = \frac{f}{x_{3f}^i} x_{1f}^i + O_x, C_y^i = \frac{f}{x_{3f}^i} x_{2f}^i + O_y, D^i = \sqrt{x_{1f}^i{}^2 + x_{2f}^i{}^2 + x_{3f}^i{}^2}, \tag{3}$$

where $O_x$ and $O_y$ are the coordinates of the image center in pixel space, f is the camera focus. The coordinates in the frame $Hog_y^i$ can be obtained in a similar manner. The process of decryption of the archaeology sites is carried out on the basis of digital DTM relief models obtained both from single view point and from different viewing points. In the second case, it is necessary to solve the problem of 3D data registration. The paper considers 3D models in the form of a point cloud. Let's $X = \{x_1, \ldots, x_n\}$ and $Y = \{y_1, \ldots, y_m\}$ – the pair of point cloud in $R^3$, respectively. In digital information processing, an ICP algorithm [22,23] is often used to solve 3D data registration task. This algorithm searches for a geometric transformation between X and Y in the following form $Rx_i + T$, where $R$ is the rotation matrix, $T$ is the vector of parallel transfer, $i = 1, \ldots, n$ and uses an incremental an approach to calculating a sparse 3D model of a scene and a bundle method to refine the camera parameters and coordinates of points. In this paper we suggested an accurate algorithm Fusion ICP for registering 3D data with a point-to-point metric for affine transformations class, in which the closed solution of the variational problem is

$$J(R^F, R^D) = \frac{\alpha \sum_{i \in A_f}^{m} \| M(R^F Hog_x^i) - M(Hog_y^i) \|^2}{|S_f|}$$
$$+ \frac{(1 - \alpha) \sum_{j \in A_d}^{n} \| R^D x_j + T - y_j \|^2}{|S_d|}, \quad (4)$$

where $R^D$ is affine rotation matrices for depth data; $T$ is a parallel transfer vector; $\alpha$ is hyperparameter, the value equal to 0.3 is used in the work; $S_f$ a set which contains the correspondences between features points; $S_d$ a set which contains the correspondences between points $x_j$ and $y_j$ in $X$ and $Y$. Let $RT^*(i)$ be the best geometric transformation for the i-th step. This algorithm has been adapted to the task solution of the registration of archaeological sites, uses two channels: data on the frame color and data on the depth in the form of a point cloud and can be presented as procedure (Algorithm 4, FICP):

- Step 1. Determine the feature points using an image matching algorithm based on DHNG (Algorithm 3);
- Step 2. Establish a correspondence between the special points $Hog_x^i$ and $Hog_y^i$ in frames using a k-d tree to improve performance. We use the obtained result of matching points as the initial values of the geometric transformation $RT^*(0)$ of the ICP;
- Step 3. Establish the correspondence between the $S_d$ points in the clouds $X$ and $Y$ using the KNN nearest neighbor method. Eliminate outliers based on the RANSAC;
- Step. 4. Solve the variational problem (4) with respect to the transformation $RT^*(i)$.

We search for the transformation until one of the stop criteria is met. We will investigate the accuracy (RMSE) and convergence rate of the proposed registration algorithm for a class of affine transformations of data in controlled (see Fig. 1) and uncontrolled conditions (see Fig. 2), which are associated with noises of various nature. To conduct tests, we will use the proposed DHNG descriptor and descriptors such as SIFT (Scale-invariant feature transform), SURF (Speeded up robust features), ORB (Oriented FAST and Rotated BRIEF) are known. From Fig. 1 it can be seen that the combination of the FCIP + DHNG registration method allows for better accuracy, but in general, under controlled conditions, the accuracy of solving the registration problem does not strongly depend on the choice of a 2D descriptor that is used to select the initial values of FCIP. In uncontrolled conditions(see Fig. 2), the FCIP + DHNG combination has significant advantages in terms of accuracy and convergence over all other combinations (FCIP + SIFT, FCIP + SURF, FCIP + ORB), our method of registering 3D data converges after the 10-th step. The processing time of algorithms based on DHNG and ORB is

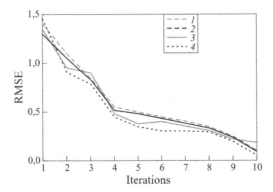

**Fig. 1.** Comparative analysis of registration methods in terms of convergence in controlled conditions (1 - SIFT; 2 - SURF; 3 - ORB; 4 - DHNG).

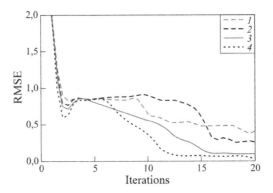

**Fig. 2.** Comparative analysis of registration methods in terms of convergence in uncontrolled conditions (1 - SIFT; 2 - SURF; 3 - ORB; 4 - DHNG).

about 2 s. The processing time of the SIFT algorithm is more then 15 s. The processing time of the SURF is 0.6 s. The method FCIP solves the problem of the dependence of the 3D registration result on the correctness of the selection of initial values R and T. FICP can be used for accurate registration of point clouds with arbitrary spatial resolution and scale relative to each other. DHNG descriptor allows to improve the convergence of the ICP.

## 3 Modified DGCNN* for Semantic Classification 3D Data

When deciphering archaeological sites, there is always a need for its 3D registration based on data obtained from different viewing points. Therefore, in order to overcome the disadvantages inherent in multi-species segmentation methods [11], a methodology (see Fig. 3) is used in which the following sequence of steps is performed:

- Step 1. 3D registration of an archaeological sites using a combined registration method based on a fusion iterative closest points algorithm (FICP);
- Step 2. 3D semantic segmentation of a point cloud associated with an archaeological site based on a modified DGCNN*;
- Step 3. Detecting objects using RGB-D data based on the YOLO detector [24];
- Step 4. Combining the results of segmentation and classification in Step 2 and Step 3 using the Bayes formula.

This methodology makes it possible to eliminate two key disadvantages of the DGCNN and RGCNN convolutional neural networks for classification and segmentation 3D data. The first disadvantage is the limitation of the dimensionality of the processed 3D data, which makes these methods inapplicable for many applied tasks related to the semantic processing of large-scale scenes, the second disadvantage is related to the dependence of these CNNs on the method of collecting data about the point cloud, for example, the quality of the registration of point clouds. The following modifications were made to the DGCNN architecture: the modality of the CNN has been expanded, the accurate solution is obtained based on a combination of various geometric and independent features of points in the cloud; the first EdgeConv layer has been replaced with a specialized layer that performs two functions: forms a multimodal feature vector consisting of point coordinates and their normalized coordinates, coordinates of normals and independent HSV color features; performs a higher discretization of the point cloud, which allows you to form a homogeneous dense point cloud and thereby get the best quality point cloud segmentation; the metric classifier that generates the output values of the DGCNN network has been replaced by a combination of two MLP network and an one RBF (Radial Basis Function) network - segmentation output. The scheme of concatenation of data from the outputs of various EdgeConv layers in the DGCNN network was also changed, which made it possible to increase the efficiency of processing local features of objects.

The introduced changes, as will be shown in Sect. 4, made it possible to obtain an accurate solution to the problem of semantic segmentation and classification for 3D large-scale DTMs, which is important when processing data on archaeological sites and objects. For some classes, the proposed 3D object classification method does not provide the required quality (at least 0.8 according to the F1-score), therefore, a combination of a classification 3D data method based on DGCNN* and a classification 2D data method based on the YOLO v.8 detector was used in the this work. The results of these methods were combined using the Bayes formula

$$p(j|m_{DGCNN*} \bigwedge m_{YOLO}) = w_j p(m_{DGCNN*}|j) \times p(m_{YOLO}|j), \qquad (5)$$

where $p(m_{DGCNN*})$ is the confidence returned by DGCNN*, $p(m_{YOLO})$ is the confidence returned by YOLO detector, $w_j$ is normalization hyperparameter, $m_{DGCNN*}$ and $m_{YOLO}$ are class labels for DGCNN* and YOLO respectively.

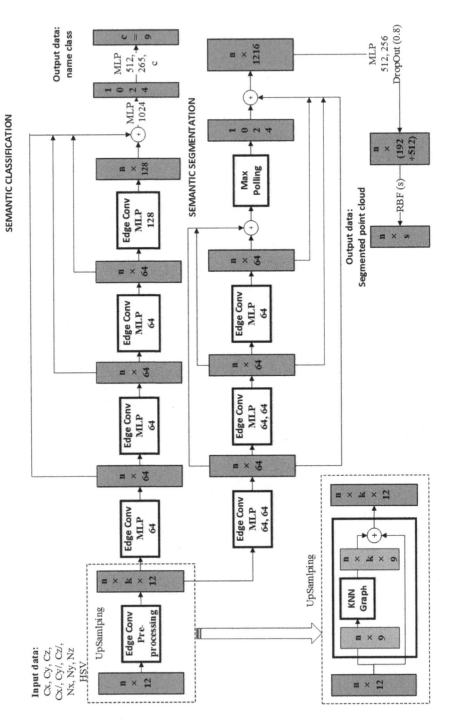

**Fig. 3.** The architecture of the modified DGCNN*.

Let's solve a variational problem of the following form in order to determine the class label with the highest probability value

$$J = \arg\max_{j} p(j|m_{DGCNN*} \bigwedge m_{YOLO}). \qquad (6)$$

Two algorithms were proposed to solve the task: Algorithm 5. Semantic segmentation algorithm based on clarifying the position of the detecting frame and Algorithm 6. A combined semantic segmentation algorithm using Algorithm 5 and DGCNN* (see Fig. 4).

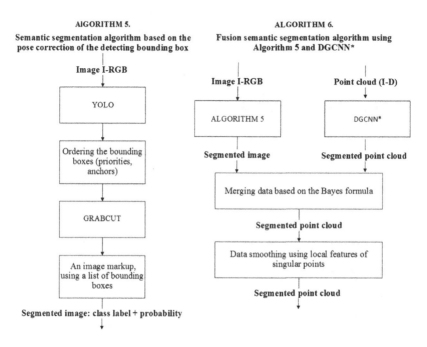

**Fig. 4.** The block diagrams of algorithms for 3D semantic classification and segmentation.

## 4 Computer Simulation

This section presents and discusses the results of computer modeling. Let's evaluate the accuracy and convergence of the proposed methodology in relation to the task of classifying archaeological objects in comparison with the known approaches to 3D classification based on MVCNN and DGCNN. In a comparative analysis, we will evaluate the quality of these methods both individually and in combination with the YOLO v8. For computer modeling, we will use the created data set about archaeological sites of the Bronze Age which consist of: aerial photography; results of remote sensing of the Resource-P, Canopus-B satellites; total station survey data obtained using the Trimble 3300 total station. We are going to expand the training subsample with examples of archaeological objects for the analysis of 2D data models by about 10 times using geometric transformations and Mask R-CNN. Archaeologists of Chelyabinsk State University have identified signs of deciphering archaeological sites which are characteristic of Bronze Age monuments in the Southern Urals [25]. The following classes of archaeological objects have been identified: Bronze Age dirt mound (on virgin soil, covered with turf) K1, Bronze Age Dirt Mound (Plowed surface) K2, A dirt mound of the early Iron Age (on virgin soil, covered with turf) K3, Stone mound of the early Iron Age or the Middle Ages (with a stone shell) K4, A dirt or stone mound of the early Middle Ages with a "mustache" K5, Burial cult complexes M1, Burial grounds with stone fences of the Middle Ages M2, Fortified settlement of the Bronze Age (with linear or concentric layout) P1, The undefended settlement of the Bronze Age P2. All evaluation tests were carried out using the following hardware platform: an Intel Core i7-based computer with a GPU, training of CNN was carried out for 150 epochs, the size of the training sample was 274 frames, the test sample was 117 frames. Table 1 presents the results of the classification of objects that relate to mounds classes.

**Table 1.** F1-score (in $10^{-3}$) of semantic classification methods for the kurgan classes

| Name methodology | K1 | K2 | K3 | K4 | K5 |
|---|---|---|---|---|---|
| ICP + DGCNN | 853 | 786 | 428 | 701 | 588 |
| MVCNN | 813 | 684 | 401 | 655 | 567 |
| ICP + DGCNN + YOLO | 868 | 845 | 355 | 812 | 477 |
| MVCNN + YOLO | 827 | 745 | 387 | 847 | 404 |
| FICP + DGCNN* | 902 | 894 | 443 | 877 | 518 |
| FICP + DGCNN* + YOLO | 893 | 922 | 432 | 902 | 511 |

Analyzing the data in Table 1, it can be concluded that the proposed methodology based on the fusion of FICP + DGCNN* + YOLO methods allows for the classification of mounds of the Early Iron Age (K2) or Middle Ages (K4) and soil mounds (K1) with high accuracy, while soil or stone mounds of the early Middle Ages with "moustaches" (K5) and Iron Age soil mounds (K3) are classified with low accuracy. Also from the Table 1 it can be seen that the use of a combination of the 3D classification method and the YOLO object detector almost always leads to an increase in the accuracy of solving the problem of 3D classifying archaeological objects. The results of computer modeling confirm the well-known fact that methods of semantic data processing based on CNN have advantages in terms of accuracy over multi-view methods such as MVCNN. On the other hand, it can be seen that using a combination of FICP + DGCNN* + YOLO almost always has accuracy advantages over FICP + DGCNN*, ICP + DGCNN + YOLO and ICP + DGCNN. When classifying a dirt mounds (K3), many errors of the first and second kind occur, in this case the object of interest may be mistakenly correlated with the background in the image or the point cloud. This is due to the presence of noise of various nature in the images, outliers in 2D and 3D data, and most importantly ambiguous signs of decryption of archaeological objects of this class. Table 2 presents the results of the classification of objects that belong to settlements and burial grounds (M1, M2) of various classes. Analyzing the data in the Table 2 it can be concluded that the proposed methodology based on the fusion of FICP + DGCNN* +YOLO methods allows for the classification of fortified (P1) and non-fortified (P2) settlements of the Bronze Age with high accuracy, while burial grounds with stone fences of the Middle Ages (M2) are classified with F1-score no more than 0.748, and funeral cult complexes (M1) are in principle poorly detected, thar it is due to their shape and signs of decryption (see Fig. 5).

**Table 2.** F1-score (in $10^{-3}$) of semantic classification methods for burial grounds and settlements

| Name methology | P1 | P2 | M1 | M2 |
|---|---|---|---|---|
| ICP + DGCNN | 804 | 825 | 388 | 698 |
| MVCNN | 682 | 626 | 261 | 414 |
| ICP + DGCNN + YOLO | 834 | 844 | 454 | 704 |
| MVCNN + YOLO | 718 | 696 | 291 | 501 |
| FICP + DGCNN* | 961 | 872 | 577 | 748 |
| FICP + DGCNN* + YOLO | 966 | 922 | 569 | 742 |

**Fig. 5.** The results of the classification of the archaeological site in the area of the village of "Levoberezhnoye according to aerial photography

For the settlement and burial grounds classes, it is also true that the accuracy of the classification process of archaeological objects is increased when using the YOLO object detector in the data processing process. The MVCNN method also showed worse accuracy than the DGCNN method and its combinations with ICP, FICP and YOLO, while the worst results were obtained for the M1 class. From the Table 2 it can be seen that using a combination of FICP + DGCNN* + YOLO methods has accuracy advantages in comparison with ICP + DGCNN, ICP + DGCNN + YOLO and FICP + DGCNN* for objects of settlement classes (P1, P2), while for objects from the burial grounds class group, accuracy was obtained similar to the FICP + DGCNN*. The results of computer simulation showed that the proposed DGCNN* architecture has advantages in F1-score compared to the classical version of DGCNN, on average, the F1-score can achieve an increase in accuracy by 0.095 points. The proposed methodology in this paper doesn't have advantages in terms of performance relative to state-of-the-art methodologies (see Table 1 and Table 2).

## Conclusions

This work is devoted to the study of methods for solving the task of segmentation and classification of 3D data of archaeological sites of the Bronze Age using deep machine learning methods. The paper proposes a new methodology for solving the 3D classification task based on a combination of the FICP registration method, a modified DGCNN* architecture and the YOLO object detector. The study suggests a new neural network architecture based on DGCNN with EdgeConv convolutional layers, which allows to obtain an accurate solution to the problem of classification and segmentation of 3D data based on a combination of geometric and color features of archaeological objects. The results of computer modeling based on data obtained as a result of the geometric survey of Bronze Age monuments in the area of the Sintashta River in the Chelyabinsk region and remote sensing data of the Earth showed advantages in accuracy of the proposed methodology in comparison with well-known approaches to classification and segmentation of 3D data for various classes of archaeological objects. The point-to-point problem in a closed form for affine transformations without initial initialization (FICP) based on a joint solution of the variational problem was solved using fusion depth data and features. The theoretical results were obtained, which allow us to evaluate the impact of features with using suggested descriptor based on histograms of oriented gradients on the accuracy and convergence of the solution of the variational ICP problem with point-to-point metric. The results obtained were used to solve the problem of constructing a digital mathematical model of an archaeological site.

**Acknolwledgments.** The work was supported by the Russian Science Foundation, project no. 23-11-20007.

# References

1. Dalal, N., Triggs, B.: Histograms of oriented gradients for human detection. Comput. Vision Pattern Recogn. **1**, 886–893 (2005)
2. Lowe, D.G.: Object recognition from local scale-invariant features. In: Proceedings of the International Conference on Computer Vision. IEEE, Kerkyra (1999)
3. Bay, H., Ess, A., Tuytelaars, T., Van Gool, L.: SURF: speeded up robust features. Comput. Vis. Image Underst. **110**(3), 346–359 (2008)
4. Calonder, M., Lepetit, V., Strecha, C., Fua, P.: BRIEF: binary robust independent elementary features. In: Daniilidis, K., Maragos, P., Paragios, N. (eds.) ECCV 2010. LNCS, vol. 6314, pp. 778–792. Springer, Heidelberg (2010). https://doi.org/10.1007/978-3-642-15561-1_56
5. Manzurv, T., Zeller, J., Serati, S.: Optical correlator based target detection, recognition, classification, and tracking. Appl. Opt. **51**, 4976–4983 (2012)
6. Ouerhani, Y., Jridi, M., Alfalou, A., Brosseau, C.: Optimized preprocessing input plane GPU implementation of an optical face recognition technique using a segmented phase only composite filter. Opt. Commun. **2013**(289), 33–44 (2013)
7. Kumar, B.V.-K.V., Mahalanobis, A., Juday, R.D.: Correlation Pattern Recognition. Cambridge University Press, Cambridge (2005)
8. Wang, P.S., Sun, C.Y., Liu, Y.: Adaptive O-CNN: a patch-based deep representation of 3D shapes. ACM Trans. Graphics **37**(6), 1–11 (2018)
9. Brock, A., Lim, T., Ritchie, J.M.: Generative and discriminative voxel modeling with convolutional neural networks. http://arxiv.org/abs/1608.04236. Accessed 08 June 2023
10. You, Y., Lou, Y., Qi, L., Tai, Y.W., Wang, W., Ma, L.: PRIN: pointwise rotation-invariant network. http://arxiv.org/abs/1811.09361. Accessed 08 June 2023
11. Su, H., Maji, S., Kalogerakis, E., Learned-Miller, E.: Multi-view convolutional neural networks for 3D shape recognition. In: Proceedings of International Conference on Computer Vision (ICCV). IEEE, Santiago (2015)
12. Maturana, D., Scherer, S.: VoxNet: a 3D convolutional neural network for real-time object recognition. In: Proceedings of International Conference on Intelligent Robots and Systems (IROS). IEEE, Hamburg (2015)
13. Li, J., Chen, B.M., Lee, G.H.: SO-net: self-organizing network for point cloud analysis. In: Proceedings Computer Vision and Pattern Recognition. IEEE, Salt Lake City (2018)
14. Lambers, K., Verschoof-van der Vaart, W.V., Bourgeois, Q.P.G.: Integrating remote sensing, machine learning, and citizen science in Dutch archaeological prospection. Remote Sens. **11**(7), 794 (2019)
15. Charles, R.Q., Su, H., Kaichun, M., Guibas, L.J.: PointNet: deep learning on point sets for 3D classification and segmentation. In: Proceedings of Conference on Computer Vision and Pattern Recognition. IEEE, Honolulu (2017)
16. Charles, R.Q., Li, Y., Hao, S., Leonidas, J. G.: PointNet++: deep hierarchical feature learning on point sets in a metric space. In: Proceedings of 31st Conference on Neural Information Processing Systems (NIPS). NeurIPS Media Kit, Long Beach (2017)
17. Zhang, Y., Rabbat, M.: A graph-CNN for 3D point cloud classification. In: Proceedings of International Conf. on Acoustics, Speech and Signal Processing. IEEE, Calgary (2018)
18. Wang, Y., Sun, Y., Liu, Z., Sarma, S.E., Bronstein, M.M., Solomon, J.M.: Dynamic graph CNN for learning on point clouds. ACM Trans. Graphics. **38**(5), 146, 1–12 (2018)

19. Te, G., Hu, W., Guo, Z., Zheng, A., Guo, Z.: RGCNN: regularized graph CNN for point cloud segmentation MM. In: Proceedings of the 26th ACM International Conference on Multimedia. ACM Digital Library, Seoul (2018)
20. Horn, B.K.P.: Closed form solution of absolute orientation using unit quaternions. J. Opt. Soc. Am. A **4**(4), 629–642 (1987)
21. Vokhmintcev, A.V., Sochenkov, I.V., Kuznetsov, V.V., Tikhonkikh, D.V.: Face recognition based on matching algorithm with recursive calculation of local oriented gradient histogram. Dokl. Math. **466**(3), 453–459 (2016)
22. Vokhmintcev, A., Timchenko, M.: The new combined method of the generation of a 3D dense map of environment based on history of camera positions and the robot's movements. Acta Polytech. Hung. **17**(8), 95–108 (2020)
23. Vokhmintcev, A.V., Melnikov, A.V., Pachganov, S.A.: Simultaneous localization and mapping method in 3D space based on the combined solution of the point-point variation problem ICP for an affine transformation. Inform. Appl. **14**(1), 101–112 (2020)
24. YOLO by Ultralytics 2023. https://github.com/ultralytics/. Accessed 08 June 2023
25. Zdanovich, G.B., Batanina, I.M., Levit, N.V., Batanin, S.A.: Step'-lesostep'. Kizil'skij rajon. Arheologicheskij atlas Chelyabinskoj oblasti 2003. https://search.rsl.ru/ru/record/01002755616?ysclid=lm3sa1zfkp696194850. Accessed 09 Jan 2023

# Network Analysis

# Visualization-Driven Graph Sampling Strategy for Exploring Large-Scale Networks

Gagik Khalafyan[(✉)], Irina Tirosyan🆔, and Varduhi Yeghiazaryan

American University of Armenia, Yerevan, Armenia
{gkhalafyan,itirosyan,vyeghiazaryan}@aua.am
http://www.aua.am

**Abstract.** Graph sampling is crucial for analyzing and understanding large-scale networks across various domains. While numerous approaches have been proposed in the existing literature, a comprehensive evaluation of these methods, with regards to both quality and execution time, is still needed. This paper addresses this gap by offering an exhaustive review of current graph sampling techniques and by introducing three distinct modifications to the Mino-centric graph sampling (MCGS) method. These modified algorithms, along with established methods, are rigorously evaluated through a quantitative analysis that encompasses two comparative iterations and multiple metrics. In addition to the quantitative analysis, we also conduct a qualitative user study, where survey participants assess the quality of the sampling from a visual perspective. Our findings indicate that one of our modified versions of the MCGS algorithm, namely Batch major CC MCGS, not only outperforms other methods in the context of visual evaluation but also significantly optimizes execution time in comparison with the original MCGS algorithm. This improvement equips researchers and practitioners with a powerful tool for exploring large-scale networks in diverse fields.

**Keywords:** Large-scale networks · Graph sampling · Network visualization · Mino-centric graph sampling

## 1 Introduction

The analysis of large-scale networks has become increasingly important due to the significant expansion and the growing complexity of real-world graphs such as social, biological, and transportation networks. These networks often contain millions or even billions of nodes and edges, presenting significant computational and visual challenges [4,13]. An example of such a network is shown in Fig. 1 (top left), representing Facebook pages and mutual connections between them, drawn from the Stanford Large Network Dataset Collection, part of the Stanford Network Analysis Project (SNAP) [14], and visualized with the Cosmograph web-based app [6].

© The Author(s), under exclusive license to Springer Nature Switzerland AG 2024
D. I. Ignatov et al. (Eds.): AIST 2023, LNCS 14486, pp. 311–324, 2024.
https://doi.org/10.1007/978-3-031-54534-4_22

Graph sampling is a technique that addresses these challenges by focusing on a smaller, representative subgraph, thereby reducing computational requirements and improving visualization quality [9]. Despite its usefulness, existing methods have limitations, creating a need for novel visualization-driven graph sampling strategies that can better support the exploration of large-scale networks [2].

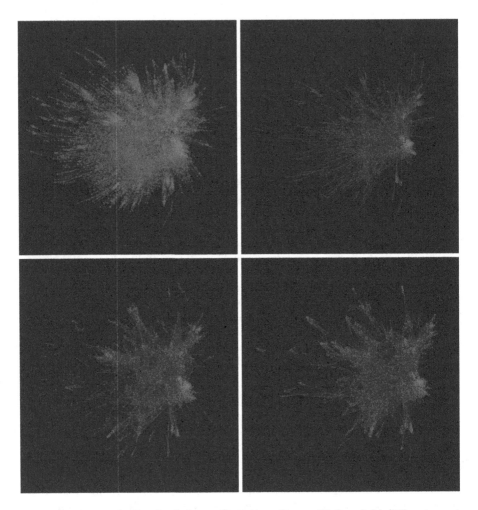

**Fig. 1.** The original "Facebook Large Page-Page Network" (top left) [20] represents connections between Facebook pages. It contains 22,470 nodes and 171,002 edges, making it challenging to analyze visually. The rest of the images are sampled versions of the graph (with a sampling rate of 0.1): MCGS (top right), Random edge–node sampler (bottom left) and Random jump sampler (bottom right). The visualizations are created with the Cosmograph web-based app [6].

This study presents a detailed examination of existing graph sampling approaches in the context of network visualization, shortlisting Random edge–node sampling, Random jump sampling, and Mino-centric graph sampling (MCGS) techniques, with the latter being the current state-of-the-art. The selection of the algorithms was made taking into account the results of the experiments in [8, 24, 26].

We then propose three modifications to MCGS and conduct an extensive quantitative analysis to evaluate their performance against the well-established algorithms. A qualitative user study is also performed to assess the visual effectiveness of the different algorithms from a user perspective.

## 2   Graph Sampling: State-of-the-Art

Over the years, a wide range of graph sampling techniques have emerged, with each technique designed to preserve different graph properties in unique ways.

**Traditional Graph Sampling Techniques.** Uniform sampling is considered one of the most primary techniques, characterized by the random selection of nodes or edges with equal probabilities. One of the most common algorithms, Random edge–node sampling, selects nodes by first choosing a random edge and then selecting its incident nodes. Nevertheless, the simplicity of this approach often falls short in capturing the structural complexities of large-scale, diverse graphs [22]. For instance, it has a tendency to drastically increase the number of connected components, as seen in the bottom left of Fig. 1.

Random walk sampling, another conventional graph sampling technique, operates by exploring nodes based on particular probabilities, generally correlated with their degrees. Although this method tends to better capture structural properties such as degree distribution and clustering coefficient compared to Uniform sampling, it exhibits bias towards high-degree nodes, limiting its ability to reflect the global structure of the graph accurately. Introducing a variation, the Random jump sampling method builds upon a random walk by allowing occasional jumps to completely random nodes at certain steps, which can mitigate the inherent bias of the conventional Random walk sampling and can ensure broader coverage of the graph's structure. However, due to its inherent randomness, the method can still overlook significant regions of the graph, focusing solely on specific segments, as illustrated in the bottom right of Fig. 1.

Importance sampling selects nodes or edges according to their importance within the graph. This importance is quantified by centrality measures like degree centrality, betweenness centrality, or eigenvector centrality [3]. Despite its superior ability to preserve fundamental graph structures, Importance sampling can be computationally expensive for large graphs and may overemphasize high-centrality nodes or edges.

Other techniques, such as Graph-based anomaly detection (GBAD), aim to identify and characterize representative substructures within large graphs.

GBAD is adept at preserving essential graph properties and detecting anomalies, however, due to the complexity of the technique in terms of computational and algorithmic processing, it may pose significant computational challenges when applied to large graphs [18].

**Spectral Sampling Techniques.** Spectral sampling is a class of graph sampling techniques that utilize the spectral properties of a graph—specifically its eigenvalues and eigenvectors—to construct representative samples. One notable spectral sampling technique is the Biconnected-components-tree-based (BC-tree-based) spectral sampling, which combines spectral sampling with a tree decomposition of the graph [8]. While effective in preserving key structural properties of the original graph, the computational complexity of spectral sampling techniques often makes them impractical for very large graphs.

**Mino-Centric Graph Sampling.** MCGS [26] tackles the challenge of preserving minority structures within a graph, a vital consideration in areas like social networks, biological networks, and fraud detection. Minority structures, in general, are those components within the graph that appear less frequently or are less connected compared to others, but can play an important role. The method combines a global sampling phase that prioritizes minority structures with a local expansion phase that accommodates majority structures. This dual focus allows for the preservation of both minority and majority structures, creating a more representative and accurate sampling of the original graph.

The selection of minority structures in the MCGS algorithm is deterministic, conducted through a systematic and efficient process, which prioritizes nodes based on certain criteria such as connectivity, node centrality, or other relevant graph properties, ensuring that the sampled graph retains key characteristics of the original, as detailed in Section 5 of [26]. Once minority selection is complete, the remaining nodes are considered for inclusion in the sample in the majority selection phase. These nodes are added to the sample one at a time, and after each addition, a series of metrics are evaluated. The node that yields the best performance according to these metrics is selected, and this process is repeated until a satisfactory number of nodes have been sampled. This one-by-one addition and evaluation can potentially slow down the sampling process, constituting a significant computational bottleneck in the algorithm.

While MCGS has shown a superior performance in preserving vital structural properties of graphs [26], as evidenced in the top right of Fig. 1, there is still room for improvement. Enhancements can be made by addressing the bottleneck issue in the majority selection process and extending the algorithm's capability to preserve a wider variety of minority structures. Lastly, the algorithm often completely leaves out small connected components in the graph. This oversight also needs to be avoided.

# 3  Proposed Algorithms

We introduce seven novel graph sampling algorithms, each addressing specific limitations observed in the original MCGS.

## 3.1  Enhanced Minor MCGS

The first algorithm, Enhanced minor MCGS, addresses the issue where MCGS occasionally misses short chain-connections between significant nodes, i.e., super pivots and huge stars. This occurs due to the existence of longer paths between the components containing these nodes, causing the shorter chains to go unnoticed.

To resolve this, we redesign the minority selection process of MCGS and additionally include all nodes in the shortest paths between the selected super pivots and huge stars. This technique ensures a more comprehensive selection of essential minority nodes and their inclusion in the final sampled graph, as demonstrated in Fig. 2.

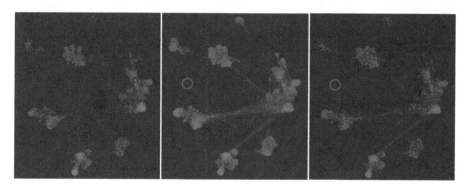

**Fig. 2.** Enhanced minor MCGS (right) successfully retrieves a crucial direct connection (highlighted with a red circle) between two components in the "Social circles: Facebook" graph (center) [15], while standard MCGS (left) fails to do so. (Color figure online)

## 3.2  Batch Major MCGS

The second algorithm, Batch major MCGS, addresses the long execution time of MCGS's majority selection step. This stems from the greedy selection algorithm that selects a single node at each step. This becomes computationally expensive when repeated until the desired sample size is achieved.

We address this by implementing a batch node selection mechanism. Instead of selecting a single node in each step, we choose a batch of nodes during each iteration. Through extensive testing, we found a batch size of 10 nodes to be optimal (see relevant experiments and discussion in Sect. 4.2).

### 3.3   Connected Component (CC) MCGS

The third proposed algorithm, Connected component (CC) MCGS, addresses the issue of under-representation of some connected components during sampling.

We modify the algorithm to apply the MCGS approach to each connected component individually, considering them as separate entities. The individual samples obtained from each component are then combined to form the final sampled graph, which ensures that all connected components, regardless of their size, are adequately represented, as illustrated in Fig. 3.

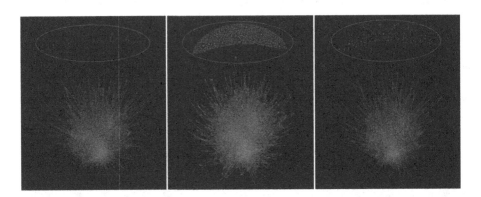

**Fig. 3.** CC MCGS (right) retrieves more connected components (highlighted with a red oval) from the original "Condense Matter collaboration network" (center) [17] than standard MCGS (left). (Color figure online)

### 3.4   Ensemble Algorithms

Further, we propose four ensemble algorithms, each integrating a combination of the three modifications. These ensemble algorithms, namely, Batch major CC MCGS, Batch major enhanced minor MCGS, Enhanced minor CC MCGS, and Batch major enhanced minor CC MCGS, aim to tackle multiple limitations simultaneously, thereby enhancing the efficiency and the representativeness of the graph sampling process.

## 4   Quantitative Analysis of Existing and Proposed Algorithms

We conducted a rigorous quantitative evaluation of various graph sampling techniques through extensive experimentation. This assessment provided an in-depth understanding of the performance and the efficiency of each algorithm. The eight graphs used in the experiments were: "Cora" [16], "Social circles: Facebook" [15], "General Relativity and Quantum Cosmology collaboration network" [12], "Autonomous systems AS-733" [11], "Wikivote" [10], "LastFM

Asia" [21], "Facebook Large Page-Page Network" [20], and "Condense Matter collaboration network" [17].

All our algorithms were developed in Python, leveraging the NumPy library for numerical operations and NetworkX for graph manipulation. For performance evaluation, the algorithms were tested on a MacBook Pro 2021, powered by an Apple M1 Pro processor with 16GB RAM. Our implementations were built upon the codebase of the original MCGS algorithm, accessible at [25]. This approach ensured uniform testing conditions throughout all evaluations and iterations.

### 4.1 Metrics and Measurements for Quantitative Evaluation

Following the example of $[1, 7, 19, 23]$, for a thorough quantitative evaluation of the graph sampling algorithms, we selected a group of metrics, each quantifying a different aspect of network analysis.

**Average and Global Clustering Coefficients:** The average clustering coefficient ($C_{avg}$) and the global clustering coefficient ($C_{global}$) [7] measure the extent to which nodes in a graph cluster together.

The average clustering coefficient is the mean of the local clustering coefficients of all nodes in the graph, where $n$ denotes the total number of nodes in the graph and $C_i$ is the clustering coefficient of the $i$-th node:

$$C_{avg} = \frac{1}{n} \sum_{i=1}^{n} C_i \tag{1}$$

The global clustering coefficient is the ratio of the number of closed triplets (or triangles) to the number of all triplets (connected triples of nodes):

$$C_{global} = \frac{3 \times \text{number of triangles}}{\text{number of connected triples}} \tag{2}$$

**Number of Connected Components:** This metric [23] provides a measure of the graph's connectivity and indicates the extent to which the connectivity of the original graph is preserved in the sampled graph.

**Kolmogorov–Smirnov Distance:** The Kolmogorov–Smirnov distance ($D_{KS}$) quantifies the difference between two probability distributions [19] (degree distributions in the scope of graphs). It is the maximum distance between the cumulative distribution functions $F_1(x)$ and $F_2(x)$ of the two distributions:

$$D_{KS} = \max_x |F_1(x) - F_2(x)| \tag{3}$$

**Skew Divergence Distance:** The skew divergence distance $(D_{SD})$ [1] provides a symmetric divergence measure between two probability distributions. It is a smoothed version of the Kullback–Leibler (KL) divergence, where $P$ and $Q$ represent the two probability distributions, $\|$ denotes the divergence of one distribution from another in the context of the KL divergence, and $\alpha$ is a smoothing factor typically set between 0 and 1. It is given by

$$D_{SD}(P,Q) = \frac{1}{2}\left(D_{KL}\Big(P\|(\alpha P+(1-\alpha)Q)\Big) + D_{KL}\Big(Q\|(\alpha Q+(1-\alpha)P)\Big)\right) \quad (4)$$

### 4.2  Optimal Batch Size Selection for Batch Major MCGS

The process of optimal batch size selection is critical for effective and efficient execution of the Batch major MCGS algorithm. For this purpose, we performed extensive experimentation on the eight diverse graphs with varying batch sizes. The primary scope of our experiments spanned batch sizes from 1 to 100, with extended testing conducted at increments of 100 up to 1000.

During these experiments, we noticed that the metrics used for evaluating the accuracy of graph sampling did not vary significantly across different batch sizes. However, we found substantial differences in execution times, with larger batch sizes yielding faster execution. Figure 4 shows the execution times for different batch sizes for the largest graph in our set, the "Condense Matter collaboration network" (the visualization is capped at a major batch size of 50 for clearer illustration). Results for sampling rates 0.1, 0.2, 0.3, 0.4 and 0.5 are included.

Across all tested networks improvements in execution time were marginal beyond a certain batch size. To identify the point where further increases in batch size yielded diminishing returns in execution time, we employed the elbow method [5], marking the point from which the improvements were marginal. We determined that a batch size of 10 is an optimal choice for our graph sampling experiments, ensuring a balance between computational efficiency and resource utilization. The selected batch size reduces the average running time of the whole algorithm 7.87 times over standard MCGS (tested with the eight graphs using sampling rates 0.1, 0.2, 0.3, 0.4 and 0.5).

### 4.3  Comparison of Algorithms Using Numerical Metrics

The evaluation of different algorithms was performed in two stages, employing six different sampling rates (0.05, 0.1, 0.2, 0.3, 0.4, 0.5) across the eight distinct graphs. For a comprehensive analysis, each algorithm was executed four times due to the implicit randomness of the process.

The first iteration of our evaluation involved analyzing the performance of our seven newly proposed algorithms using the five key metrics delineated in Sect. 4.1. For each of our eight benchmark graphs, samples were produced at the predefined six sampling rates using the algorithms under investigation. This approach resulted in a total of $8 \times 7 \times 4 = 224$ possible outcomes for each sampling rate for a single metric, considering all combinations of graphs, algorithms, and their executions.

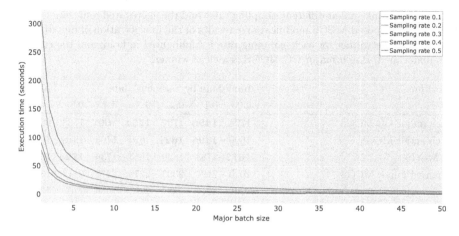

**Fig. 4.** Execution times in seconds of the Batch major MCGS algorithm for different batch sizes (range 1–50) and sampling rates (0.1–0.5) on the "Condense Matter collaboration network" [17]. As the batch size increases, a trend of reducing execution times can clearly be seen.

Given that all our metrics serve as distance metrics, the algorithm achieving the shortest distance was assigned the top rank (1), whereas the algorithm with the longest distance got the bottom rank (224). In cases where multiple algorithms showcased identical performance, they were assigned the same rank. This ranking procedure was executed sequentially for each of the five metrics, and the resulting ranks for each metric were summed, providing a composite rank for each algorithm at a particular sampling rate.

Subsequently, the ranks obtained across the four executions of an algorithm were aggregated to deduce its overall performance rank for that particular sampling rate. These ranks were further consolidated across all sampling rates to determine the final performance rank of each algorithm. An algorithm with a lower rank value was deemed superior, reflecting its consistent top-tier performance. According to this comprehensive ranking methodology, Batch major CC MCGS emerged as the standout algorithm among our proposed methods, as highlighted in Table 1.

Upon analyzing the results from the first iteration of the evaluation, a notable observation was the underperformance of the Enhanced minor MCGS. A probable reason for this might be its overemphasis on the shortest paths between super pivots and huge stars. While this strategy was intended to capture essential connections, it possibly led to a neglect of other vital components and interconnections in the graph. This overfocus inadvertently affected the overall performance of the ensemble of all modifications, as the inclusion of the suboptimal Enhanced minor MCGS pulled down the ensemble's aggregated result. Future refinements should consider a balanced approach to node selection, ensuring that the importance of diverse graph structures is acknowledged and represented.

**Table 1.** The rank sum at different sampling rates and the aggregated rank sum (Aggr) of the seven proposed MCGS modifications: results of the first iteration of quantitative comparison. The winner for each sampling rate is highlighted in bold, and the runner-up is underlined. Batch major CC MCGS is a clear winner.

| Algorithm | Rank Sum by Sampling Rate | | | | | | Aggr ↓ |
|---|---|---|---|---|---|---|---|
| | 0.05 | 0.1 | 0.2 | 0.3 | 0.4 | 0.5 | |
| Batch major CC MCGS | 1728 | **1496** | <u>1707</u> | **1554** | **1697** | <u>1733</u> | 9915 |
| Batch major MCGS | <u>1690</u> | **1496** | 1621 | <u>1629</u> | 1765 | 1918 | 10119 |
| CC MCGS | **1672** | 1753 | 1829 | 1684 | <u>1736</u> | **1724** | 10398 |
| Enhanced minor MCGS | 2015 | 2100 | 2363 | 2279 | 2277 | 2118 | 13152 |
| Batch major enhanced minor MCGS | 2206 | 2212 | 2307 | 2257 | 1986 | 2205 | 13173 |
| Enhanced minor CC MCGS | 2456 | 2566 | 2228 | 2360 | 2151 | 1979 | 13740 |
| Batch major enhanced minor CC MCGS | 2383 | 2664 | 2185 | 2229 | 2243 | 2250 | 13954 |

In the second iteration, we compared the performance of Batch major CC MCGS, the best-scoring proposed algorithm, with three existing methods, using the same methodology and metrics. In this setting MCGS outperformed Batch major CC MCGS with a slight margin. However, Batch major CC MCGS had a tenfold improvement in time complexity for majority selection due to the introduction of the batch size parameter. The full ranking statistics are in Table 2.

## 5    Qualitative Analysis of Algorithms

In addition to the quantitative analysis, we carried out a qualitative analysis to gain further insights into the performance of the proposed and existing graph sampling algorithms.

We created an environment where real survey-takers could evaluate the visual quality of graph sampling. Each participant was given 16 triplets of images, with each triplet shown on a separate webpage on the screen. The middle image in each triplet represented the original graph, while the images on the sides displayed two samples generated by two of the four shortlisted algorithms with the same sampling rate (Fig. 5).

For each of the eight graphs, we generated two triplets with sampling rates randomly chosen from 0.1, 0.2, 0.3, and 0.4. Subsequently, we randomly selected two algorithms and used their best-scoring run among their four executions for the chosen graph and sampling rate. The survey-takers were tasked with choosing one of the graphs as a better representation of the original; in this case,

**Table 2.** The rank sum at different sampling rates and the aggregated rank sum (Aggr) of the three existing graph sampling algorithms and the proposed Batch major CC MCGS: results of the second iteration of quantitative comparison. The winner for each sampling rate is highlighted in bold, and the runner-up is underlined. While Batch major CC MCGS is slightly behind MCGS in terms of rank sum, it offers a substantial computational advantage over MCGS.

| Algorithm | Rank Sum by Sampling Rate | | | | | | Aggr ↓ |
|---|---|---|---|---|---|---|---|
| | 0.05 | 0.1 | 0.2 | 0.3 | 0.4 | 0.5 | |
| MCGS | **955** | **877** | **935** | **919** | **911** | **936** | 5533 |
| Batch major CC MCGS | <u>1086</u> | <u>915</u> | <u>939</u> | <u>952</u> | <u>1007</u> | <u>953</u> | 5852 |
| Random jump sampler | 1338 | 1438 | 1317 | 1355 | 1298 | 1303 | 8049 |
| Random edge–node sampler | 1228 | 1381 | 1424 | 1373 | 1380 | 1397 | 8183 |

the selected graph was awarded 2 points. Alternatively, if the participant deemed both samples to be equally good, each sample received 1 point.

**Table 3.** Total points for the three existing graph sampling algorithms and the proposed Batch major CC MCGS based on the 100 responses to the survey. Batch major CC MCGS is the leader.

| Algorithm | Points ↑ |
|---|---|
| Batch major CC MCGS | 914 |
| Random edge–node sampler | 877 |
| MCGS | 736 |
| Random jump sampler | 595 |

The results achieved after surveying 100 people are presented in Table 3. The proposed algorithm, Batch major CC MCGS, outperforms the other contenders. With a solid advantage over the runner-up Random edge–node sampler, the proposed algorithm surpasses current state-of-the-art in the context of visual evaluation.

The qualitative analysis, in conjunction with the quantitative evaluation, provides a comprehensive understanding of the performance of the proposed graph sampling algorithm. The survey results indicate that our algorithm not only performs well in terms of numerical measurements but is also perceived as superior by real participants evaluating the visual quality of the samples. This combination of strong quantitative performance and positive qualitative feedback positions our algorithm as a robust and effective solution for visualization-driven graph sampling tasks.

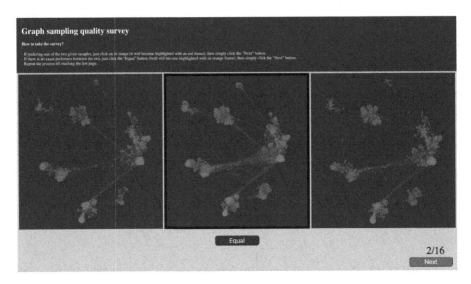

**Fig. 5.** Sample webpage of the survey environment. The middle image depicts the original graph, and sampled graphs are presented on the sides.

Additionally, the proposed sampler shows a considerable advantage over the standard MCGS in terms of time complexity for the majority selection process. These findings highlight the potential of our algorithm as a valuable tool for researchers and practitioners working with graph data.

## 6   Conclusions

We introduced several novel graph sampling algorithms. The best performer among these was rigorously evaluated against existing methods. Throughout all our experiments we employed multiple metrics to provide a comprehensive understanding of each algorithm's strengths and weaknesses.

By comparing all the proposed algorithms with each other, we identified Batch major CC MCGS as the best-performing algorithm among them. Moreover, we demonstrated that our algorithm not only performs well in terms of numerical measurements but also offers superior visual quality, as confirmed by survey participants in the qualitative analysis.

Thus, our research contributes to the field of graph sampling by introducing a robust and effective algorithm, which showcases significant advantages both in performance and in efficiency. We hope that this work provides valuable insights for researchers and practitioners working with graph data, and contributes to future studies in graph sampling.

The proposed algorithm has shown promising results, however, additional components or strategies could further enhance its performance. Investigating new techniques or more advanced graph-theoretic concepts could lead to improvements in the algorithm's ability to generate high-quality samples. It would also be beneficial to test the algorithm on a larger set of graphs to better understand the generalizability of the method. Finally, investigating how the proposed algorithm performs across diverse real-life domains and adapting it to specific requirements or constraints could lead to new insights and further validate its usefulness in real-world applications.

**Acknowledgements.** We thank the survey participants for their time and contribution to our research.

# References

1. Ahmed, N., Neville, J., Kompella, R.R.: Network sampling via edge-based node selection with graph induction. Technical report, Purdue University (2011)
2. Ahmed, N.K., Neville, J., Kompella, R.: Network sampling: from static to streaming graphs. ACM Trans. Knowl. Discov. Data **8**(2), 1–56 (2013)
3. Ahuja, R.K., Magnanti, T.L., Orlin, J.B.: Network flows (1988)
4. Barabási, A.L.: Network science. Philos. Trans. Roy. Soc. A: Math. Phys. Eng. Sci. **371**(1987), 20120375 (2013)
5. Bholowalia, P., Kumar, A.: EBK-means: a clustering technique based on elbow method and K-means in WSN. Int. J. Comput. Appl. **105**(9), 17–24 (2014)
6. Cosmograph: Interactive network visualization (2023). https://cosmograph.app. Accessed 19 July 2023
7. Hardiman, S.J., Katzir, L.: Estimating clustering coefficients and size of social networks via random walk. In: Proceedings of the 22nd International Conference on World Wide Web, pp. 539–550 (2013)
8. Hu, J., et al.: Bc tree-based spectral sampling for big complex network visualization. Appl. Netw. Sci. **6**(1), 60 (2021)
9. Lee, S.H., Kim, P.J., Jeong, H.: Statistical properties of sampled networks. Phys. Rev. E **73**(1), 016102 (2006)
10. Leskovec, J., Huttenlocher, D., Kleinberg, J.: Signed networks in social media. In: Proceedings of the SIGCHI Conference on Human Factors in Computing Systems, pp. 1361–1370 (2010)
11. Leskovec, J., Kleinberg, J., Faloutsos, C.: Graphs over time: densification laws, shrinking diameters and possible explanations. In: Proceedings of the Eleventh ACM SIGKDD International Conference on Knowledge Discovery in Data Mining, pp. 177–187 (2005)
12. Leskovec, J., Kleinberg, J., Faloutsos, C.: Graph evolution: densification and shrinking diameters. ACM Trans. Knowl. Discov. Data **1**(1), 2-es (2007)
13. Leskovec, J., Rajaraman, A., Ullman, J.D.: Mining of Massive Data Sets. Cambridge University Press, Cambridge (2020)
14. Leskovec, J., Sosič, R.: SNAP: a general-purpose network analysis and graph-mining library. ACM Trans. Intell. Syst. Technol. **8**(1), 1–20 (2016)
15. McAuley, J., Leskovec, J.: Learning to discover social circles in ego networks. In: Advances in Neural Information Processing Systems, vol. 25, pp. 539–547 (2012)
16. McCallum, A.K., Nigam, K., Rennie, J., Seymore, K.: The Cora dataset. https://paperswithcode.com/dataset/cora. Accessed 19 July 2023
17. Newman, M.E.J.: The structure of scientific collaboration networks. Proc. Natl. Acad. Sci. **98**(2), 404–409 (2001)

18. Noble, C.C., Cook, D.J.: Graph-based anomaly detection. In: Proceedings of the Ninth ACM SIGKDD International Conference on Knowledge Discovery and Data Mining, pp. 631–636 (2003)
19. Rezvanian, A., Rahmati, M., Meybodi, M.R.: Sampling from complex networks using distributed learning automata. Phys. A **396**, 224–234 (2014)
20. Rozemberczki, B., Allen, C., Sarkar, R.: Multi-scale attributed node embedding. J. Complex Netw. **9**(2), cnab014 (2021)
21. Rozemberczki, B., Sarkar, R.: Characteristic functions on graphs: birds of a feather, from statistical descriptors to parametric models. In: Proceedings of the 29th ACM International Conference on Information & Knowledge Management, pp. 1325–1334 (2020)
22. Stumpf, M.P.H., Wiuf, C.: Sampling properties of random graphs: the degree distribution. Phys. Rev. E **72**(3), 036118 (2005)
23. Uehara, R.: The number of connected components in graphs and its applications. Technical report, Komazawa University (1999)
24. Wu, Y., Cao, N., Archambault, D., Shen, Q., Qu, H., Cui, W.: Evaluation of graph sampling: a visualization perspective. IEEE Trans. Visual Comput. Graphics **23**(1), 401–410 (2016)
25. Zhao, Y., et al.: Mino-centric graph sampling (MCGS). GitHub repository (2020). https://github.com/csuvis/MCGS. Accessed 19 July 2023
26. Zhao, Y., et al.: Preserving minority structures in graph sampling. IEEE Trans. Visual Comput. Graphics **27**(2), 1698–1708 (2020)

# Limit Distributions of Friendship Index in Scale-Free Networks

Sergei Sidorov$^{(\boxtimes)}$ (iD), Sergei Mironov (iD), and Alexey Grigoriev (iD)

Saratov State University, Saratov 410012, Russia
sidorovsp@sgu.ru

**Abstract.** The friendship index measures a node's popularity relative to its friends on a social network. The friendship index is calculated by dividing the average degree of a node's friends by its own degree, i.e. it is the ratio of the sum of the degrees of its neighbors to the square of the degree of the node itself. Under some assumptions, the numerator of this fraction can be viewed as the sum of some random variables distributed according to the cumulative degree distribution function in the given network. It is known that for the vast majority of real complex networks, their degree distributions follow a power-law with some exponent $\gamma$. We examine the dependence of the average value of the friendship index among nodes of the same degree $k$ in the network on $k$. We will explore scale-free networks with degree-degree neutral mixing and find the limit distributions of the friendship index with the network size tending to infinity in the configuration model. Moreover, we compare our findings with the behavior of empirical friendship index distributions for several real networks.

**Keywords:** Social network analysis · Complex networks · Friendship index · Friendship paradox · Configuration model

## 1 Introduction

The friendship index is a significant measure used to determine the popularity of a node in comparison to its friends on a social network. Popularity is typically determined by the number of contacts a node has, known as its degree. The friendship index is calculated by dividing the average degree of a node's friends by its own degree [12]. This measure plays a vital role in social network analysis [1,2,5,6,8,9,11]. If a node's friendship index is greater than 1, it means that its friends are, on average, more popular than the node itself. Conversely, a value less than 1 indicates that the node is more popular than its neighbors.

Numerous studies on real social networks have consistently shown that the most nodes have the index greater than 1. This phenomenon, known as the friendship paradox in sociological sciences, essentially means that, on average, your friends are most likely more popular than you. The friendship paradox

The work was supported by the Russian Science Foundation, project 23-21-00148.

D. I. Ignatov et al. (Eds.): AIST 2023, LNCS 14486, pp. 325–337, 2024.
https://doi.org/10.1007/978-3-031-54534-4_23

has been extensively explored in various research papers, including the study by [12], which examined the aggregated friendship index value across entire networks, both real and model-generated. Additionally, researchers have also explored other characteristics that compare an element's popularity within a network to that of its neighbors [4,5,9,11].

Paper [12] focuses on examining the uneven distribution of node degrees at the local level. To measure this, the authors utilize the friendship index. Additionally, another study by [16] investigates the characteristics of the friendship paradox in networks created through the triadic closure model. It also compares these findings with quantitative estimates of the friendship index in actual networks. Paper [4] extends the concept of the friendship paradox to encompass arbitrary node characteristics in complex networks. The researchers analyze two networks: one based on co-authorship in Physical Review journals and another using Google Scholar profiles. The study determines that the generalized friendship paradox holds true at both individual and network levels across various characteristics, such as the number of coauthors, citations, and publications.

Previous studies have confirmed the presence of the friendship paradox in networks generated using the Barabási–Albert model [16,17]. This could be due to the scale-free network structure, where most nodes are connected to hubs, resulting in a high average degree among their neighbors. Thus, the model accurately represents the static properties of networks. It should be noted that many real social networks are scale-free, which, for example, refers to networks with a power-law degree distribution, for which the second moment of the distribution takes quite large values.

The friendship index for a node is equal to the ratio of the sum of the degrees of its neighbors to the square of its own degree. In many cases, we can consider the numerator of this ratio as the sum of random variables distributed according to the cumulative degree distribution function of the network. Real complex networks often exhibit a power-law degree distribution. We investigate scale-free networks with degree-degree neutral mixing and determine the limit distributions of the average value of the friendship index among nodes of the same degree as the network size approaches infinity. Additionally, we compare our findings to the observed distribution of the averaged friendship index values in several real networks.

## 2    Friendship Index Distribution in the Configuration Model

### 2.1    Notations

A graph with $n$ vertices is denoted by $G_n = (V_n, E_n)$. The vertices are labelled by integers as $V_n = \{1, 2, \ldots, n\}$. $E_n$ is the set of (undirected) edges. Let $d_i$ denote the degree of vertex $i$ and let $D_n = \{d_1, d_2, \ldots, d_n\}$ be the degree sequence of graph $G_n$. We have that total degree is $\sum_{i=1}^{n} d_i = 2|E_n|$.

Let $(i, j)$ denote the (directed) half-edge from node $i$ to node $j$. The undirected edge $e_{ij}$ is the sum of two half-edges $(i, j)$ and $(j, i)$. Let $|E_{ij}|$ be the number of all half-edges from node $i$ to node $j$. If $G_n$ is multi-graph then $|E_{ij}|$ may be greater than 1. If graph $G_n$ is undirected then $|E_{ii}| = |E_{ji}|$. Self-loops of vertex $i$ should be counted twice. Then we have $d_i = \sum_{j=1}^{n} |E_{ij}|$.

Let $\xi$ be a positive integer-valued random variable with the probability density function $f_\xi(k) = p(\xi = k)$ and the cumulative distribution function $F_\xi(k) = \sum_{j \leq k} f_\xi(k)$.

We assume that the degree sequences $D_n$ are obtained as $n$ independent and identically distributed samples of random variable $\xi$. If $\sum_{i=1}^{n} d_i$ turns out to be odd then we set $d_1 = d_1 + 1$ without the loss of generality, since the addition of 1 to the total degree of graph $G_n$ does not affect its asymptotic properties when $n$ is tending to infinity.

Denote $V_n(k) = \{i : d_i = k\} \subset V_n$ the set of vertices of graph $G_n$ having the degree $k$.

Let

$$f_n(k) = \frac{1}{n} \sum_{i \in V_n(k)} 1 = \frac{|V_n(k)|}{n}, \quad k = 1, 2, \ldots,$$

be the empirical degree density function. Then the size-based empirical degree density function is

$$f_n^* = \frac{k f_n(k)}{2|E_n|}, \quad k = 1, 2, \ldots.$$

Denote

$$\beta_i = \frac{\sum_{j \in E_{ij}} d_j}{d_i^2}$$

the friendship index of node $i$. We are interested in the value of friendship indices averaged over all nodes with degree $k$:

$$\Psi_n(k) := \begin{cases} \frac{1}{k} \sum_{l=1}^{n} l P(l|k), & V(k) \neq \varnothing, \\ 0, & V(k) = \varnothing. \end{cases}$$

where $P(l|k)$ is the probability that a $k$-degree node is joining to a $l$-degree vertex.

If we denote the empirical joint degree distribution on both sides of an arbitrary chosen edge from $E_n$ as follows

$$h_n(k,l) = \frac{1}{2|E_n|} \sum_{i<j} \begin{cases} 1, & d_i = k \text{ and } d_j = l, \\ 0, & \text{otherwise}, \end{cases}$$

then

$$\Psi_n(k) := \begin{cases} \frac{1}{k} \frac{\sum_{j=1}^{n} l h_n(k,l)}{f^*(k)}, & V(k) \neq \varnothing, \\ 0, & V(k) = \varnothing. \end{cases}$$

To calculate this value for a specific network of size $n$, we may use the relation

$$\Psi_n(k) = \frac{1}{|V_n(k)|} \sum_{i \in V_n(k)} \beta_i,$$

where the sum is taken over all vertices of degree $k$.

## 2.2   Configuration Model

The configuration model (CM) [3] is a graph generation model that creates graphs of a given size $n$ and with a given degree sequence. It is important for analyzing degree-degree correlations in complex networks and allows one to construct graphs $G_n$ with a specific degree distribution $D_n$. By assigning stubs to each node and randomly pairing them, one could obtain a multi-graph with the desired degree sequence.

If one has a positive integer-valued random variable $\xi$ with a probability density function $f$, then the degree sequence $D_n = (d_1, d_2, \ldots, d_n)$ can be obtained as $n$ independent and identically distributed samples of random variable $\xi$. It is known that the CM generates random multi-graphs in such a way that the empirical degree distribution $f_n$ converges to $f$ as $n$ approaches infinity.

However, the CM can be modified to generate simple graphs through the repeated configuration model (RCM) or the erased configuration model (ECM) depending on the properties of the degree distribution.

The RCM approach is based on the repeating the wiring process until a simple graph is obtained. The RCM is successful in generating a graph with desired size and a given degree sequence only if the probability of obtaining a simple graph converges to a non-zero value as $n$ approaches infinity. It is known [7] that this condition holds true if and only if $\mathbb{E}\xi^2 < \infty$, i.e. random variable $\xi$ has a finite second moment.

The generation of a simple graph through the erased configuration model is as follows: all self-loops should be removed, and multiple edges between nodes $i$ and $j$ must be replaced with a single edge. It is known [7] that the ECM generates a simple graph with the correct asymptotic degree distribution when the degree sequence $D_n$ are $n$ independent and identically distributed samples of random variable $\xi$ in the case when random variable $\mathbb{E}\xi < \infty$, i.e. $\xi$ has a finite mean.

To estimate degree-degree correlation in complex networks, the Average Nearest Neighbor Degree (ANND) [10,15,19] is used, which is defined as follows:

$$\Phi_n(k) := \begin{cases} \sum_{l=1}^{n} lP(l|k), & V(k) \neq \varnothing, \\ 0, & V(k) = \varnothing. \end{cases}$$

In networks built using the configuration model, the value of $\Phi_n(k)$ does not depend on $k$ and is a constant. This means that such networks exhibit neutral mixing.

We will use the properties of the ANND distribution established in paper [19] to obtain similar results related to the quantity of $\Psi_n(k)$.

## 2.3   Limiting Distributions of $\Psi_n(k)$ in the Configuration Model

In this section we assume that the network degree distribution of underlying random variable $\xi$ follows the power law with parameter $\gamma$, i.e. $f(k) = p(\xi = k) = c(\gamma)k^{-\gamma-1}$, where $c(\gamma) = \frac{1}{\zeta(\gamma+1)}$, where $\zeta(s) := \sum_{k\geq 1} k^{-s}$ is the Riemann zeta function. We show that the limiting behavior of $\Psi_n(k)$ as $n \to \infty$ depends on $\gamma$ and is different for the following two cases: $\gamma > 2$ and $1 < \gamma \le 2$. The case when the values of $\gamma$ less than 1 is of no interest since in this instance the first moment of the $\xi$-distribution does not exists.

We show that the limiting distribution of $\Psi_t(k)$ over $k$ depends on the value of the exponent of the power law for the underlying degree distribution $P(\xi > k) \sim ck^{-\gamma-1}$ of the network. If the exponent $\gamma$ is greater than 2, then for the network obtained as a result of the configuration model, the limit distribution will exist. However, if the exponent is between 1 and 2, then the distribution converges to a random variable. The results rely on the properties that were proved for ANND under some assumptions in [19].

Let $S_\alpha(\sigma, \beta, \mu)$ denote an $\alpha$-stable random variable $\eta$ with parameters $\alpha, \sigma, \mu$ as it given in [14]. The function $l_0(x)$ is said to be slowly varying if $\lim_{x \to \infty} \frac{f(ax)}{f(x)} = 1$ for any $a > 0$.

**Theorem 1.** *Let $\{G_n\}$, $n > 1$, be the sequence of multi-graphs obtained by the configuration model, where each of the degree sequences $D_n$ is the result of $n$ independent and identically distributed samples of power-law distributed random variable $\xi$ with exponent $\gamma > 2$. Then*

- *if $\gamma > 2$ then $\Psi_n(k) \to \frac{\nu_2}{k\nu_1}$ in distribution;*
- *if $1 < \gamma \le 2$ then there exists a slowly varying function $l_0(n)$ such that $\frac{\Psi_n(k)}{l_0(n)n^{\frac{2}{\gamma}-1}} \to \frac{1}{k}S_{\frac{\gamma}{2}}(1, 1, 0)$ in distribution.*

*Proof.* The first case implies the existence of the first and the second moments of $\xi$, $\nu_1 = \mathbb{E}\xi = c(\gamma)\sum_k k^{-\gamma} = \frac{\zeta(\gamma)}{\zeta(\gamma+1)} < \infty$, $\nu_2 = \mathbb{E}\xi^2 = c(\gamma)\sum_k k^{-\gamma+1} = \frac{\zeta(\gamma-1)}{\zeta(\gamma+1)} < \infty$.

It was shown in [19](Theorem 5.1) that if the underlying degree distribution of the network has a finite variance $\nu_2 < \infty$ then $\Phi_n(k)$ converges uniformly to a constant in probability. The limit of $\Psi_n(k)$ as $n \to \infty$ is

$$\Psi(k) = \frac{1}{k}\frac{\sum_{l \ge 1} h(k,l)l}{f^*(k)}, \tag{1}$$

where $f^*(k) = \frac{kf(k)}{\mathbb{E}\xi}$. Since the configuration model exhibit the neutral mixing property we have $h(k,l) = f^*(k)f^*(l)$ for both directed and undirected graphs. It follows from (1) that

$$\Psi(k) = \frac{1}{k}\frac{\sum_{l \ge 1} lf^*(k)f^*(l)}{f^*(k)} = \frac{1}{k}\sum_{l \ge 1}\frac{l^2 f(l)}{\nu_1} = \frac{\nu_2}{k\nu_1}. \tag{2}$$

It follows from [19](Theorem 5.1) that $\Psi_n(k)$ converges to $\frac{\nu_2}{k\nu_1}$ in probability uniformly in $k$ as $n$ tends to infinity.

In the case of $1 < \gamma \le 2$ there exists the first moment of $\xi$, while the second one is diverging. Then it follows from [19](Theorem 5.3 and 5.7) and a generalized central limit theorem [18] that if $\xi$ has infinite variance then $\Psi_n(k)$ converges to a random variable than has a stable distribution and infinite variance. More precisely, there exists a slowly varying function $l_0(n)$ such that $\dfrac{\Psi_n(k)}{l_0(n)n^{\frac{2}{\gamma}-1}} \rightarrow$ $\frac{1}{k}S_{\frac{\gamma}{2}}(1,1,0)$, where $\eta \sim S_\alpha(\sigma,\beta,\mu)$ denotes an $\alpha$-stable random variable $\eta$ with parameters $\alpha,\sigma,\mu$ as it is given in [14]. The proof of the fact is based on the stable law central limit theorem [18].                                        $\square$

*Remark 1.* While the statement of Theorem 1 is formulated for multi-graphs obtained on the basis of the configuration model, the same result also holds for simple graphs obtained using the erased configuration model (ECM). This model is known to effectively build simple graphs even if the original power distributions have infinite variance. It should be noted that ECM removes some semi-edges in the process of constructing a graph with a given degree sequence. Therefore, the resulting sequence of degrees may differ from the original one, and therefore the values of the friendship indices averaged over $k$-degree nodes in the constructed graph may be different than the corresponding values of $\Psi_n(k)$ in the multi-graph built on the basis of the configuration model. However, as it was shown in paper [19], the limiting distributions of $\Psi_n(k)$ coincide in both models.

### 2.4    Numerical Experiments on Simulated Networks with a Finite Size

We illustrate the behavior of $\Psi_n(k)$-distributions for networks simulated by the configuration model. First we generate the degree sequences $D_n$ using the Pareto-distributed random variable $\xi$ with parameter $\gamma > 1$, $p(\xi = k) \sim \gamma m^\gamma k^{-\gamma-1}$, $p(\xi > t) = m^\gamma t^{-\gamma}$ for $t \geq 1$. We set as an input parameter when generating random variables distributed according to the Pareto law, a scaling factor $m$ equal to 5.

Then we construct graphs using the degree sequences $D_n$ obtained for different $\gamma = 1.5$, $\gamma = 2$ and $\gamma = 2.5$. All networks are of the same size $n = 300,000$. In addition, we build a network based on the Barabási–Albert model with the linear preferential attachment mechanism and the number of attached links at each iteration equal to 5.

Table 1 shows some characteristics of the simulated networks: $|V|$ is the number of nodes, $|E|$ is the number of links, $\alpha$ is the exponent of the power law degree distribution, $\kappa$ is the share of network nodes for which their friendship index is more than 1.

**Table 1.** The features of simulated networks

| Model | $|V|$ | $|E|$ | $\alpha$ | $\kappa$ |
|---|---|---|---|---|
| CM, $\gamma = 1.5$ | 300000 | 2204918 | 2.5 | 0.997 |
| CM, $\gamma = 2.0$ | 300000 | 1493771 | 3 | 0.919 |
| CM, $\gamma = 2.5$ | 300000 | 1246494 | 3.5 | 0.840 |
| BA | 300000 | 1499985 | 3 | 0.929 |

We calculate the values of $\Psi_n(k)$ for each $k$, $V_n(k) \neq \varnothing$. Moreover, we are interested in estimating the spread of friendship index values among nodes of the same degree. For this purpose, we have found the empirical variances

$$\Sigma_n^2(k) := \frac{1}{|V_n(k)|} \sum_{i \in V_n(k)} (\beta_i - \Psi_n(k))^2$$

Further, we have found the values of the coefficient of variation $CV_n(k)$ for each $k$, $V_n(k) \neq \varnothing$,

$$CV_n(k) = \frac{\Sigma_n(k)}{\Psi_n(k)}.$$

The coefficient of variation is a dimensionless quantity which makes it possible to better understand the distribution of the friendship index among vertices of the same degree.

Figure 1 presents log-log plots of dependencies $\Psi_n(k)$, $\Sigma_n^2(k)$ and $CV_n(k)$ on $k$ for networks obtained by the configuration model with $\gamma = 1.5$ (the first row), $\gamma = 2$ (the second row) and $\gamma = 2.5$ (the third row).

It is known that in the Barabási–Albert networks the node degree distribution is a power law $p_k \sim k^{-\alpha}$ with its exponent equal to $\alpha = 3$. Therefore, it is appropriate to compare them with networks built according to the configuration model with parameter $\gamma = 2$. Figure 1 shows that their behavior in terms of the $\Psi_n(k)$-distribution is almost identical. This is to be quite expected as the Barabási–Albert networks exhibit neutral mixing.

Figures 1(a-d) show that in the simulated networks the distribution of $\Psi_n(k)$ (the average value of the friendship index among nodes of the same degree $k$) over $k$ follows a power law with an exponent close to $-1$ in value. This behavior coincides with the theoretical results obtained in Theorem 1.

Figures 1(a-d) also clearly show that $\Sigma_n^2(k)$ (the variances of the values of the friendship index among nodes of the same degree $k$) in the simulated networks decrease over $k$. The dependence of the dispersion on the degree is a power law with an exponent close to $-3$.

It should be noted that the values of the coefficient of variation decrease for all networks, which means that the relative value of the scatter in the values of the friendship index for nodes with small degrees is much larger than the similar scatter for nodes with large degrees.

# 3   The Distribution of Friendship Index in Real Networks

In this section, we examine the distribution of the friendship index in four real networks:

- *Amazon recommendation network.* Amazon, a popular global e-commerce platform, enables users to purchase products from different merchants and post reviews. The recommendations from various buyers create a bipartite network, as documented in [13].

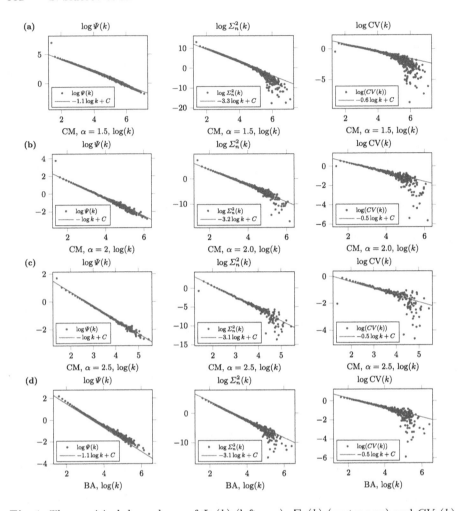

**Fig. 1.** The empirical dependence of $\Psi_n(k)$ (left row), $\Sigma_n(k)$ (center row) and $CV_n(k)$ (the third row) on $k$ in four networks of the same size $n = 300,000$ simulated by (a) the configuration model with $\gamma = 1.5$, (b) the configuration model with $\gamma = 2.0$, (c) the configuration model with $\gamma = 2.5$, (d) the Barabási–Albert model.

- *Yahoo email network.* This is a social network where users are connected if they exchanged emails. We obtained the dataset from [13], which focuses on users of the Yahoo email platform.
- *Enron mail network.* The social network is based on Enron employee interactions. Two employees are linked if they exchange corporate email messages. Compared to the Yahoo network, this social network has a higher density of connections. Additionally, it exhibits a notable presence of nodes with high and extremely high degrees.

– *SuperUser network.* We examine data from the SuperUser platform, where users can ask computer and technology-related questions and provide answers. The network is constructed using a dataset that captures user interactions on the platform. Nodes in this network represent users, and they are connected if one user answers a question asked by another. Moreover, connections are also established if a user comments on a question or answer from another user.

Table 2 shows their main features: $|V|$ is the number of nodes, $|E|$ is the number of links, $\alpha$ is the exponent of the power law degree distribution, $\kappa$ is the share of network nodes for which their friendship index is more than 1.

**Table 2.** Real network statistics

| Real network | $|V|$ | $|E|$ | $\alpha$ | $\kappa$ | Other traits |
|---|---|---|---|---|---|
| *Amazon* products | 2.100.000 | 5.800.000 | 3.96 | 0.9662 | Bipartite network |
| *Enron* mails | 87.100 | 1.100.000 | 2.09 | 0.8425 | Close community |
| *Yahoo* messages | 100.000 | 3.200.000 | 2.31 | 0.9743 | - |
| *SuperUser* reactions | 194.000 | 1.400.000 | 3.38 | 0.9798 | - |

Let us denote the sample mean of the degrees for a network $\mu_n = \frac{1}{n}\sum_{i=1}^{n} d_i$, the sample variance $\sigma_n^2 = \frac{1}{n}\sum_{i=1}^{n}(d_i - \mu_n)^2$. We assume that

– the network is degree-degree uncorrelated, i.e. $P(l|k)$ does not depends on $l$ and $k$;
– $d_i$ and $d_j$ are independent for all $i,j$;
– the underlying degree distribution follows the power law with exponent $\nu$.

Then the empirical conditional mean of $\beta_i$ is

$$\mathbb{E}(\beta_i|d_i = k) = \frac{\sum_{\{j:\ (j,i)\in E_n\}} \mathbb{E}(d_j)}{k^2} = \frac{k\mu_n}{k^2} = \frac{\mu_n}{k}.$$

Then

$$\Psi_n(k) = \frac{1}{|V_n(k)|} \sum_{i\in V_n(k)} \mathbb{E}(\beta_i|d_i = k) = \frac{1}{|V_n(k)|}\frac{\mu_n|V_n(k)|}{k} = \frac{\mu_n}{k}.$$

Therefore, the distribution of $\Psi_n(k)$ over $k$ should follow the power law with the exponent equal to $-1$.

We have constructed empirical distributions of $\Psi_n(k)$ for the five real networks as follows. For each $k = 1, 2, \ldots, V_n(k) \neq \varnothing$, we formed the set of vertices $V_n(k)$ of the network with degree $k$. For each of the vertices from the set $V_n(k)$, we found the value of its friendship index, and then calculated their average value over all vertices of the $V_n(k)$ set, i.e. we found the empirical value of $\Psi_n(k)$. Having done the corresponding procedure for $k = 1, 2, \ldots$, we obtained the corresponding values of $\Psi_n(k)$. The log-log plots of $\Psi_n(k)$ versus $k$ are shown in the left-sided parts of Figs. 2(a-e) for five real networks. To find the parameters of the linear regression dependence, the log-binning procedure was used.

The plots show that there is a visible linear dependence of $\log \Psi_n(k)$ on $\log k$, or, in the other words, the dependence of $\Psi_n(k)$ on $k$ follows a power law, and the value of the exponent of this power dependence (the linear regression parameter $\log \Psi_n(k)$ on $\log k$) is close to -1 for all five real networks. Small deviations of the exponent values from -1 for real networks are apparently associated with the presence of degree-degree correlations in real networks.

Let us now estimate the variance of friendship index values over vertices with degree $k$. We will use the fact that for independent (uncorrelated) random variables, the variance of the sum is equal to the sum of its variances. We can write down

$$\mathrm{Var}(\beta_i | d_i = k) = \frac{\sum_{j:\ (j,i)\in E_n} \mathrm{Var}(d_j | d_i = k)}{k^4} = \frac{\sum_{j:\ (j,i)\in E_n} \mathrm{Var}(d_j)}{k^4} = \frac{\sigma_n^2}{k^3}.$$

Therefore $\Sigma_n^2(k) = \frac{\sigma_n^2}{k^3}$. Thus, under the assumptions, the dependence of $\Sigma_n^2(k)$ on $k$ should satisfy a power law with exponent $-3$.

Let's see how this dependence looks in real networks. We found the empirical values of $\Sigma_n^2(k)$ for $k = 1, 2, \ldots$ and plotted the dependence of $\Sigma_n^2(k)$ on $k$ on log-log scale, which can be seen in center parts of Figs. 2(a-e). We notice that the exponent varies from $-2.35$ to $-4.2$, i.e. it is around $-3$. We may suggest that the assumption of independence for neighbor node degree variances on the node degree does not entirely hold for real networks and, as the consequence, it distorts the value of the exponent.

Further, we have found the values of the coefficient of variation $CV_n(k)$ for each $k$, $V_n(k) \neq \varnothing$. Dependencies of $CV_n(k)$ on $k$ for four real networks are shown in the right-hand side plots of Figs. 2(a-d).

For all networks, we observe a decrease in the coefficient of variation for $\Psi_n(k)$ as $k$ increases. This means that the friendship index of nodes with a higher degree has a smaller relative spread around the mean value.

It is interesting to note that the value of the coefficient of variation in scale-free networks with $1 < \gamma < 2$ significantly exceeds the corresponding values in networks with $\gamma > 2$. In real networks, the value of this coefficient is greater than 1 for almost all degrees.

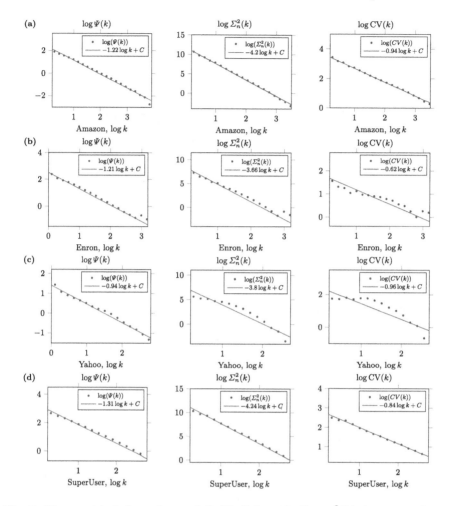

**Fig. 2.** The empirical dependence of $\Psi_n(k)$ (left row), $Sigma_n^2(k)$ (center row) and $CV_n(k)$ (the third row) on $k$ in four networks (a) user product recommendations of Amazon service, (b) network of email communications within Yahoo email service, (c) Enron mail network, (d) user interactions on the SuperUser platform.

# 4   Conclusion

This study aims to investigate the distribution of the friendship index in both real and synthetic networks. Unlike previous works, we specifically focus on how the characteristics of scale-free networks impact the distribution of the friendship index.

We consider the value of the friendship index averaged over all vertices of the same degree $k$ and study how this value depends on $k$. We examine the limiting value of this quantity in networks with neutral mixing and network sizes tending to infinity. We show that in networks with a power-law degree distribution with a finite second moment, this value tends to a constant divided by $k$. However, if the second moment of the distribution of powers is not bounded, this quantity converges to a stable distributed random variable divided by $k$. With the growth of the network, the distribution of $\Psi_t(k)$ over $k$ is scaling with the growth of the network, just as it happens for the value of average nearest neighbor degrees (ANND) [19]. Both $\Psi_t(k)$ and ANND values are scaled relative to the size $t$ of the network, which should lead to this effect. Therefore, $\Psi_n(k)$ is not appropriate measure for comparison networks of different sizes. Therefore, it would be useful to propose a measure for the friendship paradox evaluation, which does not depend on the size of network.

# References

1. Alipourfard, N., Nettasinghe, B., Abeliuk, A., Krishnamurthy, V., Lerman, K.: Friendship paradox biases perceptions in directed networks. Nat. Commun. **11**(1), 707 (2020). https://doi.org/10.1038/s41467-020-14394-x
2. Bollen, J., Gonçalves, B., van de Leemput, I., Ruan, G.: The happiness paradox: your friends are happier than you. EPJ Data Sci. **6**(1), 1–17 (2017). https://doi.org/10.1140/epjds/s13688-017-0100-1
3. Chen, N., Olvera-Cravioto, M.: Directed random graphs with given degree distributions. Stochast. Syst. **3**(1), 147–186 (2013). https://doi.org/10.1214/12-SSY076
4. Eom, Y.H., Jo, H.H.: Generalized friendship paradox in complex networks: the case of scientific collaboration. Sci. Rep. **4**, 4603 (2014). https://doi.org/10.1038/srep04603
5. Fotouhi, B., Momeni, N., Rabbat, M.G.: Generalized friendship paradox: an analytical approach. In: Aiello, L.M., McFarland, D. (eds.) SocInfo 2014. LNCS, vol. 8852, pp. 339–352. Springer, Cham (2015). https://doi.org/10.1007/978-3-319-15168-7_43
6. Higham, D.J.: Centrality-friendship paradoxes: when our friends are more important than us. J. Complex Netw. **7**(4), 515–528 (2018). https://doi.org/10.1093/comnet/cny029
7. Hofstad, R.V.D.: Random Graphs and Complex Networks. Cambridge Series in Statistical and Probabilistic Mathematics. Cambridge University Press, Cambridge (2016). https://doi.org/10.1017/9781316779422
8. Jackson, M.O.: The friendship paradox and systematic biases in perceptions and social norms. J. Polit. Econ. **127**(2), 777–818 (2019). https://doi.org/10.1086/701031
9. Lee, E., Lee, S., Eom, Y.H., Holme, P., Jo, H.H.: Impact of perception models on friendship paradox and opinion formation. Phys. Rev. E **99**(5), 052302 (2019). https://doi.org/10.1103/PhysRevE.99.052302
10. Litvak, N., van der Hofstad, R.: Uncovering disassortativity in large scale-free networks. Phys. Rev. E **87**, 022801 (2013). https://doi.org/10.1103/PhysRevE.87.022801
11. Momeni, N., Rabbat, M.: Qualities and inequalities in online social networks through the lens of the generalized friendship paradox. PLoS ONE **11**(2), e0143633 (2016). https://doi.org/10.1371/journal.pone.0143633
12. Pal, S., Yu, F., Novick, Y., Bar-Noy, A.: A study on the friendship paradox – quantitative analysis and relationship with assortative mixing. Appl. Netw. Sci. **4**(1), 71 (2019). https://doi.org/10.1007/s41109-019-0190-8

13. Paranjape, A., Benson, A.R., Leskovec, J.: Motifs in temporal networks. In: Proceedings of the Tenth ACM International Conference on Web Search and Data Mining. ACM (2017). https://doi.org/10.1145/3018661.3018731
14. Samorodnitsky, G., Taqqu, M.S.: Stable Non-Gaussian Random Processes. Routledge (2017). https://doi.org/10.1201/9780203738818
15. Sidorov, S., Mironov, S., Grigoriev, A.: Measuring the variability of local characteristics in complex networks: empirical and analytical analysis. Chaos: Interdisc. J. Nonlinear Sci. **33**(6), 063106 (2023). https://doi.org/10.1063/5.0148803
16. Sidorov, S.P., Mironov, S.V., Grigoriev, A.A.: Friendship paradox in growth networks: analytical and empirical analysis. Appl. Netw. Sci. **6**, 51 (2021). https://doi.org/10.1007/s41109-021-00391-6
17. Sidorov, S., Mironov, S., Malinskii, I., Kadomtsev, D.: Local degree asymmetry for preferential attachment model. In: Benito, R.M., Cherifi, C., Cherifi, H., Moro, E., Rocha, L.M., Sales-Pardo, M. (eds.) COMPLEX NETWORKS 2020 2020. SCI, vol. 944, pp. 450–461. Springer, Cham (2021). https://doi.org/10.1007/978-3-030-65351-4_36
18. Whitt, W.: Stochastic-Process Limits. Springer, New York (2002). https://doi.org/10.1007/b97479
19. Yao, D., van der Hoorn, P., Litvak, N.: Average nearest neighbor degrees in scale-free networks. Internet math. **2018**, 1–38 (2018). https://doi.org/10.24166/im.02.2018 https://doi.org/10.24166/im.02.2018

# Theoretical Machine Learning
# and Optimization

# The Problem of Finding Several Given Diameter Spanning Trees of Maximum Total Weight in a Complete Graph

E. Kh. Gimadi[1,2(✉)] [ID] and A. A. Shtepa[2] [ID]

[1] Sobolev Institute of Mathematics, prosp. Akad. Koptyuga, 4,
630090 Novosibirsk, Russia
[2] Novosibirsk State University, Pirogova, 2, 630090 Novosibirsk, Russia
gimadi@math.nsc.ru
http://www.math.nsc.ru/

**Abstract.** We consider the following $NP$-hard problem. Given an undirected complete edge-weighted graph and positive integers $m$, $D$, the goal is to find $m$ edge-disjoint spanning trees with diameter $D$ of maximum total weight in complete undirected graph. We propose an $\mathcal{O}(n^2)$-time approximation algorithm for the problem, where $n$ is the number of vertices in the input graph. For the case when edge weights are randomly uniformly chosen from the interval $(0; 1)$, we prove sufficient conditions under which the proposed algorithm gives asymptotically optimal solutions.

**Keywords:** given diameter Spanning Tree Problem · approximation algorithm · probabilistic analysis · asymptotic optimality

## 1 Introduction

The Minimum Spanning Tree (MST) problem is the well-known discrete optimization problem. It consists of finding a spanning tree (connected acyclic subgraph, which covers all the vertices) of minimum total edge weight in a given edge-weighted undirected graph $G = (V, E)$. The problem is solvable in polynomial time, for example, using classical algorithms by Boruvka (1926), Kruskal (1956), and Prim (1957). These algorithms have complexities $\mathcal{O}(u \log n)$, $\mathcal{O}(u \log u)$, and $\mathcal{O}(n^2)$ respectively, where $u = |E|$ and $n = |V|$. It is interesting that the expected value for an MST's weight in a graph with edge weights being randomly uniformly chosen from the interval $(0; 1)$ is close to 2.02 with a high probability [7]. Similar results are obtained in [1,5].

One of the possible generalizations of this classic problem is a problem of finding the MST with bounded diameter. The diameter of a graph is the number of edges in the longest simple path within a graph connecting a pair of vertices. This problem is $NP$-hard in cases when the diameter is bounded from

The study was carried out within the framework of the state contract of the Sobolev Institute of Mathematics (project FWNF-2022-0019).

D. I. Ignatov et al. (Eds.): AIST 2023, LNCS 14486, pp. 341–348, 2024.
https://doi.org/10.1007/978-3-031-54534-4_24

above [8,16] or from below [12] by a given number. The problem of finding MST with diameter bounded from above has several applications in wireless ad-hoc networks [4], network design [2], in development of data compression algorithm [3] and distributed mutual exclusion algorithm [18].

We assume that weights of graph edges are independent identically distributed random reals with uniform distribution on the interval $(0; 1)$. For short, further, we will say that such weights belong to the class UNI$(0; 1)$.

In previous papers [10–13], we have studied the problem of finding MST with diameter bounded from above or from below in directed and undirected graphs with edge weights from UNI$(a_n; b_n)$. UNI$(a_n; b_n)$ is the class of independent identically distributed random reals with uniform distribution on the interval $(a_n; b_n)$. In the case of $a_n > 0$, an asymptotically optimal approach was implemented for these problems.

In current paper, we consider the problem of finding $m$ spanning trees of total maximum weight with diameters equal to a given integer $D$ in a complete undirected graph. We will refer to this problem as $m$-$D$-UMaxST. An ideologically close setting was considered in papers [6,19], where an upper bound was imposed on the radius $R$ of the graph. Approximation $\mathcal{O}(n^2)$-time algorithms were presented with performance ratios $\min\{\frac{2}{5}, \frac{4}{R+7}, \frac{2}{R+1}\}$ for the problem with triangle inequality in [6] and $\frac{1}{R}$ for general case problem in [19].

In [14] an asymptotically optimal approach was proposed for finding the only one spanning tree of maximum total weight with a given diameter $D$ in complete undirected graph with edge weights from UNI$(0; 1)$. In current work, we generalize this approach to find several edge-disjoint spanning trees, solving $m$-$D$-UMaxST.

Note that our approach can also be transformed to solve the problem of finding maximum-weight spanning tree with bounded diameter from below or above.

**Statement 1** [14]. *Let $G = (V, E)$, $G' = (V, E')$ be graphs with set of vertices $V$, $(w_e) : E \to \mathbb{N} \cup \{0\}$ and $(w'_e) : E' \to \mathbb{N} \cup \{0\}$ be two weight functions on $G$ and $G'$ respectively. $T$ is a tree with total weight $W'(T) = \sum_{e \in T} w'_e$ constructed in a graph $G' = (V, E')$ with weight function $(w'_e)$, where $w'_e = 1 - w_e$. Then and only then $T$ is a tree with total weight $W(T) = n - 1 - W'(T)$, constructed in origin graph $G$ with weight function $(w_e)$.*

We are going to construct approximation algorithm for the $m$-$D$-UMaxST in graph $G$ by solving analogous problem for the several edge-disjoint spanning trees of minimum total weight in graph $G'$. According to statement 1, it is the required solution for the $m$-$D$-UMaxST.

## 2  Description of Approximation Algorithms $\mathcal{A}'$ and $\mathcal{A}$

In this section we construct approximation Algorithm $\mathcal{A}'$ for problem of finding $m$ edge-disjoint spanning trees of minimum total weight with given diameter $d$ in complete undirected graph. On the base of this algorithm we implement approximation Algorithm $\mathcal{A}$ for the $m$-$D$-UMaxST.

## The description of Algorithm $\mathcal{A}'$

---

**Input of Algorithm $\mathcal{A}'$:** edge-weighted undirected complete graph $G' = (V, E')$ ($|V| = n$), positive integers $d$, $m$ such that $m(d+1) \leq n$.

**Output of Algorithm $\mathcal{A}'$:** $m$ edge-disjoint spanning trees $T_1, \ldots, T_m$ of minimum total weight with diameter of each tree is equal to $d$.

---

**Preliminary Step 0.** In graph $G'$, choose an arbitrary $(n - m(d+1))$-vertex subset $V'$, and split the remaining vertices into $m$ disjoint subsets $V_1, \ldots, V_m$

**Step 1.** For each $l = 1, \ldots, m$, starting at an arbitrary vertex of the subgraph $G(V_l)$, construct a Hamiltonian path $P_l$ of length $d = d_n$, using the approach "go to the nearest unvisited vertex". After that set $T_l = P_l$.

**Step 2.** Without loss of generality, let $d$ be odd. Consider each pair of paths $P_i$ and $P_j$, $1 \leq i < j \leq m$. Connect them by the set $E_{ij}$ of $2(d+1)$ edges, so that the constructed subgraph was composed of two $2(d+1)$-vertex edge-disjoint subtrees with a diameter equals $d$. For that, represent each path as two halves (first and second) $P_l^1$ and $P_l^2$, $1 \leq l \leq m$. Each half contains one *end* vertex and $\frac{d-1}{2}$ *inner* vertices, $\frac{d+1}{2}$ vertices totally. For each $i$ and $j$ carry out the construction of $E_{ij}$ in the following way.

2.1. (2.2.) Connect each inner vertex of $P_i^1$ ( $P_i^2$ ) by the shortest edge to the inner vertex of $P_j^1$ ($P_j^2$). Add this edge to $T_j$.

2.3. (2.4.) Connect each inner vertex of $P_j^1$ ( $P_j^2$) by the shortest edge to the inner vertex of $P_i^2$ ( $P_i^1$). Add this edge to $T_i$.

2.5. (2.6.) Connect each end vertex of the path $P_i$ ($P_j$) by the shortest edge to the inner vertex of the path $P_j$ ($P_i$.). Add this edge to $T_j$ ($T_i$).

**Step 3.** For $l = 1, \ldots, m$ connect each vertex of the subgraph $G(V')$ by the shortest edge to the inner vertex of the path $P_l$. Thus, add this edge to corresponding $T_l$.

Description of Algorithm $\mathcal{A}'$ is completed.

Let us formulate two statements concerning Algorithm $\mathcal{A}'$, they are proved in [15].

**Statement 2** [15]. *Algorithm $\mathcal{A}'$ constructs feasible solution for the problem of finding several edge-disjoint spanning trees of minimum total weight with a given diameter in complete undirected graph.*

**Statement 3** [15]. *Time complexity of Algorithm $\mathcal{A}'$ is $\mathcal{O}(n^2)$.*

## The description of Algorithm $\mathcal{A}$

---

**Input of Algorithm $\mathcal{A}$:** edge-weighted undirected complete graph $G = (V, E)$ ($|V| = n$), positive integers $D$, $m$ such that $m(D+1) \leq n$.

**Output of Algorithm $\mathcal{A}$:** $m$ edge-disjoint spanning trees $T_1, \ldots, T_m$ of maximum total weight with diameter of each tree is equal to $D$.

---

**Step 1.** Change weight function $(w_e)$ of graph $G$ to weight function $(w'_e)$ obtaining graph $G'$ such that $w'_e = 1 - w_e$. Set $d = D$.

**Step 2.** Apply Algorithm $\mathcal{A}'$ to the graph $G'$ with weight function $(w'_e)$, obtain spanning trees $T_1, \ldots, T_m$.

**Step 3.** The constructed spanning trees $T_1, \ldots, T_m$ are solution for the $m$-$D$-UMaxST.

Description of Algorithm $\mathcal{A}$ is completed.

It is not hard to understand that Algorithm $\mathcal{A}$ finds feasible solution for the $m$-$D$-UMaxST according to statements 1 and 2. Also time complexity of Algorithm $\mathcal{A}$ is equal to $\mathcal{O}(n^2)$ since changing of weight function in $n$-vertex graph can be done in $\mathcal{O}(n^2)$ time.

## 3  Probabilistic Analysis of Algorithm $\mathcal{A}$

We perform the probabilistic analysis of Algorithm $\mathcal{A}$ under conditions that weights of graph edges are random variables $\xi$ from the class UNI$(0; 1)$.

By $W_A(I)$ and $OPT(I)$ we denote respectively the approximate (obtained by some approximation algorithm $A$) and the optimum value of the objective function of the problem on the input $I$.

Algorithm $A$ is said to have *performance guarantees* $(\varepsilon_n, \delta_n)$ on the set of random inputs of the $n$-sized problem (where $n$ is amount of input data required to describe the problem, see [8]), if for each input $I$

$$\mathbb{P}\big\{|F_A(I) - OPT(I)| > \varepsilon_n OPT(I)\big\} \le \delta_n, \tag{1}$$

where $\varepsilon_n = \varepsilon_A(n)$ is an estimation of *the relative error* of the solution obtained by algorithm $A$, $\delta_n = \delta_A(n)$ is an estimation of *the failure probability* of the algorithm, which is equal to the proportion of cases when the algorithm does not hold the relative error $\varepsilon_n$ or does not produce any answer at all.

Following [9] we say that approximation algorithm $A$ is *asymptotically optimal* on the class of input data of the problem, if there exist such performance guarantees that for all inputs $I$ of size $n$

$$\varepsilon_n \to 0 \text{ and } \delta_n \to 0 \text{ as } n \to \infty.$$

We denote random variable equal to minimum over $k$ variables from the class UNI$(0; 1)$ by $\xi_k$. Let $W'_{\mathcal{A}'}$ be the total weight of trees $T_1, \ldots, T_m$ constructed by Algorithm $\mathcal{A}'$. According to the description of Algorithm $\mathcal{A}'$ the weight $W_{\mathcal{A}'}$ is a random variable equal to $W'_1 + W'_2 + W'_3$, where random variables $W'_1, W'_2, W'_3$ correspond to Steps of Algorithm $A'$.

$W'_1 = \sum_{i=1}^{m} \sum_{k=1}^{d} \xi_k$, since we construct the path $P_i$ consisting of $d$ edges and repeat this construction $m$ times during Step 1.

$W'_2 = C_m^2\big(4\frac{d-1}{2}\xi_{(d-1)/2} + 4\xi_{(d-1)}\big)$, because for each pair of paths (totally, $C_m^2 = \frac{m(m-1)}{2}$ such pairs) we connect each inner vertex of a one half of a path ($\frac{d-1}{2}$ such vertices) by the shortest edge to the inner vertex of a half of another path on the Steps 2.1–2.4, this process is modeled by random variable $\xi_{(d-1)/2}$. Since we connect first and second halves of a path to first and second halves of another path, there is a factor 4 in the first term. Finally, we connect each end vertex in each pair of paths by the shortest edge to the inner vertex of path on the Steps 2.5–2.6. We find this shortest edge by looking over all $(d-1)$ inner vertices of the corresponding path. It is modeled by random variable $\xi_{(d-1)}$. Since every pair of paths has 4 end vertices, there is a factor 4 in the second term.

$W_3' = m(n - m(d+1))\xi_{(d-1)}$, since we connect each vertex from $G(V')$ (where $|V'| = n - m(d + 1)$) by the shortest edge to the inner vertex of the path $P_l$ (there are $(d-1)$ such vertices so random variable $\xi_{(d-1)}$ arises), $1 \le l \le m$, and we repeat this construction $m$ times.

**Lemma 1.** *The Algorithm $\mathcal{A}$ for the $m$-$D$-UMaxST on $n$-vertex complete graph with weights of edges from $UNI(0;1)$ has the following estimates of the relative error $\varepsilon_n$ and the failure probability $\delta_n$:*

$$\varepsilon_n = \frac{2\widehat{\mathbb{E}W'_{\mathcal{A}'}}}{m(n-1)}, \tag{2}$$

$$\delta_n = \mathbb{P}\left\{ \widetilde{W}'_{\mathcal{A}'} > \widehat{\mathbb{E}W'_{\mathcal{A}'}} \right\}, \tag{3}$$

*where $\widetilde{W}'_{\mathcal{A}'} = W'_{\mathcal{A}'} - \mathbb{E}W'_{\mathcal{A}'}$, $\widehat{\mathbb{E}W'_{\mathcal{A}'}}$ is some upper bound for expectation $\mathbb{E}W'_{\mathcal{A}'}$.*

*Proof.* For the case of $m$-$D$-UMaxST, we have

$$\mathbb{P}\left\{ W'_{\mathcal{A}} < (1 - \varepsilon_n)OPT \right\} \le \mathbb{P}\left\{ W'_{\mathcal{A}} < (1 - \varepsilon_n)m(n - 1) \right\} =$$

$$= \mathbb{P}\left\{ m(n - 1) - W'_{\mathcal{A}'} < (1 - \varepsilon_n)m(n - 1) \right\} = \mathbb{P}\left\{ W'_{\mathcal{A}'} > \varepsilon_n m(n - 1) \right\} =$$

$$= \mathbb{P}\left\{ W'_{\mathcal{A}'} - \mathbb{E}W'_{\mathcal{A}'} > \varepsilon_n m(n - 1) - \mathbb{E}W'_{\mathcal{A}'} \right\} \le$$

$$\le \mathbb{P}\left\{ \widetilde{W}'_{\mathcal{A}'} > 2\widehat{\mathbb{E}W'_{\mathcal{A}'}} - \mathbb{E}W'_{\mathcal{A}'} \right\} = \mathbb{P}\left\{ \widetilde{W}'_{\mathcal{A}'} > \widehat{\mathbb{E}W'_{\mathcal{A}'}} \right\} = \delta_n.$$

The lemma is proved.

Further analysis of the performance guarantees of Algorithm $\mathcal{A}$ will be based on the following theorem.

**Petrov's Theorem** [17]. *Consider independent random variables $X_1, \ldots, X_N$. Let there be positive constants $T$ and $h_1, \ldots, h_N$ such that for all $k = 1, \ldots, N$ and $0 \le t \le T$ the following inequalities hold:*

$$\mathbb{E}e^{tX_k} \le e^{\frac{h_k t^2}{2}}. \tag{4}$$

Set $S = \sum_{k=1}^{N} X_k$ and $H = \sum_{k=1}^{N} h_k$. Then

$$\mathbb{P}\{S > x\} \le \begin{cases} \exp\{-\frac{x^2}{2H}\}, & \text{if } 0 \le x \le HT, \\ \exp\{-\frac{Tx}{2}\}, & \text{if } x \ge HT. \end{cases}$$

We will also need the following lemmas from [15].

**Lemma 2** [15]. *Let $\xi_k$ be random variable equal to minimum over $k$ independent random variables from $UNI(0;1)$. Given constants $T = 1$ and $h_k = \frac{1}{(k+1)^2}$. Then for the biased random variables $\widetilde{\xi}_k = \xi_k - \mathbb{E}\xi_k$ condition (4) of Petrov's Theorem holds for each $t \le T$ and $1 \le k < n$.*

**Lemma 3** [15]. *In the case $D = D_n \ge \ln n$, the following upper bound is true*

$$H \le \frac{mn}{D}$$

*where $H = \sum_{k=1}^{n} h_k$ with $h_k = \frac{1}{(k+1)^2}$.*

**Lemma 4** [15]. *For $\mathbb{E}W'_{\mathcal{A}'}$ the following inequality holds*

$$\widehat{\mathbb{E}W'_{\mathcal{A}'}} = 2m\ln n + \frac{2mn}{D} \ge \mathbb{E}W'_{\mathcal{A}'}.$$

**Theorem 1.** *Let $D = D_n \ge \ln n$. Then we get the following failure probability and the relative error for Algorithm $\mathcal{A}$:*

$$\delta_n = \frac{1}{n^m}, \tag{5}$$

$$\varepsilon_n = \mathcal{O}\left(\frac{\ln n}{n} + \frac{1}{D}\right), \tag{6}$$

*which tend to zero as $n \to \infty$.*

*Proof.* First, we note that in the courses of Algorithm $\mathcal{A}'$ and Algorithm $\mathcal{A}$ we are dealing with random variables only of the type $\xi_k$, $1 \le k \le D$. These variables satisfy the conditions of the Petrov's Theorem for constants $T = 1$ and $h_k = \frac{1}{(k+1)^2}$ according to lemma 2.

Using lemmas 3 and 4 and setting the threshold to be equal to $x = \widehat{\mathbb{E}W'_{\mathcal{A}'}} = 2m\ln n + \frac{2mn}{D}$, we have

$$TH \le \frac{mn}{D} < 2m\ln n + \frac{2mn}{D} = x$$

By Petrov's Theorem, taking into account the inequality $x \ge 2m\ln n$, we have an estimate for the failure probability

$$\delta_n = \mathbb{P}\{\widetilde{W}'_{\mathcal{A}'} > x\} \le \exp\left\{-\frac{Tx}{2}\right\} \le \exp\left\{-m\ln n\right\} = \frac{1}{n^m} \to 0 \text{ as } n \to \infty.$$

Then, according to lemma 4 and the formula (2), for the relative error we have

$$\varepsilon_n = \frac{2\widehat{\mathbb{E}W'_{\mathcal{A}'}}}{m(n-1)} = \frac{4n}{(n-1)} \cdot \left(\frac{\ln n}{n} + \frac{1}{D}\right) = \mathcal{O}\left(\frac{\ln n}{n} + \frac{1}{D}\right) \to 0 \text{ as } n \to \infty,$$

because $D = D_n \geq \ln n$.

Therefore, we obtain asymptotically optimal solution for the $m$-$D$-UMaxST with weights of edges from UNI$(0;1)$.

Theorem 1 is proved.

**Remark 1.** Results obtained for the considered UNI$(0;1)$ entries will also hold for the case of UNI$(a_n;b_n)$ entries, where $0 < a_n < b_n$.

**Remark 2.** In contrast to the case of minimization problem from [10–13], here there is no need to impose additional condition on the scatter of edge weights like $\frac{b_n}{a_n} = o\left(\frac{n}{\ln n}\right)$.

# 4   Conclusion

In this work, we have considered the problem of finding several edge-disjoint spanning trees of maximum total edge with given diameter in edge-weighted complete undirected graph $G$. We have used $\mathcal{O}(n^2)$-time algorithm from [15] for the minimization problem and applied this known algorithm for the graph $G'$ with modified weight function. For the uniform distribution of edge weights on interval $(0;1)$ we have carried out the probabilistic analysis of the algorithm and obtain sufficient conditions of its asymptotic optimality.

# References

1. Angel, O., Flaxman, A.D., Wilson, D.B.: A sharp threshold for minimum bounded-depth and bounded-diameter spanning trees and Steiner trees in random networks. Combinatorica **32**(1), 1–33 (2012). https://doi.org/10.1007/s00493-012-2552-z
2. Bala, K., Petropoulos, K., Stern, T.E.: Multicasting in a linear Lightwave network. In: Proceedings of the IEEE INFOCOM 1993, pp. 1350–1358 (1993). https://doi.org/10.1109/INFCOM.1993.253399
3. Bookstein, A., Klein, S.T.: Compression of correlated bit-vectors. Inform. Syst. **16**(4), 387–400 (1991)
4. Clementi, A.E.F., Ianni, M.D., Monti, A., Rossi, G., Silvestri, R.: Experimental analysis of practically efficient algorithms for bounded-hop accumulation in ad-hoc wireless networks. In: Proceedings of the 19th IEEE International Parallel Distributed Processing Symposium (IPDPS 2005), pp. 8–16 (2005). https://doi.org/10.1109/IPDPS.2005.210
5. Cooper, C., Frieze, A., Ince, N., Janson, S., Spencer, J.: On the length of a random minimum spanning tree. Comb. Probab. Comput. **25**(1), 89–107 (2016). https://doi.org/10.1017/S0963548315000024

6. Erzin, A.I.: The problem of constructing a spanning tree of maximal weight with a bounded radius. Upravlyaemye Sistemy, iss. **27**, 70–78 (1987). (in Russian)

7. Frieze, A.: On the value of a random MST problem. Discret. Appl. Math. **10**(1), 47–56 (1985). https://doi.org/10.1016/0166-218X(85)90058-7

8. Garey, M.R., Johnson, D.S.: Computers and Intractability, p. 340. Freeman, San Francisco (1979)

9. Gimadi, E.K., Glebov, N.I., Perepelitsa, V.A.: Algorithms with estimates for discrete optimization problems. Problemy Kibernetiki, iss. **31**, 35–42 (1975). (in Russian)

10. Gimadi, E. K., Istomin, A.M., Shin, E.Y.: On algorithm for the minimum spanning tree problem bounded below. In: Proceedings of the DOOR 2016, Vladivostok, Russia, 19–23 September 2016, CEUR-WS, 1623, pp. 11–17 (2016)

11. Gimadi, E.K., Istomin, A.M., Shin, E.Y.: On given diameter MST problem on random instances. In: CEUR Workshop Proceedings, pp. 159–168 (2019)

12. Gimadi, E.K., Serdyukov, A.I.: A probabilistic analysis of an approximation algorithm for the minimum weight spanning tree problem with bounded from below diameter. In: Inderfurth, K., Schwödiauer, G., Domschke, W., Juhnke, F., Kleinschmidt, P., Wäscher, G. (eds.) Operations Research Proceedings 1999. ORP, vol. 1999, pp. 63–68. Springer, Heidelberg (2000). https://doi.org/10.1007/978-3-642-58300-1_12

13. Gimadi, E.K., Shevyakov, A.S., Shin, E.Y.: Asymptotically optimal approach to a given diameter undirected MST problem on random instances. In: Proceedings of 15-th International Asian School-Seminar OPCS-2019, Publisher: IEEE Xplore, pp. 48–52 (2019)

14. Gimadi, E.K., Shtepa, A.A.: Asymptotically optimal approach for the maximum spanning tree problem with given diameter in a complete undirected graph on UNI(0; 1)-entries. Problems Inform. **57**(4), 53–62 (2022)

15. Gimadi, E.K., Shtepa, A.A.: On asymptotically optimal approach for finding of the minimum total weight of edge-disjoint spanning trees with a given diameter. Autom. Remote Control, **84**(7), 872–888 (2023). https://doi.org/10.25728/arcRAS.2023.42.85.001 https://doi.org/10.25728/arcRAS.2023.42.85.001

16. Nadiradze, G.: Bounded Diameter Minimum Spanning Tree, Master thesis, Central European University (2013)

17. Petrov, V.V.: Limit Theorems of Probability Theory. Sequences of Independent Random Variables, p. 304. Clarendon Press, Oxford (1995)

18. Raymond, K.: A tree-based algorithm for distributed mutual exclusion. ACM Trans. Comput. Syst. **7**(1), 61–77 (1989). https://doi.org/10.1145/58564.59295

19. Serdyukov, A.I.: On problem of maximal spanning tree with bounded radius. Diskretn. Anal. Issled. Oper. Ser. 1, **5**(3), 64–69 (1998). (in Russian)

# Is Canfield Right? On the Asymptotic Coefficients for the Maximum Antichain of Partitions and Related Counting Inequalities

Dmitry I. Ignatov$^{(\boxtimes)}$ ⓘ

National Research University Higher School of Economics, Moscow, Russia
dignatov@hse.ru
http://www.hse.ru

**Abstract.** This paper dates back to the asymptotic solutions of Rota's problem on the size of maximum antichain in the set partition lattice by Canfield and Harper and others. The knowledge of asymptotic coefficients could pave the way to the asymptotic solutions of such problems as (maximal) antichain counting in partition lattices. In addition to our attempt to reduce uncertainty in the values of these coefficients, we provide some inequalities for the discrepancy between the number of antichains and maximal antichains in partition lattices and give alternative proof for the number of maximal antichains obtained by us recently and recorded in the Online Encyclopaedia of Integer Sequences (https://oeis.org/A358041).

**Keywords:** partition lattice · maximal antichains · concept lattices · Formal Concept Analysis · enumerative combinatorics · asymptotic analysis

## 1 Introduction

More than half a century ago G.C. Rota formulated one of his famous problems related to maximum antichains in lattices [1], which asks whether a particular graded poset has its maximum antichain composed of all elements of the same rank, i.e. it is a level set. In its turn, this problem dates back for almost a century ago to the Sperner theorem, which states that a Boolean lattice on $n$ elements has its maximum antichain consisting of all subsets of size $\lfloor n/2 \rfloor$. Many interesting results related o Sperner theorem have been obtained since then [2]. Thus, our focus is on one of the fundamental lattices, $\mathcal{P}_n$ the partition lattice of $n$ elements.

In [3], R.L. Graham summarised the sate-of-the-art by that time as follows: "It was shown that any antichain in $\mathcal{P}_n$ had at most $\max_{k \leq n}\{{n \atop k}\}$ elements for $n \leq 20$ (and even this was not completely trivial because of the size $\{{n \atop k}\}$,

D. I. Ignatov et al. (Eds.): AIST 2023, LNCS 14486, pp. 349–361, 2024.
https://doi.org/10.1007/978-3-031-54534-4_25

e.g., $\{^{20}_{8}\}$=15170932662679) and it was generally believed this would continue to hold for all $n$"[1].

Here, the size of the level sets of the partition lattice is given by Stirling numbers of the second kind, $\{^{n}_{k}\}$.

However, it turned out the maximal value of the Stirling number of the second kind, $\max_{k\leq n}\{^{n}_{k}\}$, is not always equal to the size of the maximum antichain in $\mathcal{P}_n$. Graham continued, "it was quite unexpected when Canfield showed that there are antichains in $\mathcal{P}_n$ having many more than $\{^{n}_{k}\}$ elements when $n$ becomes very large" and "Canfield estimates that his techniques will start working at about $n = 6.526 \cdot 10^{26}$". The mentioned techniques rely on replacement of some level subset of partitions, for example, taken the level of maximal cardinality $k$, by a subset of incomparable partitions of another level, e.g., $k - 1$.

Later on, the bound where the discrepancy arises was lowered to $3.4 \cdot 10^6$ in [5] as well the asymptotic lower bound for the size of maximum antichain divided by $\max_{k\leq n}\{^{n}_{k}\}$ was established in [6]. Finally, the asymptotic upper bound was also found [7] and surprisingly for its discoverers, its functional form coincides with the lower bound up to the unknown constants.

Thus, from [7], we know that the connection between these numbers is non-linear with unknown constants [7]:

$$d(\mathcal{P}_n) = \max_{k\leq n}\{^{n}_{k}\}\Theta(n^a(\ln n)^{-a-1/4}), \text{ where}$$

$d(\mathcal{P}_n)$ is is the size of the maximal antichain in $\mathcal{P}_n$ and $a = \frac{2-e\ln 2}{4} \approx 0.02895765$. In the original paper, Canfield formulates the inequalities for $n > 1$ and writes "The symbols $c_1, c_2, \ldots$, denote positive real constants; it would be possible but distracting to replace these by appropriate explicit values." Similarly to [8], this sentence raises the main question to answer here: Is Canfield right?

We assume that the interval $1 \leq n \leq 20$ is discrepancy-free, and take into account that the discrepancy arises when $n \geq 3.4 \cdot 10^6$ [5,6]. Based on this, we can decrease uncertainty about the unknown constants. Moreover, Canfield uses the substitution $re^r = n$ and performs several "crudifications" going back to $n$ from $r$ in the last steps of his analyses. Using the principal branch of Lambert $W_0$ function, we can also avoid these steps.

Another part of this paper, is devoted to two related problems, the number of antichains and maximal antichains in the partition lattices. Note that the number of antichains in the Boolean lattice on a set of $n$ elements is known as $n$-th Dedekind number [9].

Recently, with the help of Ganter's Next Closure algorithm [10,11] from Formal Concept Analysis (an applied branch of modern Lattice Theory; cf. Galois lattices) [12], we have confirmed the results on the number of antichains in the partition lattices[2], $acp(n)$, and obtained new results for maximal antichains[3],

---

[1] See also [4] for $n < 20$.

[2] OEIS A302250.

[3] OEIS A358041.

$macp(n)$, also up to $n = 5$. Here, since $macp(n) < acp(n)$ for $n \geq 1$, we would like to strength this inequality by considering level antichains and between-level antichains. In addition, we provide another proof for $acp(4)$ and $macp(4)$ based on pattern mining with concept lattices and bipartite graphs.

## 2  Basic Definitions

Formal Concept Analysis is an applied branch of modern lattice theory aimed at data analysis, knowledge representation and processing with the help of (formal) concepts and their hierarchies. Here we recall basic definitions from [12,13] and a related tutorial [14].

First, we recall several notions related to lattices and partitions.

**Definition 1.** *A partition of a nonempty set $A$ is a set of its nonempty subsets $\sigma = \{B \mid B \subseteq A\}$ such that $\bigcup\limits_{B \in \sigma} B = A$ and $B \cap C = \emptyset$ for all $B, C \in \sigma$. Every element of $\sigma$ is called* block.

**Definition 2.** *A poset $\mathbf{L} = (L, \leq)$ is a **lattice**, if for any two elements $a$ and $b$ in $L$ the supremum $a \vee b$ and the infimum $a \wedge b$ always exist. $\mathbf{L}$ is called a **complete lattice**, if the supremum $\bigvee X$ and the infimum $\bigwedge X$ exist for any subset $A$ of $L$. For every complete lattice $\mathbf{L}$ there exists its largest element, $\bigvee L$, called the **unit element** of the lattice, denoted by $\mathbf{1}_L$. Dually, the smallest element $\mathbf{0}_L$ is called the **zero element**.*

**Definition 3.** *A partition lattice of set $A$ is an ordered set $(Part(A), \vee, \wedge)$ where $Part(A)$ is a set of all possible partitions of $A$ and for all partitions $\sigma$ and $\rho$ supremum and infimum are defined as follows:*

$$\sigma \vee \rho = \left\{ \bigcup conn_{\sigma,\rho}(B) \mid \forall B \in \sigma \right\},$$

$$\sigma \wedge \rho = \{B \cap C \mid \exists B \in \sigma, \exists C \in \rho : B \cap C \neq \emptyset\}, \text{where}$$

*$conn_{\sigma,\rho}(B)$ is the connected component to which $B$ belongs to in the bipartite graph $(\sigma, \rho, E)$ such that $(B, C) \in E$ iff $C \cap B \neq \emptyset$.*

**Definition 4.** *Let $A$ be a set and let $\rho, \sigma \in Part(A)$. The partition $\rho$ is finer than the partition $\sigma$ if every block $B$ of $\sigma$ is a union of blocks of $\rho$, that is $\rho \leq \sigma$.*

Equivalently one can use the traditional connection between supremum, infimum and partial order in the lattice: $\rho \leq \sigma$ iff $\rho \vee \sigma = \sigma$ ($\rho \wedge \sigma = \rho$).

**Definition 5.** *A **formal context** $\mathbb{K} = (G, M, I)$ consists of two sets $G$ and $M$ and a relation $I$ between $G$ and $M$. The elements of $G$ are called the **objects** and the elements of $M$ are called the **attributes** of the context. The notation $gIm$ or $(g, m) \in I$ means that the object $g$ has attribute $m$.*

**Definition 6.** *For $A \subseteq G$, let*

$$A' := \{m \in M \mid (g, m) \in I \text{ for all } g \in A\}$$

*and, for $B \subseteq M$, let*

$$B' := \{g \in G \mid (g, m) \in I \text{ for all } m \in B\}.$$

*These operators are called **derivation operators** or **concept-forming operators** for $\mathbb{K} = (G, M, I)$.*

Let $(G, M, I)$ be a context, one can prove that operators

$$(\cdot)'' \colon 2^G \to 2^G, \ (\cdot)'' \colon 2^M \to 2^M$$

are closure operators (i.e. idempotent, extensive, and monotone).

**Definition 7.** *A **formal concept** of a formal context $\mathbb{K} = (G, M, I)$ is a pair $(A, B)$ with $A \subseteq G$, $B \subseteq M$, $A' = B$ and $B' = A$. The sets $A$ and $B$ are called the **extent** and the **intent** of the formal concept $(A, B)$, respectively. The **subconcept-superconcept relation** is given by $(A_1, B_1) \leq (A_2, B_2)$ iff $A_1 \subseteq A_2$ $(B_2 \subseteq B_1)$.*

This definition implies that every formal concept has two constituent parts, namely, its extent and intent.

**Definition 8.** *The set of all formal concepts of a context $\mathbb{K}$ together with the order relation $\leq$ forms a complete lattice, called the **concept lattice** of $\mathbb{K}$ and denoted by $\underline{\mathfrak{B}}(\mathbb{K})$.*

**Definition 9.** *For every two formal concepts $(A_1, B_1)$ and $(A_2, B_2)$ of a certain formal context their **greatest common subconcept** is defined as follows:*

$$(A_1, B_1) \wedge (A_2, B_2) = (A_1 \cap A_2, (B_1 \cup B_2)'').$$

*The **least common superconcept** of $(A_1, B_1)$ and $(A_2, B_2)$ is given as*

$$(A_1, B_1) \vee (A_2, B_2) = ((A_1 \cup A_2)'', B_1 \cap B_2).$$

We say supremum instead of "least common superconcept", and instead of "greatest common subconcept" we use the term infimum.

In Fig. 1, one can see the context whose concept lattice is isomorphic to the partition lattice of a four-element set and the line (or Hasse) diagram of its concept lattice.

**Definition 10.** *A subset $X$ of the intent $B$ of a formal concept $(A, B)$ of a certain context $\mathbb{K}$ such that $X' = B$ (equivalently, $X'' = A$) is called generator of $B$. By $gen(B)$ we denote the set of all generators of $B$.*

Dually, the notion of generator can be defined for extents.

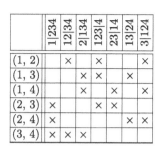
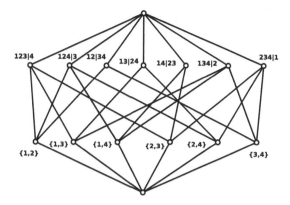

**Fig. 1.** The formal context (left) and the line diagram of the concept lattice (right) which is isomorphic to $\mathcal{P}_4$.

**Theorem 1.** *(Ganter&Wille [12]) For a given partially ordered set $\mathfrak{P} = (P, \leq)$ the concept lattice of the formal context $\mathbb{K} = (J(P), M(P), \leq)$ is isomorphic to the Dedekind–MacNeille completion of $\mathfrak{P}$, where $J(P)$ and $M(P)$ are sets of join-irreducible and meet-irreducible elements of $\mathfrak{P}$, respectively.*

A join-irreducible[4] lattice element cannot be represented as the supremum of strictly smaller elements; dually, for meet-irreducible elements. If $(P, \leq)$ is a lattice, then $\mathbb{K} = (J(P), M(P), \leq)$ is called its **standard context**.

**Theorem 2.** *(Bocharov et al. [15]) For a given partition lattice $\mathfrak{L} = (Part(A), \vee, \wedge)$ there exist a formal context $\mathbb{K} = (P_2, A_2, I)$, where $P_2 = \{\{a, b\} \mid a, b \in A \text{ and } a \neq b\}$, $A_2 = \{\sigma \mid \sigma \in Part(A) \text{ and } |\sigma| = 2\}$ and $\{a, b\}I\sigma$ when a and b belong to the same block of $\sigma$. The concept lattice $\mathfrak{B}(P_2, A_2, I)$ is isomorphic to the initial lattice $(Part(A), \vee, \wedge)$.*

There is a natural bijection between elements of $\mathfrak{L} = (Part(A), \vee, \wedge)$ and formal concepts of $\mathfrak{B}(P_2, A_2, I)$. Every $(A, B) \in \mathfrak{B}(P_2, A_2, I)$ corresponds to $\sigma = \bigwedge B$ and every pair $\{i, j\}$ from $A$ is in one of $\sigma$ blocks, where $\sigma \in Part(A)$. Every $(A, B) \in \mathfrak{B}(J(\mathfrak{L}), M(\mathfrak{L}), \leq)$ corresponds to $\sigma = \bigwedge B = \bigvee A$.

## 3 Problem Statement

Let us denote the partition lattice of set $[n] = \{1, \ldots, n\}$ by $\mathcal{P}_n = (Part([n]), \leq)$, where $Part([n])$ is the set of all partitions of $[n]$.

Three related problems, which we are going to consider are as follows.

**Problem 1. (MaxACPS)** *What is the size of the largest antichain of $\mathcal{P}_n = (Part([n]), \leq)$ for a given $n \in \mathbb{N}$?*

---

[4] join- and meet-irreducible elements are also called supremum- and infimum-irreducible elements, respectively.

**Problem 2. (#ACP)** *Count* $acp(n)$*, the number of antichains of* $\mathcal{P}_n$ *for a given* $n \in \mathbb{N}$.

**Problem 3. (#MaxACP)** *Count* $macp(n)$*, the number of maximal antichains of* $\mathcal{P}_n$ *for a given* $n \in \mathbb{N}$.

# 4    The Size of the Maximum Antichain

Let us consider Problem 1 and say that an interval $[n_1, n_2]$ has *zero-discrepancy* if $d(\mathcal{P}_n) = \max_{k \le n}\{\binom{n}{k}\}$ for all $n \in [n_1, n_2]$.

**Proposition 1.** *If the interval* $[1, 3.4 \cdot 10^6]$ *has zero-discrepancy, then for* $n > 1$

$$c_1 n^a (\ln n)^{-a-1/4} \le \frac{d(\mathcal{P}_n)}{\max_{k \le n}\{\binom{n}{k}\}} \le c_2 n^a (\ln n)^{-a-1/4}, \; where$$

$c_1 \le \frac{(ln2)^{a+1/4}}{2^a} \approx 0.884871$ *and* $c_2 \ge \frac{(ln15265)^{a+1/4}}{15265^a} \approx 1.423252$.

*Proof.* On the interval $[2, 3.4 \cdot 10^6)$, Canfield's function $f(x) = x^a (\ln x)^{-a-1/4}$ takes its maximum value at $n_{max} = 2$ and its minimum value at $x_{min} = e^{(a+1/4)/a} = e^{1+1/\ln\frac{e^2}{2e}} \approx 15264.667991$ or $n_{min} = 15265$ in integers.

However, we do not know whether the smallest $n_C$ where the discrepancy is non-zero arises before $3.4 \cdot 10^6$. So, what if a little bird [16] or an oracle has told us where $n_C$ is located? Then, there are the following options: a) $n_C \le n_{min}$ and b) $n_C > n_{min}$. Case a) affects only $c_2$ implying $c_2 \ge 1/f(n_C - 1)$. Case b) has three possible subcases: 1) $f(n_{min}) \le f(n_C - 1) \le f(2)$; 2) $f(n_C - 1) \ge f(2)$; 3) $f(n_C - 1) \ge f(n_{min})$.

The first subcase does not affect on the ranges for $c_1 \le 1/f(2)$ and $c_2 \ge 1/f(n_{min})$. Subcase 2) forces $c_1 \le 1/f(n_C - 1)$, while subcase 3) ensures $c_2 \ge 1/f(n_C - 1)$. However, 2) and 3) do not take place.

**Proposition 2.** *Since the interval* $[1, 20]$ *has zero-discrepancy, then for* $n > 1$

$$c_1 n^a (\ln n)^{-a-1/4} \le \frac{d(\mathcal{P}_n)}{\max_{k \le n}\{\binom{n}{k}\}} \le c_2 n^a (\ln n)^{-a-1/4}, \; where$$

$c_1 \le \frac{(ln2)^{a+1/4}}{2^a} \approx 0.884871$ *and* $c_2 \ge \frac{(ln20)^{a+1/4}}{20^a} \approx 1.24523$.

Let us have a look at the original inequalities in terms of $re^r = n$ substituion, where $r = W_0(n)$. Since we deal with integer positive values $n$, then we can use $W$ instead of $W_0$ without ambiguity.

**Proposition 3.** *If the interval* $[1, 3.4 \cdot 10^6)$ *has zero-discrepancy, then for* $n > 1$

$$\tilde{c}_1 e^{ar} r^{-1/4} \le \frac{d(\mathcal{P}_n)}{\max_{k \le n}\{\binom{n}{k}\}} \le \tilde{c}_2 e^{ar} r^{-1/4}, \; where$$

$\tilde{c}_1 \le \frac{W(2)^{1/4}}{e^{aW(2)}} \approx 0.93749$, $\tilde{c}_2 \ge \frac{r_{min}^{1/4}}{e^{ar_{min}}} \approx 1.33497$, *and* $r_{min} = \frac{1}{2-e\ln 2}$.

*Proof.* Let us consider inequalities for the lower and upper bounds obtained by Canfield and Harper.

1) Lower bound. In [17], the following inequality is obtained

$$d(\mathcal{P}_n) \geq c\frac{n^{1/2}/r}{n^{e\ln 2/4}/r^{\beta/2}} \max_{k\leq n}\{{n \atop k}\},$$

where $c > 0$, $\beta = \frac{1+e\ln 2}{2}$. Substituting $n = re^r$, we get $d(\mathcal{P}_n) \geq ce^{ar}r^{-1/4} \max_{k\leq n}\{{n \atop k}\}$.

On the interval $[2, 3.4 \cdot 10^6]$, $f(r) = e^{ar}r^{-1/4}$ takes maximum value at $r = W(2)$, which defines $\tilde{c}_1 \leq 1/f(W(2))$.

2) Upper bound. In [7], the following inequality takes place

$$d(\mathcal{P}_n) \leq \left(\frac{\varepsilon_0}{M^{1/2}} + 2n^{-1}\right)\omega_n,$$

where $M > c_5 2^{er/2}r^{-1/2}$, $\varepsilon_0$ and $c_5$ are some positive constants, $\omega_n$ is the $n$-th Bell number. Doing the same substitution for $n$ and taking into account $\omega_n = O(n^{1/2}/r)\max_{k\leq n}\{{n \atop k}\}$, for some $c > 0$ we get

$$d(\mathcal{P}_n) \leq ce^{ar}r^{-1/4} \max_{k\leq n}\{{n \atop k}\}.$$

Since $f(r) = e^{ar}r^{-1/4}$ takes its minimum at $r_{min} = 1/(2 - e\ln 2)$, we have $\tilde{c}_2 \geq 1/f(r_{min})$.

*Remark 1.* Proposition 3 can be formulated even for $n \geq 1$ by changing $\tilde{c}_1 \leq \frac{W(1)^{1/4}}{e^{aW(1)}} \approx 0.85367$. The function $f(r)$ takes its minimum in integers at $n = 48481$ since $x_{min} = r_{min}e^{r_{min}} \approx 48480.77$.

*Remark 2.* Similar proposition can be formulated for the zero-discrepancy interval as in Proposition 2, but in terms of $r = W(n)$. Then $\tilde{c}_2 \geq 1/f(W(20)) = 1.14320$.

# 5 The Number of (Maximal) Antichains

## 5.1 Known and Recent Results

The results for #ACP problem were published in OEIS and we have validated them with the FCA-based approach, while our results on #MAXACP were published later by Oct 29 2022. They are summarised for $n$ up to 5 in Table 1.

**Table 1.** The confirmed (the first row) and the obtained (the last row) results

| $n$ | 1 | 2 | 3 | 4 | 5 |
|---|---|---|---|---|---|
| #ACP, OEIS A302250 | 2 | 3 | 10 | 347 | 79814832 |
| #MaxACP, OEIS A358041 | 1 | 2 | 3 | 32 | 14094 |

These results were either confirmed or obtained with the help of approach to count (maximal) antichains used by K. Reuter [18] and learned by him from R. Wille.

All the contexts and codes are available on GitHub.

## 5.2   Inequalities

Thus, a simple lower bound for #ACP problem is given by $2^{\max\limits_{k\leq n}\{^n_k\}} \leq 2^{d(\mathcal{P}_n)}$, and was further improved by considering other the partition lattice levels (see Proposition 4).

**Proposition 4.** $acp(\mathcal{P}_n) \geq \sum\limits_{k=1}^{n} 2^{\{^n_k\}} - n + 1$ for $n \geq 1$.

As for enhancement of the evident inequality $macp(n) < acp(n)$ for $n \geq 1$, Proposition 5 was obtained.

**Proposition 5.** $macp(\mathcal{P}_n) \leq acp(\mathcal{P}_n) - \sum\limits_{k=1}^{n} 2^{\{^n_k\}} + 2n - 1$ for $n \geq 1$.

Now, we are going to slightly improve them by considering between-level antichains of size 2.

**Proposition 6.** $acp(\mathcal{P}_n) \geq \sum\limits_{k=1}^{n} \left(2^{\{^n_k\}} + \{^n_k\}\left(\{_{k-1}^{\;n}\} - \binom{k}{2}\right)\right) - n + 1$ for $n \geq 1$.

*Proof.* Let us consider two levels $k$ and $k - 1$, then the number of between level antichains is $\{^n_k\}\{_{k-1}^{\;n}\}$ minus the number of between level chains (i.e. edges between levels). Since every partition of level $k$ can be further coarsened by $\binom{k}{2}$ ways, we have the total number of between-level antichains of size $2 \{^n_k\} \left(\{_{k-1}^{\;n}\} - \binom{k}{2}\right)$.

**Proposition 7.** $macp(\mathcal{P}_n) \leq acp(\mathcal{P}_n) - \sum\limits_{k=1}^{n} \left(2^{\{^n_k\}} + \{^n_k\}\left(\{_{k-1}^{\;n}\} - \binom{k}{2}\right)\right) + 2n - 1$ for $n \geq 1$.

*Proof.* It follows from that every between-level antichain of size 2 in $\mathcal{P}_n$ is not maximal.

## 5.3  Case-Study for $acp(4)$ and $macp(4)$

We can consider the difference between known values of $acp(n)$ and $macp(n)$ and compare the enhancements.

Let us use $\Delta(n)$ for $acp(n) - macp(n)$, $D_l(n)$ for the decrement by levels $\sum_{k=1}^{n} 2^{\{{n \atop k}\}} - 2n + 1$, and, $D_{l+}$ for the decrement by levels and between-level antichains of size 2, i.e. $\sum_{k=1}^{n} \left( 2^{\{{n \atop k}\}} + \{{n \atop k}\} \left( \{{n \atop k-1}\} - \binom{k}{2} \right) \right) - 2n + 1$. In Table 2, it is shown that for the first three values $\Delta(n)$, $D_l(n)$, and $D_{l+}(n)$ coincide, but later the antichains different from the level antichain's subsets appear.

**Table 2.** The signed relative errors $\frac{\Delta(n) - D_l(n)}{\Delta(n)}$

| $n$ | 1 | 2 | 3 | 4 | 5 |
|---|---|---|---|---|---|
| $\Delta(n)$ | 1 | 1 | 7 | 315 | 79800738 |
| $D_l(n)$ | 1 | 1 | 7 | 189 | 33588219 |
| $D_{l+}(n)$ | 1 | 1 | 7 | 213 | 33588709 |
| $\frac{\Delta(n) - D_l(n)}{\Delta(n)}$ | 0 | 0 | 0 | 0.4 | $\approx 0.5791$ |
| $\frac{\Delta(n) - D_{l+}(n)}{\Delta(n)}$ | 0 | 0 | 0 | $\approx 0.3238$ | $\approx 0.5791$ |

Taking into account between-level antichains helped us to to decrease relative error for $n = 4$ almost by 0.08 but this improvement is already negligible for $n = 5$. This means that between-level antichains of larger sizes should be taken into account and not only for consecutive levels of $\mathcal{P}_n$.

However, $\mathcal{P}_4$ has only two consecutive levels different from its zero and unit elements, this means that among $\Delta(4) - D_{l+}(4) = 102$ antichains all are between-level antichains of size greater than 2.

Let us find all of the remaining antichains by considering the partial order $(\{{[4] \atop 2}\} \cup \{{[4] \atop 3}\}, \leq)$ on all the partitions on four elements into two and three blocks. It is clear, that this partial order forms a bipartite graph where $\sigma \in \{{[4] \atop 2}\}$ is connected by an edge with $\rho \in \{{[4] \atop 3}\}$ if $\rho$ refines $\sigma$.

Let us takes the complemented order $(\{{[4] \atop 2}\} \cup \{{[4] \atop 3}\}, \not\leq)$ and its bipartite graph induced by $\not\leq$ relation. It is clear than every biclique in this graph with at least one non-empty component forms an antichain of $(\{{[4] \atop 2}\} \cup \{{[4] \atop 3}\}, \leq)$ and of $\mathcal{P}_4$, while its maximal bicliques form maximal antichains.

It is known that formal concepts of a certain formal context biuniquely correspond to maximal bicliques of the bipartite graph on its objects and attributes with the same incidence relation. So, let us consider the concept lattice $\mathfrak{B}(\{{[4] \atop 2}\}, \{{[4] \atop 3}\}, \not\leq)$, which is simply the Dedekind-MacNeille completion of $(\{{[4] \atop 2}\} \cup \{{[4] \atop 3}\}, \not\leq)$ or $(J(\mathcal{P}_4), M(\mathcal{P}_4), \not\leq)$ (cf. the dual Dedekind-MacNeille completion [19]). We obtain that $|\mathfrak{B}(\{{[4] \atop 2}\}, \{{[4] \atop 3}\}, \not\leq)| = 30$ concepts, maximal bicliques or maximal antichains, and if we take into account zero and unit of $\mathcal{P}_4$, we obtain $macp(4) = 32$.

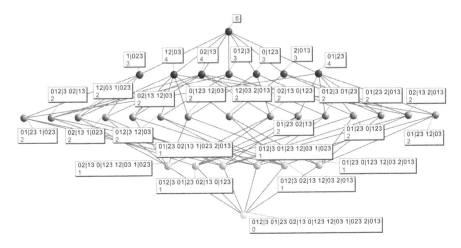

**Fig. 2.** A line diagram of $\underline{\mathfrak{B}}(\{{}^{[4]}_2\}, \{{}^{[4]}_3\}, \not\leq)$ with full labeling of intents (partitions) and the size of extent

The line diagram in Fig. 2 is obtained with Lattice Miner[5] [20].

**Table 3.** Patterns of antichains in $\mathcal{P}_4$ as concepts of $\underline{\mathfrak{B}}(\{{}^{[4]}_2\}, \{{}^{[4]}_3\}, \not\leq)$ with non-empty extent and intent

| Rank | Concept type | Size of extent generators | Size of intent generators | Total |
|------|-------------|--------------------------|---------------------------|-------|
| 3 | $K_{3,1}$ | 3 | 1 | |
| Cardinality | 4 | 4 | 4 | 4 |
| 3 | $K_{4,1}$ | 3, 4 | 1 | |
| Cardinality | 3 | $4 \cdot 3, 3$ | 3 | 15 |
| 2 | $K_{2,2}$ | 2 | 2 | |
| Cardinality | 15 | 15 | 15 | 15 |
| 1 | $K_{1,4}$ | 1 | 2, 3, 4 | |
| Cardinality | 6 | 6 | $6, 4 \cdot 6, 6$ | 36 |

In Table 3, the top and bottom concepts are not shown, since they have either empty extent or intent, i.e. their corresponding subgraph types are $K_{6,0}$ and $K_{0,7}$, respectively. They correspond to two level antichains, and every subset of their non-empty extent and intent forms an antichain in $\mathcal{P}_4$, in total $2^6 + 2^7 - 2$ non-empty antichains.

Each concept $(A, B)$ in Table 3 is given with its type as a biclique $K_a, b$, where $a = |A|$ and $b = |B|$. The generators are found using the line diagram, but their number number can be also found with the algorithm for counting

[5] https://github.com/LarimUQO/lattice-miner.

concept stability from [21] or with the Möbious function of the concept lattice $\mathfrak{B}(\{\binom{[4]}{2}\}, \{\binom{[4]}{3}\}, \not\leq)$ [22]. For example, we know that there is a unique generator of size two for each concept of rank one (with four partitions as attributes), since each of them has only five upper neighbours, while six pairs can be taken from a four-element set (cf., for example, $\{0|123, 1|023\}'' = \{02|13, 0|123, 12|03, 1|023\}$).

Another peculiar source of antichains is the middle level of $\mathfrak{B}(\{\binom{[4]}{2}\}, \{\binom{[4]}{3}\}, \not\leq)$, i.e. elements of rank two. These concepts do not have generators of their extents and intents as proper sets, but since they correspond to maximal bicliques, every their non-maximal biclique of type $K_{2,1}$ or $K_{1,2}$ also forms an antichain. In total, we have $15 \cdot 2 + 15 \cdot 2 = 60$ such antichains.

The remaining antichains are of type $K_{1,1}$ and their number is $\binom{4}{3}\left(\binom{4}{2} - \binom{3}{2}\right) = 24$.

Summing it all, we have $2^6 + 2^7 - 2 + 4 + 15 + 15 + 36 + 60 + 24 = 344$. If we take into account zero and unit of $\mathcal{P}_4$, i.e., single-element antichains of partitions $\{1234\}$ and $\{1|2|3|4\}$ along with the empty set of partitions $\emptyset$, we obtain $acp(4) = 347$.

One can require a single formula for $acp(4)$, which can be written as follows:

$$acp(4) = 2^{|J(\mathcal{P}_4)|} + 2^{|M(\mathcal{P}_4)|} + \sum_{k=1}^{3} \sum_{(A,B)\in\mathcal{L}_k} \max(|gen(A)|, |gen(B)|) +$$

$$\sum_{(A,B)\in\mathcal{L}_2} |\{(C,D)|C \times D \subset A \times B, C \neq \emptyset \text{ and } D \neq \emptyset\}| + 3 .$$

Note that $|gen(B)| = \sum_{(A,B)\leq(C,D)} 2^{|D|}\mu((C,D),(A,B))$ (dually, for $gen(A)$), where $\mu(x,y)$ is the Möbius function of the concept lattice $\mathfrak{B}(J(\mathcal{P}_4), M(\mathcal{P}_4), \not\leq)$.

# 6  Conclusion

We reduced uncertainty concerning the coefficients in the known asymptotic for the size of maximum antichain in the set partition lattice and illustrated how the values of $acp(4)$ and $macp(4)$ can be obtained from the between-level partial order of $\mathcal{P}_4$ with the help of Formal Concept Analysis along with the types of patterns obtained. This example can be used as the simplest model case for higher $n$ where antichains between non-consecutive levels appear.

**Acknowledgements.** This study was implemented in the Basic Research Program's framework at HSE University. This research was also supported in part through computational resources of HPC facilities at HSE University.

I would like to thank all the OEIS editors, especially Joerg Arndt, Michel Marcus, and N. J. A. Sloane. I also would like to thank anonymous reviewers and Jaume Baixeries for relevant suggestions, and Lev P. Shibasov and Valentina A. Goloubeva for the lasting flame of inspiration. Last but not least I would like to thank Zamira Ignatova for her care and patience.

# References

1. Rota, G.: A generalization of Sperner's theorem. J. Combin. Theory **2**, 104 (1967)
2. Engel, K.: Encyclopedia of Mathematics and its Applications Sperner Theory. Cambridge University Press, Cambridge (1997)
3. Graham, R.L.: Maximum antichains in the partition lattice. Math. Intell. **1**(2), 84–86 (1978)
4. Harper, L.H.: The morphology of partially ordered sets. J. Comb. Theory Ser. A **17**(1), 44–58 (1974)
5. Jichang, S., Kleitman, D.J.: Superantichains in the lattice of partitions of a set. Stud. Appl. Math. **71**(3), 207–241 (1984)
6. Canfield, E.R., Harper, L.H.: Large antichains in the partition lattice. Random Struct. Algorithms **6**(1), 89–104 (1995)
7. Canfield, E.R.: The size of the largest antichain in the partition lattice. J. Comb. Theory, Ser. A **83**(2), 188–201 (1998)
8. Farley, J.D.: Was Gelfand right? The many loves of lattice theory. Notices of the American Mathematical Society **69**(2) (2022)
9. Korshunov, A.D., Shmulevich, I.: The number of special monotone Boolean functions and statistical properties of stack filters. Diskretn. Anal. Issled. Oper., Ser. 1 **7**(3), 17–44 (2000)
10. Ganter, B.: Algorithmen zur formalen begriffsanalyse. In: Ganter, B., Wille, R., Wolff, K.E. (eds.) Beiträge zur Begriffsanalyse, pp. 241–254. B.I.-Wissenschaftsverlag, Mannheim (1987)
11. Ganter, B., Reuter, K.: Finding all closed sets: a general approach. Order **8**(3), 283–290 (1991)
12. Ganter, B., Wille, R.: Formal Concept Analysis: Mathematical Foundations, 1st edn. Springer-Verlag, New York Inc, Secaucus, NJ, USA (1999)
13. Aigner, M.: Combinatorial Theory. Springer, Berlin, Heidelberg (2012)
14. Ignatov, D.I.: Introduction to formal concept analysis and its applications in information retrieval and related fields. In: Braslavski, P., Karpov, N., Worring, M., Volkovich, Y., Ignatov, D.I. (eds.) RuSSIR 2014. CCIS, vol. 505, pp. 42–141. Springer, Cham (2015). https://doi.org/10.1007/978-3-319-25485-2_3
15. Bocharov, A., Gnatyshak, D., Ignatov, D.I., Mirkin, B.G., Shestakov, A.: A lattice-based consensus clustering algorithm. In: Huchard, M., Kuznetsov, S.O., (eds.) Proceedings of the Thirteenth International Conference on Concept Lattices and Their Applications, Moscow, Russia, 18–22 July 2016. Volume 1624 of CEUR Workshop Proceedings, pp. 45–56. CEUR-WS.org (2016)
16. Knuth, D.: The Art of Computer Programming, Volume 4A: Combinatorial Algorithms, Part 1. Pearson Education (2014)
17. Canfield, E., Harper, L.: A Simplified Guide to Large Antichains in the Partition (2000)
18. Reuter, K.: The jump number and the lattice of maximal antichains. Discret. Math. **88**(2), 289–307 (1991)
19. Markowsky, G., Markowsky, L.: Lattice data analytics: the poset of irreducibles and the macneille completion. In: 10th IEEE International Conference on Intelligent Data Acquisition and Advanced Computing Systems: Technology and Applications, IDAACS 2019, Metz, France, 18–21 September 2019, pp. 263–268. IEEE (2019)
20. Lahcen, B., Kwuida, L.: Lattice miner: a tool for concept lattice construction and exploration. In: Supplementary Proceeding of International Conference on Formal concept analysis (ICFCA 2010) (2010)

21. Roth, C., Obiedkov, S., Kourie, D.: Towards concise representation for taxonomies of epistemic communities. In: Yahia, S.B., Nguifo, E.M., Belohlavek, R. (eds.) CLA 2006. LNCS (LNAI), vol. 4923, pp. 240–255. Springer, Heidelberg (2008). https://doi.org/10.1007/978-3-540-78921-5_17

22. Babin, M.A., Kuznetsov, S.O.: Approximating concept stability. In: Domenach, F., Ignatov, D.I., Poelmans, J. (eds.) ICFCA 2012. LNCS (LNAI), vol. 7278, pp. 7–15. Springer, Heidelberg (2012). https://doi.org/10.1007/978-3-642-29892-9_7

# Author Index

D. I. Ignatov et al. (Eds.): AIST 2023, LNCS 14486, pp. 363–364, 2024.
https://doi.org/10.1007/978-3-031-54534-4

Printed in the United States
by Baker & Taylor Publisher Services
Printed in the United States
by Baker & Taylor Publisher Services